Rarefied Gas Dynamics

Lei Wu

Rarefied Gas Dynamics

Kinetic Modeling and Multi-Scale Simulation

Springer

Lei Wu
Southern University of Science
and Technology
Shenzhen, China

ISBN 978-981-19-2874-1 ISBN 978-981-19-2872-7 (eBook)
https://doi.org/10.1007/978-981-19-2872-7

This Springer imprint is published by the registered company Springer Nature Singapore Pte Ltd.
The registered company address is: 152 Beach Road, #21-01/04 Gateway East, Singapore 189721, Singapore

In memory of Gui-Xin Tong & Jason M. Reese

Preface

There has been three impulses in my life to publish a book of my own. The first impulse emerged when my principal supervisor Prof. Yonghao Zhang asked me to prepare the Ph.D. thesis in the spring of 2013, after I have researched in the University of Strathclyde for two and a half years. My intention then was to write a comprehensive thesis about the deterministic numerical methods for the Boltzmann equation, detailing every aspects in rarefied gas dynamics, and from the classical to quantum mechanics. Soon I abandoned this idea after realizing that my knowledge is far from being adequate, but just focused on the development and application of the fast spectral method. This turned out to be a right decision: I passed the Ph.D. viva 1 month before my scholarship expired.

The second impulse bumped in when my subsidiary Ph.D. supervisor Prof. Jason Reese invited me to attend his 50th birthday party in 2017. In order to express my gratitude to his guidance and help in the past 6 years, I decided to write a limited edition based on our joint work in rarefied gas dynamics, as a unique birthday gift. But the time is short, so eventually I compiled a book, by simply putting together our 14 joint publications and 6 of my other works in chronological order. Jason was happy after his birthday, so was I. The book is heavy; it returned to me from Jason's office after his sudden death in 2019, making it much heavier.

The third impulse revived after I joined the Southern University of Science and Technology around 2020. Two reasons gave birth to this book. First, I started to build a research group, and need some organized materials to introduce group members the recent development in rarefied gas dynamics. Second, several colleagues' postgraduates are working on the lattice Boltzmann method, a powerful method in computational fluid dynamics based on the gas kinetic theory, an in-depth knowledge in rarefied gas dynamics would be beneficial to their research. Therefore, I initiated a course entitled "Rarefied gas dynamics: from theory to application". On top of my Ph.D. thesis and the compiled book for Jason's birthday, I quickly drafted a book as the lecture material. The students' vitality and enthusiasm ignites my passion to write a complete version; my colleague Prof. Lianping Wang's multiple presence in the class also encouraged me a lot. This time I made it.

The rarefied gas dynamics is concerned with the phenomena that cannot be described by the conventional Navier-Stokes equations. It arises in gas flows when the mean free path/collision time of gas molecules is not negligible in comparison with the characteristic flow length/period. Hence, it has a wide range of applications in modern engineering problems, from the Sky to Underground, including the space shuttle's reentry into atmosphere, inertial confinement fusion, microelectromechanical systems, and shale gas extraction. In these extreme conditions, the continuum mechanics breaks down, while the molecular dynamic simulation is limited to tiny temporal and spatial scales, so the gas kinetic theory must be adopted to understand the flow physics.

The celebrated Boltzmann equation is at the heart of gas kinetic theory, which is initially proposed for dilute monatomic gas. It was extended to dense gas by Enskog, and molecular (diatomic and polyatomic) gas by Wang-Chang and Uhlenbeck. Now, kinetic models have been developed for liquid-vapor two phase flow, granular gas, plasma, radiation transport, phonon hydrodynamics, quantum gas, and social sciences, to name just a few. This fits the vast backgrounds of the postgraduates I taught, who worked on multiphase flow, cavitation, particle-laden flows, and so on. During lecturing I showed the counter-intuitive phenomena and typical numerical methods in rarefied gas dynamics, and situations where the conventional techniques of modeling and simulation are inadequate in their research fields.

This monograph is centered on the kinetic theory-based modeling and deterministic multi-scale simulation. Derivation of various Boltzmann equations and their mathematical properties, as well as the derivation of fluid-dynamic equations from the Boltzmann equation, are kept minimum, as they are elaborated in many other excellent books, such as those by Prof. Carlo Cercignani, Prof. Yoshio Sone, Prof. Henning Struchtrup, and the one entitled "An Introduction to the Boltzmann Equation and Transport Processes in Gases" by Prof. Gilberto M. Kremer. In fact, this monograph solves nearly all types of the Boltzmann equation in Kremer's book.

Major topics covered in this book include the following:

- the fast spectral method to solve the Boltzmann collision operator for dilute monatomic gas and the Enskog collision operator for dense granular gas;
- the general synthetic iterative scheme to solve the kinetic equations with the properties of fast convergence and asymptotic preserving;
- the kinetic modeling of monatomic and molecular gases, and the extraction of critical gas parameters from the experiment of Rayleigh-Brillouin scattering;
- the assessment of the fluid-dynamic equations derived from the Boltzmann equation and typical kinetic gas-surface boundary conditions;
- the applications of the fast spectral method and general synthetic iterative scheme to reveal the rarefied gas dynamics.

I tried to present the monograph in a concise manner, as more details can be found in our published papers. Nevertheless, to enhance the readers' understanding, sample MATLAB codes simulating the normal shock wave, Poiseuille flow, and thermal transpiration are provided. Anyone with basic knowledge of programming can understand and modify the codes, and solve more complicated problems on a desktop computer.

I sincerely thank my collaborators for their valuable contributions: Craig White, Jianping Meng, Tom Scanlon, Jun Zhang, Haihu Liu, Minhtuan Ho, Irina Graur, Henning Struchtrup, Lefki Germanou, Xiaojun Gu, Chang Liu, Kun Xu, Zhaoli Guo, Zhihui Li, Wei Su, Peng Wang, Lianhua Zhu, Felix Sharipov, Wei Liu, Chuang Zhang, Xingcai Pi, Weiqi Yang, Wim Ubachs, Qi Li, Yangyang Shi, Xiaowen Shan, Jianan Zeng, and Ruifeng Yuan. I acknowledge the financial supports from the University of Strathclyde and the Scottish Government, the Engineering and Physical Sciences Research Council in the United Kingdom, and the National Natural Science Foundation of China. Special thanks are given to Qi Li for the meticulous proof-reading.

I express my special thanks to my supervisor Prof. Jiefang Zhang, who led me into the academic career when I was taking the undergraduate and master's degree in the Zhejiang Normal University. His great personality, enthusiasm in research, and wide horizons have inspired and shaped me to take challenges and tackle difficulties in life and research. I am also grateful to my Ph.D. supervisors, Profs. Yonghao Zhang and Jason Reese, for giving me the freedom to choose research topics, as well as for creating a nice environment in research and collaboration in the James Weir Fluids Laboratory, especially the collaboration with Dr. Craig White allowed me not only to validate the fast spectral method, but also to develop a kinetic model for polyatomic gas.

Finally, I would like to thank my wife, Qing Tian, for her constant love, understanding, and support in the past decades starting from my undergraduate study. I also thank my children, Yutian and Liangzi, for their support; they often sat on my lap and wrote their very first fact books on my desktop; I hope one day they will have books of their own.

Shenzhen, China
March 2022

Lei Wu

Contents

Chapter 1
Introduction

Several engineering applications where the Navier-Stokes-Fourier equations fail to predict the behavior of rarefied gas flows are introduced, as well as the simple gas kinetic theory and molecular dynamics simulation.

1.1 Navier-Stokes-Fourier Equations

A macroscopic volume of gas is a system of a very large number of molecules moving constantly in a rather irregular way. This huge number of degrees of freedom makes the trace of motion of each molecule impossible. The fundamental and practical task in the study of gas dynamics is to obtain the macroscopic quantities, such as the mass density ρ, bulk velocity $\boldsymbol{u}$, temperature T, pressure tensor $\boldsymbol{p}$, and heat flux $\boldsymbol{q}$. Traditionally, the gas is treated as a continuum and the following equations are built based on the mass, momentum, and energy conservation:

$$\begin{aligned} \frac{\partial \rho}{\partial t}+\frac{\partial(\rho u_j)}{\partial x_j}&=0, \\ \frac{\partial(\rho u_i)}{\partial t}+\frac{\partial(\rho u_i u_j+p_{ij})}{\partial x_j}&=\rho a_i, \\ \frac{\partial(\rho E)}{\partial t}+\frac{\partial\left(\rho E u_j+u_i p_{ij}+q_j\right)}{\partial x_j}&=\rho a_j u_j, \end{aligned} \tag{1.1}$$

where t is the time, $\boldsymbol{x}\equiv(x_1,x_2,x_3)$ is the spatial Cartesian coordinates, $\boldsymbol{a}\equiv(a_1,a_2,a_3)$ is the external acceleration, and $E\equiv e+u_iu_i/2$ is the specific energy with e being the specific internal energy that is a function of temperature T. The subscripts $i,j,k=1,2,3$ represent the three orthogonal spatial directions, and the Einstein summation convention is used.

L. Wu, *Rarefied Gas Dynamics*,
https://doi.org/10.1007/978-981-19-2872-7_1

Equation (1.1) is not closed because expressions for the pressure tensor $\boldsymbol{p}$ and heat flux $\boldsymbol{q}$ cannot be determined from conservation laws. Usually, the phenomenological and empirical constitutive relations are adopted, e.g., Newton's law of viscosity

$$p_{ij} = \frac{\rho k_B T}{m}\delta_{ij} - \mu\left(\frac{\partial u_i}{\partial x_j} + \frac{\partial u_j}{\partial x_i} - \frac{2}{3}\frac{\partial u_k}{\partial x_k}\delta_{ij}\right) - \mu_b \frac{\partial u_k}{\partial x_k}\delta_{ij}, \tag{1.2}$$

and Fourier's law of heat conduction

$$q_i = -\kappa \frac{\partial T}{\partial x_i}, \tag{1.3}$$

where k_B is the Boltzmann constant, m is the molecular mass, δ is the Kronecker delta, μ is the shear viscosity, μ_b is the bulk viscosity, and κ is the thermal conductivity.

Together with the non-velocity-slip and non-temperature-jump conditions at the solid surface, the Navier-Stokes-Fourier (NSF) equations have found a wide range of engineering applications.

1.2 Continuum Breakdown

However, in some extreme conditions, the gas dynamics predicted by NSF equations can differ from experimental results significantly. The gas flow where the traditional NSF equations fail is called rarefied gas flow, and its dynamics is rarefied gas dynamics (RGD). It is also called non-equilibrium gas dynamics because, as we shall see in later chapters, the velocity distribution function of gas molecules departures from the Maxwell equilibrium distribution significantly. Sometimes it is also called non-continuum gas dynamics. This is because the basic assumption in continuum mechanics that "the stress at a point is related to the strain and the rate of change of strain with respect to time at the same point" is violated.

Typical examples of rarefaction effects that lead to non-intuitive phenomena beyond the description of NSF equations, with applications from the sky, earth surface, and underground, are briefly illustrated below.

1.2.1 Reentry of Space Vehicle

Figure 1.1 sketches the complex flow physics around a reentry vehicle, where the reentry speed is about 8 km/s. Due to the violent interaction with atmosphere, a strong bow shock wave is generated in front of the vehicle. Intermolecular collisions also promote energy exchanges among the translational and internal (rotational, vibrational, and electronic) modes, and eventually lead to non-equilibrium chemical reactions. Behind the shock wave, a massive amount of the kinetic energy of free

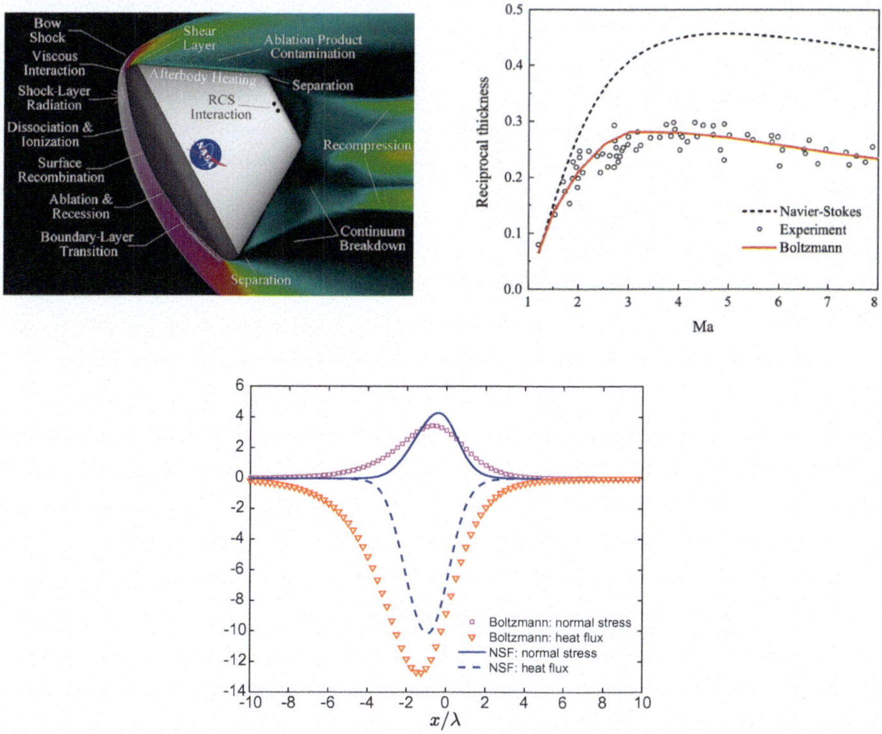

Fig. 1.1 (Left) Complicated gas dynamics around the reentry capsule (image credit: NASA). (Right) Inverse thickness of the normal shock wave of argon as a function of the Mach number [1]. (Bottom) Dimensionless normal stress and heat flux of argon shock wave with a Mach number of 4; the distance is normalized by the mean free path (λ) of the gas in upstream

stream is converted into the internal energy of surrounding gas, resulting in intense convective and radiative heating to the space vehicle. Meanwhile, supersonic expansion, flow separation, re-circulation, and re-compression take place as the flow passes through the vehicle.

Even for the simple normal shock wave, the NSF equations fail to predict its thickness when the Mach number is larger than 2. Although the thickness of shock wave is only a few mean free path (MFP, the average distance traveled by a moving molecule between collisions) of gas molecules, in high-altitude flight it is comparable to the dimension of space vehicle and hence the rarefaction effects are significant. On the contrary, the Boltzmann equation well predicts the thickness of normal shock wave over a wide range of Mach number. When the Mach number is 4, the normal stress and heat flux predicted from NSF equations are quite different from those of the Boltzmann equation, which further confirms the failure of NSF equations. In-depth knowledge of RGD is key to predicting/controlling the entry of space objects.

1.2.2 Microelectromechanical Systems

The rarefied gas flow at the micro-scale has a broad range of industrial applications and has attracted particular attention due to the rapid development of microelectromechanical systems [4]. At small length scale, and sometimes the high oscillation frequency, the NSF equations loss validity.

Figure 1.2 shows the sound resonance between the transducer and receiver, where the sound pressure at the receiver is a function of the oscillation frequency ϖ and propagation distance L. When ϖ is 0.3 times of the mean collision frequency of gas molecules $\bar{\nu}$,

$$\bar{\nu} = \frac{1}{\tau_c} = \frac{p}{\mu}, \tag{1.4}$$

where τ_c is the average time between successive collisions, the NSF equations fail to predict the sound pressure at the receiver; other high-order macroscopic equations, like the regularized 13 and 26 moment equations derived from the Boltzmann equation [5, 6], however, can well predict the sound pressure. When ϖ is increased to $1.09\bar{\nu}$, the difference between NSF predictions and experimental data further enlarges, and the regularized 13 and 26 moment equations fail too. However, the Boltzmann equation predicts the sound pressure well [3].

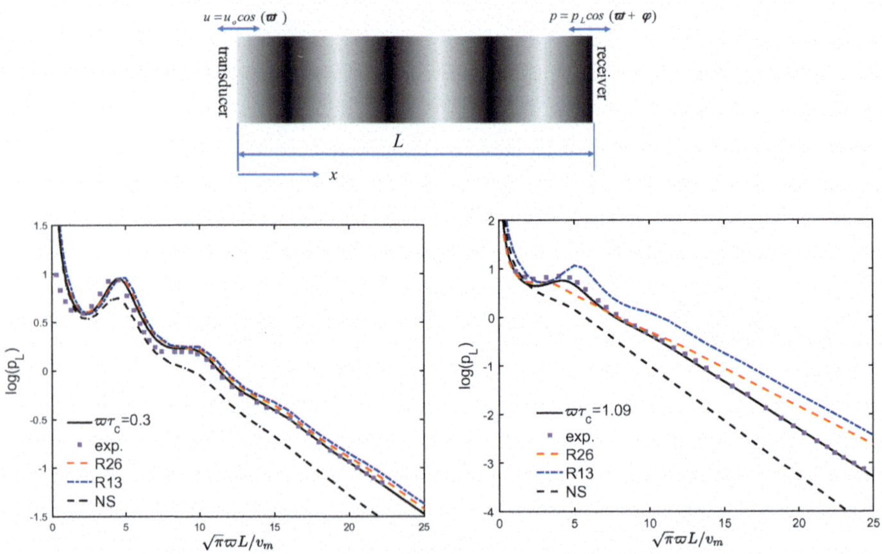

Fig. 1.2 (Top) Sound propagation between transducer and receiver. (Bottom) Dimensionless sound amplitude p_L at the receiver as a function of the dimensionless length $\sqrt{\pi}\varpi L/v_m$, where ϕ is the sound frequency and v_m is the most probable speed of gas molecules. Experimental data are collected from Ref. [2] when the gas medium is helium. Solid lines represent the numerical solutions of the Boltzmann equation with the diffuse boundary condition [3]. R13 and R26: the regularized 13 and 26 moment equations derived from the Boltzmann equation, which are more accurate than the NSF equations

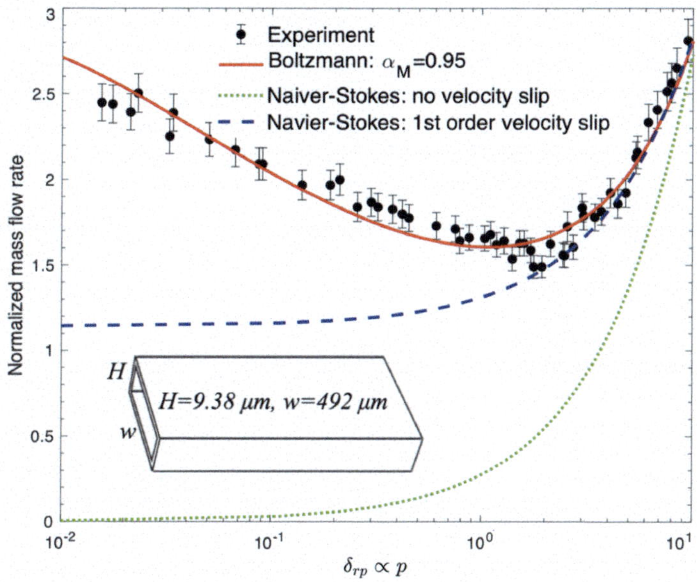

Fig. 1.3 The normalized mass flow rate as a function of the rarefaction parameter (1.12), for the gas flow through a long and narrow duct [7, 8]. In the numerical solution of the Boltzmann equation, α_M in the tangential momentum accommodation coefficient used in the Maxwell diffuse-specular boundary condition, that is, $\alpha_M = 0.95$ means that, when molecules hit the wall, 95% of them are diffusely reflected, while 5% are specularly reflected

1.2.3 Shale Gas Extraction

Unconventional gas reservoirs have attracted significant attention due to the shale gas revolution, but the accurate prediction of gas production is a grand research challenge. In shale gas exploitation, the gas pressure could reach hundred times of atmospheric pressure, but the flow passage is at the nanometer scale. Although the gas is dense [9, 10], its flow dynamics can still be rarefied.[1] Experimental data on gas flow through porous media is hard to measure and interpret, so we introduce the experiment of micro-gas flow to demonstrate the influence of rarefaction effects [7].

The duct shown in the inset of Fig. 1.3 has a cross-sectional dimension of $H \times w = 9.38\,\mu\text{m} \times 492\,\mu\text{m}$, which is connected to two large reservoirs at the ends. Given the inlet and outlet pressures (p_{in} and p_{out}, respectively), the mass flow rate (MFR, $\dot{M}$) is measured when the steady state of the flow is reached. Then the normalized MFR along the channel with length L

$$G = \frac{L\sqrt{2k_B T_0/m}}{H^2 w(p_{in} - p_{out})}\dot{M} \tag{1.5}$$

[1] The antonym of "dense gas" is "dilute gas", while the "rarefied gas" is referred to the gas in which the flow cannot be adequately described by NSF equations.

is plotted against the mean gas pressure p. At large gas pressures, the NSF equations predict the MFR well. However, when the gas pressure is reduced, the normalized MFR predicted by NSF equations always decreases, while that from the experiment first decreases and then increases, resulting in the famous phenomenon of "Knudsen minimum". At low pressures, the NSF equations could underpredict the MFR by several orders of magnitude; even with the correction of wall slip velocity, they underpredict the MFR by several times. The Boltzmann equation, however, can predict the MFR well in the whole range of gas rarefaction.

1.2.4 Global Wind Profiling

In 2018, the European Space Agency launched the satellite ADM-Aeolus to measure the global wind profile for an improved understanding of atmospheric dynamics. The ADM-Aeolus mission makes use of the Doppler wind Lidar measurement technique, and the spectrum of scattered light is important for the accurate retrieval of wind profile. Due to the high scattering frequency and short laser wavelength, as compared, respectively, to the mean collision frequency and mean free path of gas molecules, the gas dynamics is not in equilibrium, and the NSF equations have large error, see Fig. 1.4.

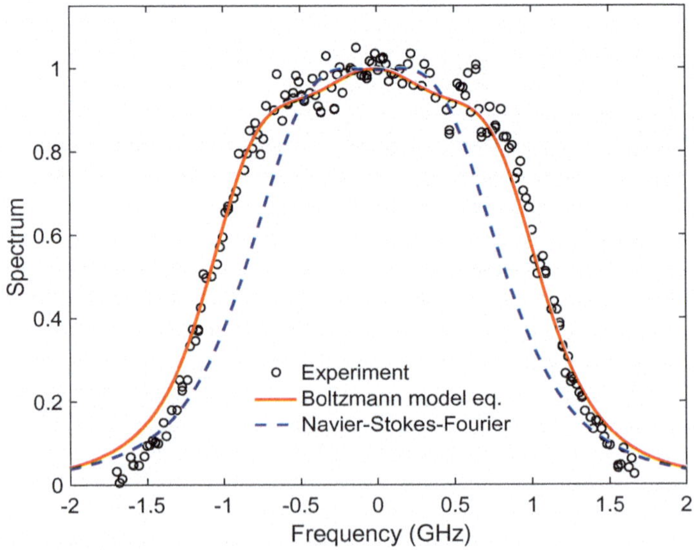

Fig. 1.4 Comparisons in the spectra of spontaneous Rayleigh-Brillouin scattering between the experimental data [11], the linearized Boltzmann equation [3] (see Chap. 15), and the NSF equations. The light is scattered from CO_2 with zero flow velocity, at a pressure of 770 mm of Hg. The Knudsen number (1.10) is $Kn = 0.14$

1.3 Simple Gas Kinetic Theory

The breakdown of NSF constitutive relations may be understood qualitatively from the simple kinetic theory, which not only allows the estimation of transport coefficients such as the shear viscosity, thermal conductivity, and self-diffusion coefficient, but also indicates the origin of rarefaction effects.

Consider a gas of hard spheres (HS) of diameter σ and number density n, traveling with the average speed $\bar{c}$. Imagine a gas molecule moves through the gas with the average relative speed $\sqrt{2}\bar{c}$, while other molecules are fixed. During the time interval Δt the moving molecule sweeps a volume of $\Delta V = \sqrt{2}\pi\sigma^2\bar{c}\Delta t$, which encounters collisions with $n\Delta V$ molecules. Hence, the average time between successive collisions is $\tau_c = 1/\sqrt{2}n\pi\sigma^2\bar{c}$, and the MFP is

$$\lambda = \bar{c}\tau_c = \frac{1}{\sqrt{2}n\pi\sigma^2}. \tag{1.6}$$

In simple kinetic theory, it is assumed that one-sixth of the molecules moves into each of the six directions of space, and within one mean free path there are no collisions among gas molecules (hence it can be derived from the ideal gas law $p = nk_BT$ that the average speed is $\bar{c} = \sqrt{3k_BT/m}$). Therefore, the number flux (the number of gas molecules per unit area per unit time) through a plane perpendicular to the x_1-axis is

$$J_1 = \frac{n(x_1-\lambda)}{6}\bar{c} - \frac{n(x_1+\lambda)}{6}\bar{c} \approx -\frac{\bar{c}}{3}\lambda\frac{dn}{dx_1}. \tag{1.7}$$

Similarly, the shear stress (momentum flux) and heat flux (due to the translational motion only) are

$$\begin{aligned}\sigma_{12} &= \frac{mn\bar{c}}{6}\left[u_2(x_1-\lambda) - u_2(x_1+\lambda)\right] \approx -\frac{mn\bar{c}}{3}\lambda\frac{\partial u_2}{\partial x_1},\\ q_1 &= \frac{mn\bar{c}}{6}\left[\frac{3}{2}\frac{k_B}{m}T(x_1-\lambda) - \frac{3}{2}\frac{k_B}{m}T(x_1+\lambda)\right] \approx -\frac{nk_B\bar{c}}{2}\lambda\frac{\partial T}{\partial x_1},\end{aligned} \tag{1.8}$$

therefore the diffusion coefficient, shear viscosity, and thermal conductivity are, respectively,

$$D = \frac{\bar{c}}{3}\lambda, \quad \mu = \frac{mn\bar{c}}{3}\lambda, \quad \kappa = \frac{nk_B\bar{c}}{2}\lambda. \tag{1.9}$$

It is clear from Eqs. (1.7) and (1.8) that these transport coefficients are derived only when the first-order Taylor expansion is used; this requires the fluid system varies slowly on a spatial scale of the size of MFP, otherwise higher order expansions are needed, and the constitutive relations will go beyond Newton's law of stress and Fourier's law of heat conduction. Therefore, the degree of rarefaction effects can be characterized by the ratio of MFP to the characteristic flow length in which the macroscopic quantities vary.

1.4 Knudsen Number

The RGD emerges in extreme conditions where the gas density is low, the characteristic flow length is small, and/or when the oscillation frequency of gas flow is high. This brings us to the introduction of dimensionless parameter, the Knudsen number, to approximately measure the degree of non-equilibrium.

1.4.1 Spatial Knudsen Number

The Knudsen number, defined as the ratio of the molecular MFP λ to the characteristic flow length L, is widely used:

$$\mathrm{Kn} = \frac{\lambda}{L}, \tag{1.10}$$

where the MFP can be defined in terms of the shear viscosity as

$$\lambda = \frac{\mu}{p}\sqrt{\frac{\pi k_B T_0}{2m}}, \tag{1.11}$$

rather than Eq. (1.6) that is derived from the simple kinetic theory based on HS molecular model. This is obvious because for other types of intermolecular potential it is difficult to determine the molecular diameter.

This Knudsen number is known as the spatial Knudsen number as spatial length scales are involved. Note that sometimes the rarefaction parameter, which is inversely proportional to the spatial Knudsen number, is used instead:

$$\delta_{rp} = \frac{pL}{\mu v_m} = \frac{\sqrt{\pi}}{2\mathrm{Kn}}, \tag{1.12}$$

where $v_m = \sqrt{2k_B T/m}$ is the most probable speed of gas molecules. This parameter appears in the kinetic model equations when the Boltzmann collision operator is simplified.

As shown in Fig. 1.5, the spatial Knudsen number is used to classify the gas flow into four regimes [12, 13]:

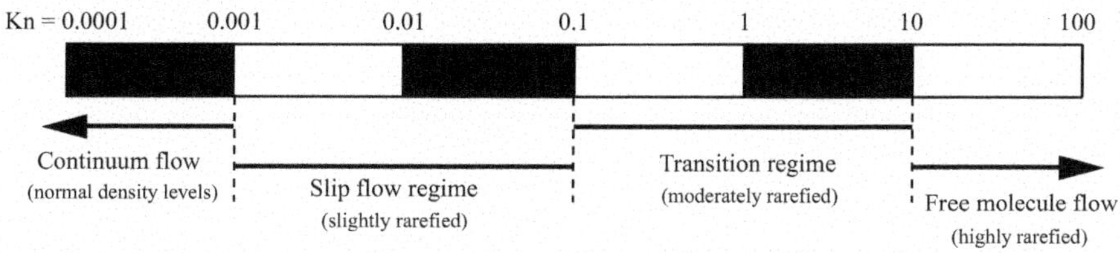

Fig. 1.5 Flow regime classification based on the spatial Knudsen number [12, 13]

- Continuum flow regime. When $\mathrm{Kn} \lesssim 0.001$, the NSF equations with the non-velocity-slip and non-temperature-jump boundary conditions are adopted to describe the gas dynamics.[2]
- Slip flow regime. When $0.001 \lesssim \mathrm{Kn} \lesssim 0.1$, weak rarefaction effects such as the velocity slip and temperature jump emerge in the Knudsen layer, i.e., a small region within a few MFP away from the solid wall. With appropriate velocity-slip and temperature-jump boundary conditions, the NSF equations may still produce reasonable results.
- Transition flow regime. When $0.1 \lesssim \mathrm{Kn} \lesssim 10$, strong rarefaction effects modify the constitutive relations as assumed in continuum mechanics so that the NSF equations break down completely, and gas kinetic equations, such as the Boltzmann equation, are used to describe rarefied gas flows [15].
- Free-molecular flow regime. When $\mathrm{Kn} \gtrsim 10$, the collision between gas molecules does not play an important role, while the collision between the gas and wall molecules (gas-surface interaction) is dominant.

Route to non-equilibrium—We take the planar Couette flow as an example to demonstrate how the rarefaction effects set in. Consider two parallel plates located at $x_2 = -L/2$ and $L/2$, moving in the x_1-direction with velocities $-u_w$ and u_w, respectively. The wall velocity is small, then the gas temperature equals the wall temperature T_w when the steady state is reached, hence the shear viscosity and heat conductivity are constant in the whole domain. This problem is effectively one dimensional (1D). According to the NSF equations, the velocity u_1 is linear and the heat flux q_1 is zero. Also, if the gas pressure remains unchanged, but the wall distance is decreased, the NSF equation predicts a shear stress:

$$p_{12} = -\mu \frac{2u_w}{L} = -\frac{4}{\sqrt{\pi}} \mathrm{Kn} \left(\frac{u_w}{v_m} p \right), \tag{1.13}$$

which eventually goes to infinity when $L \to 0$.

Figure 1.6 shows the numerical solution from the Boltzmann equation with the diffuse boundary condition. When $\mathrm{Kn} = 0.01$, the velocity profile is linear and the

---

[2] There are some exceptions, where the NSF equations do not describe the gas dynamic accurately even when Kn approaches zero. For example, the ghost effect induced by the periodical variation of wall temperature [14].

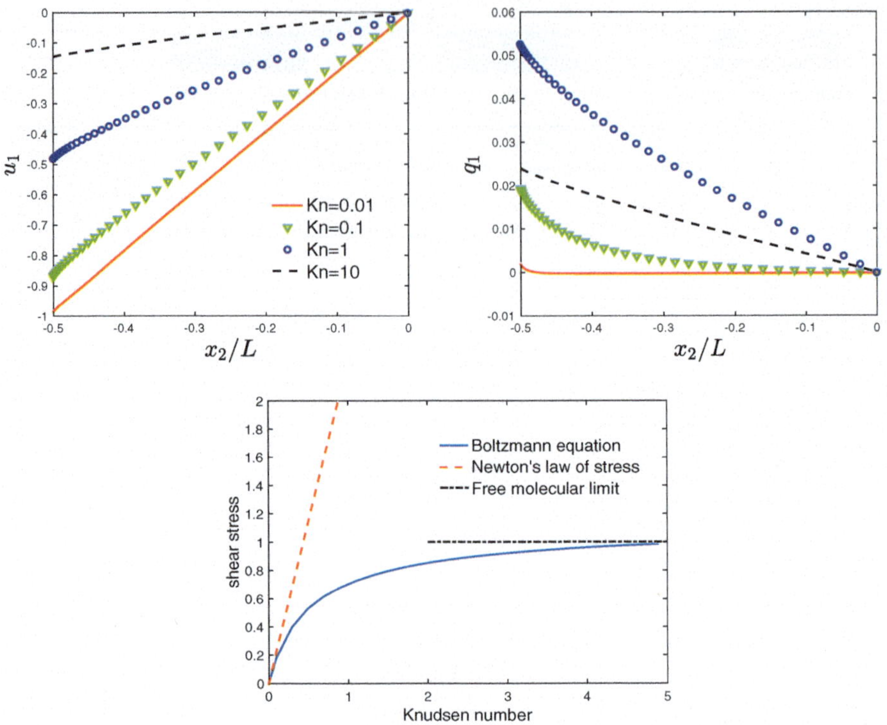

Fig. 1.6 The flow velocity (normalized by wall velocity), heat flux (normalized by the product of gas pressure and wall velocity), and shear stress (normalized by the product of gas pressure and dimensionless velocity $u_w/\sqrt{2k_BT_w/m}$) of the planar Couette flow at different values of spatial Knudsen number obtained from the linearized Boltzmann equation for HS gas with the diffuse boundary condition

flow velocity at the two plates is very close to the wall velocity. Also the heat flux is zero in the bulk region, as predicted by the NSF equations; only in the vicinity of plates can we see a very small heat flux, which is due to the rarefaction effect inside the Knudsen layer; in fact, if we zoom in the velocity profile, we shall find a tiny velocity defect there; this can be described by the Knudsen layer function in Chap. 10. When Kn $= 0.1$, the Knudsen layer stretches to the whole domain, and the heat flux is not zero, which shows that the Fourier law of heat conduction cannot describe this rarefaction effect. The flow velocity is, however, nearly linear, but there is a considerable slip at the wall. If a proper velocity-slip boundary condition is given, the velocity can be predicted by the NSF equations. When Kn $= 1$, the velocity profile is nonlinear, with increased velocity slip at the wall; also, the magnitude of heat flux increases significantly. In the free-molecular regime, the velocity and heat flux approach zero.

The shear stress at the wall is also shown. When Kn is small, Newton's law is accurate. When Kn > 0.2, large discrepancy between the Newton's law and prediction from the Boltzmann equation is clearly seen. Moreover, the shear stress does not approach infinity but saturates to a fixed value when $\text{Kn} \to \infty$, or when $L \to 0$. This phenomenon is also observed in the lid-driven cavity flow by the molecular dynamics simulation [16].

In this simple problem, if an effective shear viscosity is introduced, one may get the shear stress correct in the whole range of Knudsen number, but not the velocity profile. Moreover, one can never get the streamwise heat flux correct from the NSF equations, because there is no temperature gradient at all. To take into account these rarefaction effects, high-order constitutive relations are needed since, as we can see from Eq. (1.8), the linear constitutive relations are only valid when the fluid system varies slowly on a spatial scale of the size of MFP or, equivalently, when the spatial Knudsen number is small.

1.4.2 Temporal Knudsen Number

When the ratio of the characteristic flow frequency ϖ to the mean collision frequency of gas molecules $\bar{\nu} = 1/\tau_c$ becomes appreciable, the gas flow can also be in non-equilibrium, no matter what the spatial Knudsen number is. Thus, in addition to the spatial Knudsen number (1.10), it is necessary to define the temporal Knudsen number to measure the degree of non-equilibrium:

$$\text{Kn}_t = \varpi \tau_c = \varpi \frac{\mu}{p}. \tag{1.14}$$

Let us take the 1D oscillatory Couette flow to demonstrate the rarefaction effects. The same setting as the planar Couette flow in the previous section is used, but now the plate at $x_2 = -L/2$ oscillates in the x_1-direction with frequency ϖ and amplitude u_m. Adopting the Navier-Stokes equations with the first-order velocity-slip boundary condition, the analytical solution for the gas velocity is obtained as follows [17]:

$$\begin{aligned} u_1 =& u_w \exp(i\varpi t) \frac{\sin\left[(1+i)\delta_{rp}\sqrt{\text{Kn}_t}\left(\frac{1}{2}-\frac{x_2}{L}\right)\right]}{\sin\left[(1+i)\delta_{rp}\sqrt{\text{Kn}_t}\right]} \\ &\times \frac{1+(1+i)\sigma_P\sqrt{\text{Kn}_t}\cot\left[(1+i)\delta_{rp}\sqrt{\text{Kn}_t}\left(\frac{1}{2}-\frac{x_2}{L}\right)\right]}{1-2i\sigma_P^2\text{Kn}_t+2(1+i)\sigma_P\sqrt{\text{Kn}_t}\cot\left[(1+i)\delta_{rp}\sqrt{\text{Kn}_t}\right]}, \end{aligned} \tag{1.15}$$

where i is the imaginary unit and σ_P is the viscous slip coefficient. Here we take $\sigma_P = 1$; more sophisticated calculations will be given in Chap. 10. Note that the flow velocity is complex because it has a phase offset as compared to that of the oscillating plate.

By comparing the numerical solution of the Boltzmann equation with the analytical solution, we are able to classify the flow regime based on the temporal Knudsen number. To isolate the rarefaction effects associated with the spatial Knudsen number, we set $\text{Kn} = 0.001$. It is seen from Fig. 1.7 that the slip solution (1.15) is valid up

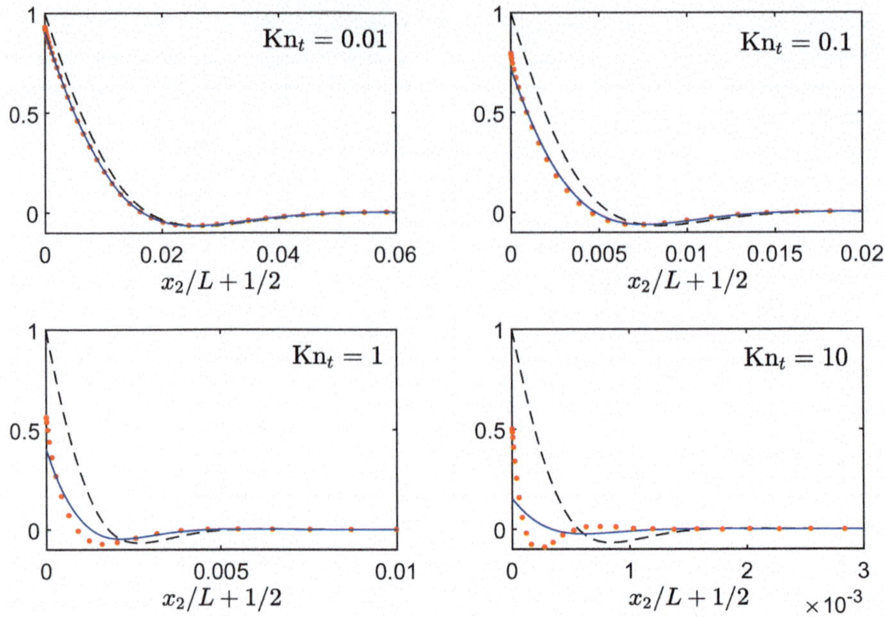

Fig. 1.7 The real part of the flow velocity (normalized by the oscillation amplitude u_w) at $t = 2m\pi/\varpi$ with $m = 0, 1, \ldots,$ at different values of the temporal Knudsen number. The spatial Knudsen number is fixed at Kn $= 0.001$. Dashed and solid lines are the solutions of the NSF equations without and with the first-order velocity-slip boundary conditions [17], respectively, while circles are solutions from the linearized Boltzmann equation for HS gas, with the diffuse boundary condition

to $\text{Kn}_t \approx 0.1$, while when $\text{Kn}_t = 10$ the velocity profile from the Boltzmann equation is very close to the collisionless limit [17], indicating that the flow has entered the free-molecular regime. Therefore, we conclude that Fig. 1.5 (Kn is replaced by Kn_t) can also be used to classify the four flow regimes in terms of the temporal Knudsen number, when the spatial Knudsen number is in the continuum flow regime. When both spatial and temporal Knudsen numbers are not small, stronger rarefaction effects are expected.

1.5 Molecular Dynamics Simulations

Alternative methods should be adopted to study the RGD due to the breakdown of NSF equations in extreme conditions. Physically, when the intermolecular potential ϕ is known, the molecular dynamics (MD) simulations, which is based on Newton's second law of motion, can be adopted to trace the movement of individual molecules:

$$m\frac{d^2\boldsymbol{x}_i}{dt^2} = \sum_j \boldsymbol{F}_{ij}, \quad \text{with } \boldsymbol{F}_{ij} = -\nabla\phi(r_{ij}), \tag{1.16}$$

where $\boldsymbol{x}_i$ is the position of the ith molecule, ∇ is the gradient operator, and r_{ij} is the distance between the ith and jth molecules.

When the position and velocity $\boldsymbol{v}$ of individual molecule are known, macroscopic quantities can be obtained through ensemble average. Suppose there are N molecules inside a volume V, the flow velocity $\boldsymbol{u}$ and temperature T are given by

$$\boldsymbol{u} = \frac{1}{N}\sum_{i=1}^{N} \boldsymbol{v}_i, \tag{1.17}$$

$$T = \frac{m}{3Nk_B}\sum_{i=1}^{N} c^2, \tag{1.18}$$

where

$$\boldsymbol{c} = \boldsymbol{v} - \boldsymbol{u} \tag{1.19}$$

is the peculiar velocity and its magnitude is denoted by c. Independent runs for transient flows or time averaging for steady flows are needed to reduce the intrinsic noise in particle-based simulations.

Transport coefficients such as the viscosity and heat conductivity that appear in the NSF equations can also be calculated in MD simulations, via the Green-Kubo formulas [18, 19]:

$$\mu = \frac{1}{V}\frac{1}{k_B T}\int_0^\infty \left\langle J^p_{xy}(0)J^p_{xy}(\tau)\right\rangle d\tau, \tag{1.20}$$

$$\frac{4}{3}\mu + \mu_b = \frac{1}{V}\frac{1}{k_B T}\int_0^\infty \left\langle J^p_{xx}(0)J^p_{xx}(\tau)\right\rangle d\tau, \tag{1.21}$$

$$\kappa = \frac{1}{V}\frac{1}{k_B T^2}\int_0^\infty \left\langle J^q_x(0)J^q_x(\tau)\right\rangle d\tau, \tag{1.22}$$

where the brackets $\langle\cdots\rangle$ denote the canonical ensemble average, $\boldsymbol{J}^p = m\boldsymbol{c}_i\boldsymbol{c}_i - pV\boldsymbol{I}$ and $\boldsymbol{J}^q = E_i\boldsymbol{c}_i$, with $\boldsymbol{I}$ being the identity matrix. For monatomic gas, $E_i = mc_i^2/2$ and κ is the translational thermal conductivity κ_{tr}. For molecular gas, suppose $E_i = \epsilon_i$ is the internal energy of the ith molecule, then Eq. (1.22) calculates the internal thermal conductivity κ_i; together with the translational thermal conductivity they form the total thermal conductivity κ.

The MD simulation is also an indispensable tool to investigate the gas-surface interaction, which, together with the gas-gas interaction, determines the gas dynamics in wall-bounded problems. However, since it is limited to small spatial and temporal scales, the MD simulation cannot be used efficiently in rarefied gas flow simulations. Therefore, the gas kinetic theory, a mesoscopic theory that bridges the physics at the microscopic level and flow dynamics (phenomenon) at the macroscopic level, is needed.

References

1. Alsmeyer A (1976) Density profiles in argon and nitrogen shock waves measured by the absorption of an electron beam. J Fluid Mech 74:497–513
2. Schotter R (1974) Rarefied gas acoustics in the noble gases. Phys Fluids 17:1163
3. Wu L, Gu XJ (2020) On the accuracy of macroscopic equations for linearized rarefied gas flows. Adv Aerodyn 2:2
4. Karniadakis G, Beskok A, Aluru NR (2005) Microflows and manoflows: fundamentals and simulations. Springer, New York
5. Struchtrup H (2005) Macroscopic transport equations for rarefied gas fows: approximation methods in kinetic theory. Springer, Heidelberg
6. Gu XJ, Emerson DR (2009) A high-order moment approach for capturing non-equilibrium phenomena in the transition regime. J Fluid Mech 636:177–216
7. Ewart T, Perrier P, Graur IA, Méolans JG (2007) Mass flow rate measurements in a microchannel, from hydrodynamic to near free molecular regimes. J Fluid Mech 584:337–356
8. Wu L, Struchtrup H (2017) Assessment and development of the gas kinetic boundary condition for the Boltzmann equation. J Fluid Mech 823:511–537
9. Wu L, Zhang YH, Reese JM (2015) Fast spectral solution of the generalized Enskog equation for dense gases. J Comput Phys 303:66–79
10. Wu L, Liu HH, Reese JM, Zhang YH (2016) Non-equilibrium dynamics of dense gas under tight confinement. J Fluid Mech 794:252–266
11. Greytak TJ, Benedek GB (1966) Spectrum of light scattered from thermal fluctuations in gases. Phys Rev Lett 17:179–182
12. Tsien HS (1946) Superaerodynamics, mechanics of rarefied gases. J Aero Sci 13:653–664
13. Gad-el-Hak M (1999) The fluid mechanics of microdevices - the Freeman Scholar lecture. J Fluids Eng 121(1):5–33
14. Sone Y (2002) Kinetic theory and fluid dynamics. Birkhauser Boston
15. Chapman S, Cowling TG (1970) The mathematical theory of non-uniform gases. Cambridge University Press, Cambridge
16. Nie XB, Robbins MO, Chen SY (2006) Resolving singular forces in cavity flow: Multiscale modeling from atomic to millimeter scales. Phys Rev Lett 94:134501
17. Sharipov F, Kalempa D (2008) Oscillatory Couette flow at arbitrary oscillation frequency over the whole range of the Knudsen number. Microfluid Nanofluid 4:363–374
18. Green MS (1954) Markoff random processes and the statistical mechanics of time-dependent phenomena. ii. irreversible processes in fluids. J Chem Phys 22:398–413
19. Kubo R (1957) Statistical-mechanical theory of irreversible processes. i. general theory and simple applications to magnetic and conduction problems. J Phys Soc Jpn 12:570–586

Chapter 2
Gas Kinetic Theory

The Boltzmann equation that describes the dynamics of dilute monatomic gas, the Wang-Chang and Uhlenbeck equation for molecular (diatomic and polyatomic) gas, as well as their heuristic extension to dense/granular gas, are introduced, together with the gas-surface boundary conditions and typical numerical methods in rarefied gas dynamics.

2.1 Velocity Distribution Function

In the gas kinetic theory, the velocity distribution function (VDF), $f(t, \boldsymbol{x}, \boldsymbol{v})$, is introduced to describe the state of the gas, where $\boldsymbol{v} = (v_1, v_2, v_3)$ is the molecular velocity. The molecular number in the phase-space volume $d\boldsymbol{x}d\boldsymbol{v}$ is

$$f(t, \boldsymbol{x}, \boldsymbol{v})d\boldsymbol{x}d\boldsymbol{v}, \tag{2.1}$$

and macroscopic quantities are defined as the moments of VDF, e.g., the number density n, bulk velocity $\boldsymbol{u}$, temperature T, pressure tensor p_{ij}, and heat flux $\boldsymbol{q}$ are given by

$$\left[n, n\boldsymbol{u}, \frac{3}{2}nk_BT, p_{ij}, \boldsymbol{q}\right] = \int \left[1, \boldsymbol{v}, \frac{m}{2}c^2, mc_ic_j, \frac{m}{2}c^2\boldsymbol{c}\right] f d\boldsymbol{v}. \tag{2.2}$$

Note that the equation of state for dilute gas is described by the ideal gas law, where the gas pressure is $p = nk_BT$. Also, the deviatoric stress tensor[1] σ_{ij} is defined as

[1] Note that while the stress in continuum mechanics is understood to be the force per unit area that the part lying on the positive side of a surface element exerts on the part lying on the negative side, in this book the stress has an opposite sign to that in continuum mechanics.

L. Wu, *Rarefied Gas Dynamics*,
https://doi.org/10.1007/978-981-19-2872-7_2

$$\sigma_{ij} = p_{ij} - p\delta_{ij} = m \int c_{\langle i}c_{j\rangle} f d\boldsymbol{v}, \tag{2.3}$$

where indices in angular brackets denote the trace-free part of a symmetric tensor; in this case, $c_{\langle i}c_{j\rangle} = \frac{1}{2}\left(c_i c_j + c_j c_i - \frac{2}{3}c^2\delta_{ij}\right) = c_i c_j - \frac{c^2}{3}\delta_{ij}$. Similarly, Eq. (1.2) can be simplified to $p_{ij} = \frac{\rho k_B T}{m}\delta_{ij} - 2\mu\frac{\partial u_{\langle i}}{\partial x_{j\rangle}} - \mu_b \frac{\partial u_k}{\partial x_k}\delta_{ij}$.

2.2 Binary Collision

Before introducing the Boltzmann collision operator (BCO), it is better to know the collision dynamics between gas molecules. Consider the collision of two molecules of equal mass m sketched in Fig. 2.1. The pre-collision velocities are $\boldsymbol{v}$ and $\boldsymbol{v}_*$, respectively, while the corresponding post-collision velocities are $\boldsymbol{v}'$ and $\boldsymbol{v}'_*$. These velocities are measured when the two molecules are too far away that their interaction potential is negligible. Conservation of momentum and energy yield the following relations:

$$\begin{aligned} \boldsymbol{v}' &= \frac{\boldsymbol{v}+\boldsymbol{v}_*}{2} + \frac{|\boldsymbol{v}-\boldsymbol{v}_*|}{2}\boldsymbol{\Omega} = \boldsymbol{v} + \frac{v_r\boldsymbol{\Omega} - \boldsymbol{v}_r}{2}, \\ \boldsymbol{v}'_* &= \frac{\boldsymbol{v}+\boldsymbol{v}_*}{2} - \frac{|\boldsymbol{v}-\boldsymbol{v}_*|}{2}\boldsymbol{\Omega} = \boldsymbol{v}_* - \frac{v_r\boldsymbol{\Omega} - \boldsymbol{v}_r}{2}, \end{aligned} \tag{2.4}$$

where $\boldsymbol{v}_r = \boldsymbol{v} - \boldsymbol{v}_*$ is the relative velocity before collision and $\boldsymbol{\Omega}$ is a unit vector in the sphere $\mathbb{S}^2$ along the relative velocity $\boldsymbol{v}' - \boldsymbol{v}'_*$ after collision. The deflection angle θ between the pre- and post-collision relative velocities satisfies $\cos\theta = \boldsymbol{\Omega}\cdot\boldsymbol{v}_r/v_r$, with $0 \le \theta \le \pi$. The post-collision velocities can also be related to the pre-collision molecular velocities through

$$\begin{aligned} \boldsymbol{v}' &= \boldsymbol{v} - (\boldsymbol{v}_r\cdot\boldsymbol{k})\boldsymbol{k}, \\ \boldsymbol{v}'_* &= \boldsymbol{v}_* + (\boldsymbol{v}_r\cdot\boldsymbol{k})\boldsymbol{k}, \end{aligned} \tag{2.5}$$

where the unit vector $\boldsymbol{k}$ is expressed as $\boldsymbol{k} = (\boldsymbol{v}_r - v_r\boldsymbol{\Omega})/|\boldsymbol{v}_r - v_r\Omega|$.

2.2.1 Deflection Angle

Provided that the intermolecular potential $\phi(r)$ is known, the defection angle of binary collision can be obtained either classically or quantum mechanically [2]. The total cross section calculated from the classical physics diverges, while that from the quantum mechanics converges. However, in gas dynamics, the important quantity is the viscosity cross section, and when the temperature is not too low, both classical and quantum calculations yield same viscosity cross section.

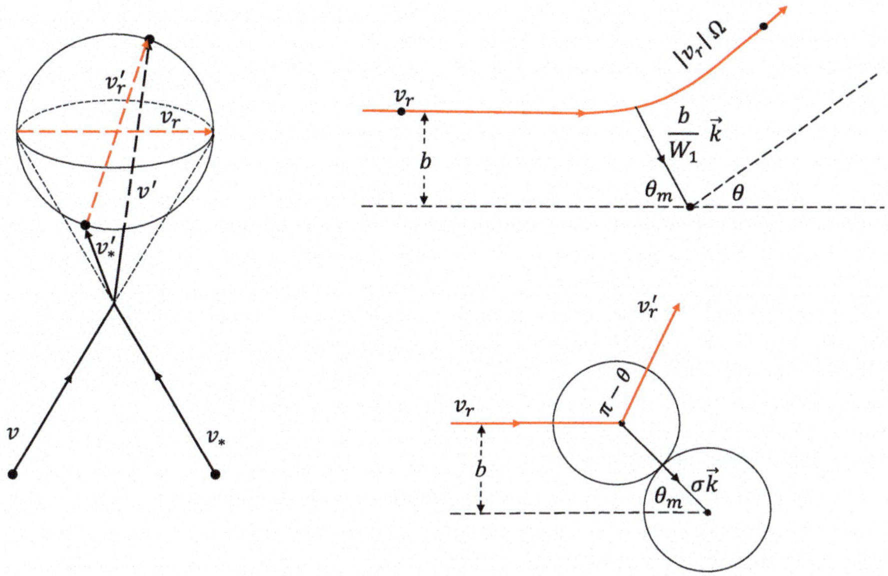

Fig. 2.1 (Left) Velocity redistribution after binary collision. Due to the conservation of momentum and energy, the pre- and post-collision relative velocities fall in a sphere and pass through the sphere center. (Top right) Classical binary collision between molecules with central force, where b/W_1 is the minimum distance between two colliding molecules. (Bottom right) Binary collision between HS molecules. b is the aiming distance, σ is the diameter of the HS, and $\boldsymbol{k}$ is a unit vector along the minimum distance between two molecules [3]

Here we adopt the classical physics to describe the collision dynamics. It is convenient to use the Binet equation in the polar coordinates system (r, θ),

$$\frac{d^2W}{d\theta^2} + W = \frac{2b}{mv_r^2W^2}\frac{d\phi}{dr}, \tag{2.6}$$

where $W = b/r$ with b being the aiming distance. At $r \to \infty$, we have $\theta \to \pi$ and $W = \sin\theta$. Therefore $dW/d\theta = 1$. On multiplying the Binet equation by $2\frac{dW}{d\theta}$ and integrating, we have $\frac{dW}{d\theta} = \sqrt{1 - W^2 - \frac{4\phi}{mv_r^2}}$. Therefore, the changed angle from $r = \infty$ to the minimum distance between two colliding particles is $\theta_m(b, v_r) = \int_0^{W_1}\left[1 - W^2 - \frac{4\phi(r)}{mv_r^2}\right]^{-1/2} dW$, where W_1 is the positive root of the term in the brackets. Hence, the deflection angle is

$$\theta(b, v_r) = \pi - 2\theta_m = \pi - 2\int_0^{W_1}\left[1 - W^2 - \frac{4\phi(r)}{mv_r^2}\right]^{-1/2} dW. \tag{2.7}$$

Note that the integrand approaches infinity when $W \to W_1$. Therefore, it is first integrated from zero to $W_1 - \epsilon$, and then from $W_1 - \epsilon$ to W_1, where ϵ is a small parameter. Introducing $y = W_1 - W$, and with the help of first-order Taylor expansion, the latter integral is integrated analytically as $2\sqrt{\epsilon}\left[2W_1 - \frac{4b}{mv_r^2 W_1^2}\frac{d\phi}{dr}|_{r=b/W_1}\right]^{-1/2}$, whereas the formal can be calculated by the adaptive numerical quadrature [1].

2.2.2 Differential Cross Section

In a real scattering experiment, information about the scatterer can be figured out from the different rates of scattering to different angles. This is related to the differential cross section σ_D, which is the fraction of the total number of scattered molecules that fall in the solid angle $d\Omega = \sin\theta d\theta d\phi$, where ϕ is the azimuthal angle. A molecule coming in with the aiming distance between b and $b + db$ will be scattered through an angle between θ and $\theta + d\theta$. Therefore, the differential cross section is

$$\sigma_D = \frac{b|db|}{\sin\theta|d\theta|}, \tag{2.8}$$

which is always non-negative and depends on the relative speed v_r and the deflection angle θ.

In gas kinetic theory, it is convenient and beneficial to study the inverse power-law potentials (where K indicates the potential strength)

$$\phi(r) = \frac{K}{\eta - 1} r^{1-\eta}, \tag{2.9}$$

although the Lennard-Jones potential is widely used (say, in the MD simulation of monatomic gas):

$$\phi(r) = 4\epsilon\left[\left(\frac{d_{LJ}}{r}\right)^{12} - \left(\frac{d_{LJ}}{r}\right)^{6}\right], \tag{2.10}$$

where ϵ is the potential depth and d_{LJ} is the distance at which the potential between two molecules is zero. When $\eta > 5$ ($\eta < 5$), the power-law potential is called hard (soft) potential. Maxwellian molecules have the potential with $\eta = 5$ and the Coulomb potential has $\eta = 2$; in reality, however, due to the Debye shielding, charged molecules interact through the shielded Coulomb potential

$$\phi(r) = \epsilon\frac{\lambda_d}{r}\exp\left(-\frac{r}{\lambda_d}\right), \tag{2.11}$$

where λ_d is the Debye shielding length. Hence, the shielded Coulomb potential has an effective $\eta > 2$ when compared to the power-law potential.

For the power-law potential, from Eq. (2.7), we know that the deflection angle is a function of

$$s = \left[\frac{m(\eta-1)}{4K}\right]^{\frac{1}{\eta-1}} b v_r^{\frac{2}{\eta-1}}. \tag{2.12}$$

That is, $\theta = \Theta(s)$. Thus, the differential cross section is

$$\sigma_D = \left[\frac{m(\eta-1)}{4K}\right]^{\frac{2}{1-\eta}} v_r^{-\frac{4}{\eta-1}} \times \frac{sds}{\sin\Theta d\Theta}. \tag{2.13}$$

In the next section, it will be shown that the collision kernel $B(\cos\theta, v_r)$, which is at the heart of the BCO, is

$$B(\cos\theta, v_r) = v_r \sigma_D. \tag{2.14}$$

Therefore, for Maxwell molecules, the collision kernel is a function of the deflection angle only. This allows easy evaluation of many problems, e.g., the exact derivation of transport coefficients through the Chapman-Enskog expansion, as well as the eigenvalues and eigenfunctions of the linearized Boltzmann equation [4].

For HS molecules with molecular diameter σ, Fig. 2.1 shows that the deflection angle satisfies the relation $b = \sigma\cos(\theta/2)$. Hence, the differential cross section and the collision kernel are, respectively,

$$\begin{aligned} \sigma_D &= \frac{\sigma^2}{4}, \\ B(\cos\theta, v_r) &= \frac{\sigma^2 v_r}{4}. \end{aligned} \tag{2.15}$$

2.2.3 *Grazing Collision*

When the aiming distance b and/or the relative speed v_r is large, the deflection angle approaches zero. In this case, the parameter s in Eq. (2.12) is large, hence the upper limit of the integral in Eq. (2.7) can be approximated as $W_1 \approx 1 - s^{1-\eta}/2$. Introducing $y = W/W_1$ and applying the first-order Taylor expansion, the deflection angle (2.7) is evaluated as follows:

$$\begin{aligned} \theta(b, v_r) =& \pi - 2W_1 \int_0^1 \left[1 - y^2 + s^{1-\eta}(y^2 - y^{\eta-1})\right]^{-1/2} dy \\ \approx& \pi - 2W_1 \int_0^1 \left[1-y^2\right]^{-1/2} \left[1 - \frac{s^{1-\eta}}{2}\frac{y^2 - y^{\eta-1}}{1-y^2}\right] dy \\ =& \frac{\pi}{2} s^{1-\eta} + s^{1-\eta} \int_0^1 \frac{y^2 - y^{\eta-1}}{(1-y^2)^{\frac{3}{2}}} dy \\ =& O(s^{1-\eta}). \end{aligned} \tag{2.16}$$

Thus, the differential cross section at small deflection angle is

$$\sigma_D \propto v_r^{-\frac{4}{\eta-1}} \theta^{\frac{2\eta}{1-\eta}} \tag{2.17}$$

and the total cross section $\sigma_T = 2\pi \int_o^\pi \sin(\theta)\sigma_D d\theta$ for inverse power potential diverges. This, however, does not pose too much problems since it will be shown in Sect. 3.2 that the viscosity cross section

$$\sigma_\mu = 2\pi \int_0^\pi \sin^3 \theta \sigma_D d\theta, \tag{2.18}$$

which determines the transport coefficients, such as the viscosity and thermal conductivity, are much more important. It can be easily shown that σ_μ is finite as long as $\eta > 2$. When $\eta = 2$ (Coulomb potential), the effective potential of charged particles will be the shielded Coulomb potential (2.11), and the corresponding transport coefficients are also finite.

The differential cross section calculated by the quantum mechanics is finite, and when the temperature is high, σ_μ calculated from the classical and quantum mechanics are very close, say, for helium when the temperature is above 100 K [5]; therefore, the classical mechanics can be safely used. At low temperature, the quantum scattering gives more accurate results [2, 6].

2.3 Boltzmann Equation

The governing equation for the VDF of a dilute gas is derived by Ludwig Boltzmann. In his description, molecules move in straight lines with fixed velocities until they encounter elastic collisions with other molecules. This is justified by the fact that, at standard temperature and pressure, the MFP of gas molecules is hundreds times the nominal atomic diameter and about 30 times the average molecular separation.

The molecular streaming and collision change the VDF as per the Boltzmann equation [4, 7]:

$$\frac{\partial f}{\partial t} + \boldsymbol{v} \cdot \frac{\partial f}{\partial \boldsymbol{x}} + \boldsymbol{a} \cdot \frac{\partial f}{\partial \boldsymbol{v}} = Q(f, f_*), \tag{2.19}$$

where the first term on the left-hand side describes the change of VDF with respect to time, the second term is the convective change, and the third term represents the change of VDF induced by external acceleration (suppose it is independent of the molecular velocity). They together describe the streaming of gas molecules.

The quadratic collision operator $Q(f, f_*)$ describes the change of molecular numbers per unit phase-space volume $d\boldsymbol{x}d\boldsymbol{v}$ and per unit time. This change consists of two effects. First, when the molecule with velocity $\boldsymbol{v}$ collides with another molecule with velocity $\boldsymbol{v}_*$, its velocity becomes $\boldsymbol{v}'$, which contributes to the loss of molecules with the very velocity $\boldsymbol{v}$ in the phase space. During time interval Δt, there are

$$f(t, \boldsymbol{x}, \boldsymbol{v}_*)v_r \Delta t b d b d\phi d\boldsymbol{v}_* \tag{2.20}$$

such collisions. Therefore, the number of molecule lost in the binary collision per unit phase-space volume and per time is

$$\begin{aligned} Q_{\text{loss}} &= \int_{\mathbb{R}^3}\int_{\mathbb{S}^2} \frac{f(t, \boldsymbol{x}, \boldsymbol{v})d\boldsymbol{x}d\boldsymbol{v} \times f(t, \boldsymbol{x}, \boldsymbol{v}_*)v_r \Delta t b d b d\phi d\boldsymbol{v}_*}{d\boldsymbol{x}d\boldsymbol{v}\Delta t} \\ &= \int_{\mathbb{R}^3}\int_{\mathbb{S}^2} \frac{f(t, \boldsymbol{x}, \boldsymbol{v})d\boldsymbol{x}d\boldsymbol{v} \times f(t, \boldsymbol{x}, \boldsymbol{v}_*)v_r \Delta t \sigma_D \sin\theta d\theta d\phi d\boldsymbol{v}_*}{d\boldsymbol{x}d\boldsymbol{v}\Delta t} \\ &= \int_{\mathbb{R}^3}\int_{\mathbb{S}^2} B(\cos\theta, v_r) f(t, \boldsymbol{x}, \boldsymbol{v}_*) f(t, \boldsymbol{x}, \boldsymbol{v}) d\Omega d\boldsymbol{v}_*. \end{aligned} \tag{2.21}$$

Second, when the molecule with velocity $\boldsymbol{v}'$ collides with another molecule with the velocity $\boldsymbol{v}'_*$, its velocity becomes $\boldsymbol{v}$, which contributes to the gain of molecules with the very velocity $\boldsymbol{v}$ in the phase space. Therefore, with the facts that $d\boldsymbol{v}d\boldsymbol{v}_* = d\boldsymbol{v}'d\boldsymbol{v}'_*$, $v_r = v'_r$, and the collision kernel is only determined by the relative collision speed and aiming distance, the gain part of the BCO is

$$\begin{aligned} Q_{\text{gain}} &= \int_{\mathbb{R}^3}\int_{\mathbb{S}^2} \frac{f(t, \boldsymbol{x}, \boldsymbol{v}')d\boldsymbol{x}d\boldsymbol{v}' \times f(t, \boldsymbol{x}, \boldsymbol{v}'_*)v'_r \Delta t b d b d\phi d\boldsymbol{v}'_*}{d\boldsymbol{x}d\boldsymbol{v}\Delta t} \\ &= \int_{\mathbb{R}^3}\int_{\mathbb{S}^2} B(\cos\theta, v_r) f(t, \boldsymbol{x}, \boldsymbol{v}'_*) f(t, \boldsymbol{x}, \boldsymbol{v}') d\Omega d\boldsymbol{v}_*. \end{aligned} \tag{2.22}$$

Finally, the BCO is written in the following form:

$$Q(f, f_*) = \int_{\mathbb{R}^3}\int_{\mathbb{S}^2} B(\cos\theta, v_r)[f(\boldsymbol{v}'_*)f(\boldsymbol{v}') - f(\boldsymbol{v}_*)f(\boldsymbol{v})]d\Omega d\boldsymbol{v}_*, \tag{2.23}$$

where t and $\boldsymbol{x}$ are omitted since the binary collision is local in time and space. Note that the "Stosszahlansatz" or assumption of molecular chaos was used implicitly, that is, the value of VDF for different velocities is independent.

2.3.1 H-Theorem

The BCO has the following symmetry property[2] for arbitrary function $\Psi(\boldsymbol{v})$ of molecular velocity:

[2] Consider the integral $S = \int \Psi(\boldsymbol{v})Q(\boldsymbol{v})d\boldsymbol{v}$. If we exchange the variables $\boldsymbol{v} \Longleftrightarrow \boldsymbol{v}_*$ and $\boldsymbol{v}' \Longleftrightarrow \boldsymbol{v}'_*$, we have $S = \int \Psi(\boldsymbol{v}_*)Q(\boldsymbol{v})d\boldsymbol{v}$. If we exchange the variables $(\boldsymbol{v}, \boldsymbol{v}_*) \Longleftrightarrow (\boldsymbol{v}', \boldsymbol{v}'_*)$, and take into account the relation $d\boldsymbol{v}d\boldsymbol{v}_* = d\boldsymbol{v}'d\boldsymbol{v}'_*$, we have $S = -\int \Psi(\boldsymbol{v}')Q(\boldsymbol{v})d\boldsymbol{v}$. If we exchange the variables $(\boldsymbol{v}, \boldsymbol{v}_*) \Longleftrightarrow (\boldsymbol{v}'_*, \boldsymbol{v}')$, we have $S = -\int \Psi(\boldsymbol{v}'_*)Q(\boldsymbol{v})d\boldsymbol{v}$. The arithmetic average of the four S leads to Eq. (2.24). Similarly, we have $\int \Psi(\boldsymbol{v})Q(\boldsymbol{v})d\boldsymbol{v} = \iiint \left[\Psi(\boldsymbol{v}') - \Psi(\boldsymbol{v})\right] B(\cos\theta, v_r) f(\boldsymbol{v}_*) f(\boldsymbol{v}) d\Omega d\boldsymbol{v}_* d\boldsymbol{v}$.

$$\int \Psi(\boldsymbol{v})Q(\boldsymbol{v})d\boldsymbol{v} = \frac{1}{4}\iiint d\Omega d\boldsymbol{v}_* d\boldsymbol{v}\left[\Psi(\boldsymbol{v}_*) + \Psi(\boldsymbol{v}) - \Psi(\boldsymbol{v}'_*) - \Psi(\boldsymbol{v}')\right]$$
$$\times B(\cos\theta, v_r)\left[f(\boldsymbol{v}'_*)f(\boldsymbol{v}') - f(\boldsymbol{v}_*)f(\boldsymbol{v})\right]. \tag{2.24}$$

Therefore, if $\Psi = 1, \boldsymbol{v}, v^2$, with the help of Eq. (2.4) we know $\int \Psi(\boldsymbol{v})Qd\boldsymbol{v} = 0$. Any function Ψ satisfying this equation is called collisional invariant. Clearly, 1, $\boldsymbol{v}$ and v^2 are the five fundamental collisional invariants, which corresponds to the three conservation laws; linear combinations of them are also collisional invariants.

Let H be the complete integral (i.e., the integral over all values of the molecular velocity and spatial coordinates) defined by the equation

$$H = -\int f \ln f d\boldsymbol{v}d\boldsymbol{x}, \tag{2.25}$$

which is a scalar related to the entropy of gas system. Without external force, the time derivative of H satisfies

$$\begin{aligned}\frac{dH}{dt} &= -\int (1 + \ln f)\frac{\partial f}{\partial t}d\boldsymbol{v}d\boldsymbol{x}\\ &= \int \boldsymbol{v}\cdot\frac{\partial f \ln f}{\partial \boldsymbol{x}}d\boldsymbol{v}d\boldsymbol{x} - \int (1 + \ln f)Qd\boldsymbol{v}d\boldsymbol{x}\\ &= \oint \boldsymbol{v}\cdot\boldsymbol{n}f \ln f d\boldsymbol{v}dS - \int Q \ln f d\boldsymbol{v}d\boldsymbol{x},\end{aligned} \tag{2.26}$$

where $\boldsymbol{n}$ is the outer normal of the system surface S. For an isolated system, the first term in the right-hand side vanishes. The second term, as per Eq. (2.24), can be expressed as

$$\frac{1}{4}\iiint d\Omega d\boldsymbol{v}_* d\boldsymbol{v} B(\cos\theta, v_r)\left[f(\boldsymbol{v}'_*)f(\boldsymbol{v}') - f(\boldsymbol{v}_*)f(\boldsymbol{v})\right]$$
$$\times\left\{\ln[f(\boldsymbol{v}'_*)f(\boldsymbol{v}')] - \ln[f(\boldsymbol{v}_*)f(\boldsymbol{v})]\right\}. \tag{2.27}$$

Since the collision kernel $B(\cos\theta, v_r)$ is non-negative, while for any two positive constants a and b the inequality $(a - b)(\ln a - \ln b) \geq 0$ always holds, we have

$$\frac{dH}{dt} \geq 0, \tag{2.28}$$

which states that H of an isolated system can never decrease. This is known as Boltzmann's H-theorem.

Since H is bounded[3] it cannot increase indefinitely, but must tend to a limit, corresponding to the equilibrium state in which the equality in Eq. (2.28) holds. That is, when $\ln[f(\boldsymbol{v}_*)f(\boldsymbol{v})] = \ln[f(\boldsymbol{v}'_*)f(\boldsymbol{v}')]$, or

$$\ln f(\boldsymbol{v}_*) + \ln f(\boldsymbol{v}) = \ln f(\boldsymbol{v}'_*) + \ln f(\boldsymbol{v}'). \tag{2.29}$$

Therefore, $\ln f$ is also a collisional invariant, which can be expressed by the linear combination of five fundamental collisional invariants as

$$\ln f = \alpha_1 + \boldsymbol{\alpha}_2 \cdot \boldsymbol{v} + \alpha_3 v^2, \tag{2.30}$$

where α_1 and α_3 are scalars, and $\boldsymbol{\alpha}_2$ is a vector; they can be determined according to the density, velocity, and temperature of a gas system. Eventually, it is found that the equilibrium VDF is $f = F_{eq}$, where

$$F_{eq} = n\left(\frac{m}{2\pi k_B T}\right)^{3/2} \exp\left(-\frac{mc^2}{2k_B T}\right). \tag{2.31}$$

For a system in global equilibrium but subject to external force, the VDF is still given by F_{eq}, but here the number density is a function of the spatial coordinates. Substituting $f = F_{eq}$ into the Boltzmann equation, we have

$$f = n_0\left(\frac{m}{2\pi k_B T}\right)^{3/2} \exp\left(-\frac{\frac{mv^2}{2} + U(\boldsymbol{x})}{k_B T}\right), \tag{2.32}$$

where $U(\boldsymbol{x}) = -m\int \boldsymbol{a}\cdot d\boldsymbol{x}$ is the potential energy and n_0 is the reference molecular number density at $U(\boldsymbol{x}) = 0$.

2.3.2 *Equilibrium Collision Frequency*

If the total cross section is finite (either through the cutoff of aiming distance or from the quantum mechanics calculation), the BCO can be separated into a gain term Q^+ and a loss term νf as

$$Q(f, f_*) = Q^+ - \nu(v)f, \tag{2.33}$$

where the collision frequency is

[3] "The integral $\int f(mc^2/2)d\boldsymbol{v}d\boldsymbol{x}$ that represents the total kinetic energy of an isolated system certainly converges. Hence if $\int f \ln f d\boldsymbol{v}d\boldsymbol{x}$ were to diverge, then $-\ln f$ would have to tend to infinity more rapidly than c^2 as $c \to \infty$. But this would imply that the VDF tends to zero more rapidly than $\exp(-c^2)$, in which case $\int f \ln f d\boldsymbol{v}d\boldsymbol{x}$ certainly converges" [4].

$$\nu(v) = \int_{\mathbb{R}^3}\int_{\mathbb{S}^2} B(\cos\theta, |\boldsymbol{v} - \boldsymbol{v}_*|) f(\boldsymbol{v}_*) d\Omega d\boldsymbol{v}_*. \tag{2.34}$$

For inverse power-law potentials, the equilibrium collision frequency corresponding to the differential cross section (2.13) and equilibrium VDF is[4]

$$\begin{aligned}
\nu_{eq}(v) &= \left[\frac{m(\eta-1)}{4K}\right]^{\frac{2}{1-\eta}} \int |\boldsymbol{v} - \boldsymbol{v}_*|^{\frac{\eta-5}{\eta-1}} F_{eq}(\boldsymbol{v}_*) s ds d\phi d\boldsymbol{v}_* \\
&= \nu_\eta^0 \int \exp(-\xi_*^2)\xi_*^2 \sqrt{X(\theta)}^{\frac{\eta-5}{\eta-1}} \sin\theta d\theta d\zeta d\xi_* \\
&= 2\pi \frac{\eta-1}{3\eta-7}\nu_\eta^0 \int_0^\infty \frac{\xi_*}{\xi}\exp(-\xi_*^2)\left[(\xi+\xi_*)^{\frac{3\eta-7}{\eta-1}} - |\xi-\xi_*|^{\frac{3\eta-7}{\eta-1}}\right] d\xi_*,
\end{aligned} \tag{2.35}$$

where

$$\begin{aligned}
\nu_\eta^0 &= \left[\frac{m(\eta-1)}{4K}\right]^{\frac{2}{1-\eta}} \frac{2n}{\sqrt{\pi}} \sqrt{\frac{2k_BT}{m}}^{\frac{\eta-5}{\eta-1}} \int s ds, \\
\nu_{HS}^0 &= n\sqrt{\frac{2k_BT}{m\pi}}\sigma^2.
\end{aligned} \tag{2.36}$$

Specifically, for Maxwellian molecules with $\eta = 5$, the collision frequency is independent of the molecular velocity and temperature:

$$\nu_{eq} = 4\pi\nu_5^0 \int_0^\infty \exp(-\xi_*^2)\xi_*^2 d\xi_* = \pi^{3/2}\nu_5^0, \tag{2.37}$$

while for HS molecules,

$$\nu_{eq}(v) = n\sqrt{\frac{2\pi k_BT}{m}}\sigma^2 \left\{\exp\left(-\xi^2\right) + \frac{\sqrt{\pi}}{2}\left(\frac{1}{\xi} + 2\xi\right)\operatorname{erf}(\xi)\right\}, \tag{2.38}$$

where $\operatorname{erf}(x)$ is the Gauss error function.

Equilibrium collision frequencies for the inverse power-law potential are shown in Fig. 2.2. For hard-potentials ($\eta > 5$), the collision frequency increases with the relative collision speed, while for soft-potentials ($\eta < 5$) the collision frequency decreases with the increasing relative speed.

[4] To simplify the calculation, the following transforms are introduced [8]: $\boldsymbol{v} = (0, 0, 1)v_m\xi$, $\boldsymbol{v}_* = (\sin\phi\sin\theta, \cos\phi\sin\theta, \cos\theta)v_m\xi_*$, and $X(\theta) = \xi^2 + \xi_*^2 - 2\xi\xi_*\cos\theta$, so that $dX = 2\xi\xi_*\sin\theta d\theta$.

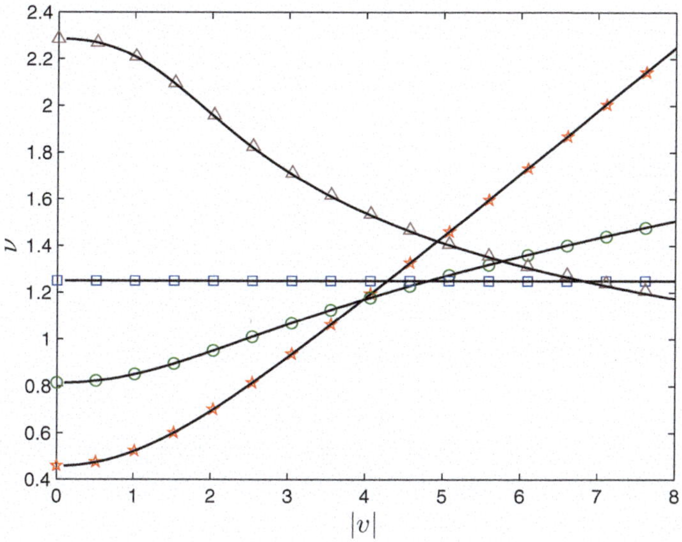

Fig. 2.2 Equilibrium collision frequency of HS molecules (stars), argon with $\eta = 7.45$ (circles), Maxwellian molecules (squares), and soft-potential molecules with $\eta = 3.86$ (triangles) as a function of the peculiar speed normalized by the most probable speed [9]. All gases have the same value of shear viscosity (i.e., $\text{Kn} = \sqrt{\pi}$). Lines are analytical results, while symbols are obtained from the fast spectral method in Chap. 4

2.3.3 Linearized Boltzmann Equation

The linearized Boltzmann equation (LBE) plays an important role in the gas kinetic theory. First, the eigenvalues and eigenfunctions of the linearized BCO are not only important in the derivation of NSF equations (and the transport coefficients such as the shear viscosity and thermal conductivity), but also form a fundamental basis in the development of simplified gas kinetic models. Second, in most experiments involving the gas flow inside microelectromechanical systems, the pressure/temperature gradients and driving velocity are so small that the Boltzmann equation, when linearized, allows efficient and accurate simulation. Third, while the Boltzmann equation is an average equation for the deterministic evolution of the one-particle distribution function, the fluctuations can be taken into account in the LBE formally, e.g., in the Rayleigh-Brillouin scattering, where the light is scattered by the spontaneous density fluctuations in gas [11].

In the LBE, the VDF is expressed as

$$f(t, \boldsymbol{x}, \boldsymbol{v}) = F_{eq} \left[1 + \phi(t, \boldsymbol{x}, \boldsymbol{v})\right], \tag{2.39}$$

where the perturbation ϕ satisfying $|\phi| \ll 1$, and the equilibrium number density and temperature in F_{eq} are n_0 and T_0, respectively. In the absence of external force, the

Boltzmann equation is linearized to

$$\frac{\partial \phi}{\partial t} + \boldsymbol{v} \cdot \frac{\partial \phi}{\partial \boldsymbol{x}} = \mathcal{J}(\phi), \tag{2.40}$$

where the linearized BCO is

$$\mathcal{J}(\phi) = \iint B(\cos\theta, v_r)[\phi(\boldsymbol{v}_*') + \phi(\boldsymbol{v}') - \phi(\boldsymbol{v}_*) - \phi(\boldsymbol{v})]F_{eq}(\boldsymbol{v}_*)d\Omega d\boldsymbol{v}_*. \tag{2.41}$$

2.3.3.1 Maxwellian Gas: Eigenvalues and Eigenfunctions

For molecules interacting with the Maxwell force law $F = K/r^5$, the collision kernel becomes $B(\cos\theta, v_r) = \sqrt{2K/m}B(\theta)$, which is a function of the deflection angle only. In this case, the eigenvalues and eigenfunctions of $\mathcal{J}$ can be given exactly [12]:

$$\begin{aligned} \mathcal{J}(\Phi_{rlm}) &= n\lambda_{rl}\Phi_{rlm}, \\ \Phi_{rlm} = g_{rl}(\xi)Y_l^m(\hat{\xi}) &= \sqrt{\frac{2\pi^{3/2}r!}{(r+l+\frac{1}{2})!}}S_{l+\frac{1}{2}}^{(r)}(\xi^2)\xi^l Y_l^m(\hat{\xi}), \end{aligned} \tag{2.42}$$

with

$$\begin{aligned} \lambda_{rl} =& 2\pi\sqrt{\frac{2K}{m}}\int_0^\pi d\theta \sin\theta B(\theta) \\ &\times \left[-1 + \cos^{2r+l}\frac{\theta}{2}P_l\left(\cos\frac{\theta}{2}\right) - \delta_{r0}\delta_{l0} + \sin^{2r+l}\frac{\theta}{2}P_l\left(\sin\frac{\theta}{2}\right)\right], \end{aligned} \tag{2.43}$$

where $P_l(x)$ is the Legendre polynomial, $S_{l+\frac{1}{2}}^{(r)}(\xi)$ is the Sonine polynomial, and $Y_l^m(\hat{\xi})$ is the spherical harmonics given on the direction of $\boldsymbol{\xi}$, see Appendix A. The dimensionless molecular velocity is

$$\boldsymbol{\xi} = \frac{\boldsymbol{c}}{v_m}, \tag{2.44}$$

with

$$v_m = \sqrt{\frac{2k_B T_0}{m}} \tag{2.45}$$

being the most probable speed.

The first few eigenvalues are $\lambda_{00} = \lambda_{01} = \lambda_{10} = 0$,

$$\lambda_{02} = 2\pi\sqrt{\frac{2K}{m}}\int_0^\pi d\theta \sin^3\theta B(\theta) \times \left(-\frac{3}{4}\right) \equiv -\frac{3}{4}A_2,$$
$$\lambda_{11} = -\frac{1}{2}A_2, \tag{2.46}$$

and $\lambda_{03} = -9A_2/8$, while the first several Φ_{rlm} are

$$\Phi_{000} = 1, \quad \Phi_{010} = \sqrt{2}\xi_z, \quad \Phi_{100} = \sqrt{\frac{2}{3}}\left(\frac{3}{2} - \xi^2\right),$$
$$\Phi_{020} = \frac{1}{3\sqrt{3}}\left(\xi_z^2 - \frac{\xi^2}{3}\right), \quad \Phi_{110} = \frac{2}{\sqrt{5}}\left(\frac{5}{2} - \xi^2\right)\xi_z, \tag{2.47}$$

when the dimensionless velocity along z-axis is $\xi_z = \xi\cos\theta$.

Finally, it is interesting to note that these eigenfunctions are closely related to the perturbed quantities such as the density ϱ, velocity $\boldsymbol{u}$, temperature T, stress $\boldsymbol{\sigma}$, and heat flux $\boldsymbol{q}$:

$$\varrho = \int F_{eq}\phi d\boldsymbol{v}, \quad \boldsymbol{u} = \frac{1}{n_0}\int \boldsymbol{v}F_{eq}\phi d\boldsymbol{v}, \quad \sigma_{ij} = m\int v_{\langle i}v_{j\rangle}F_{eq}\phi d\boldsymbol{v},$$
$$T = \frac{2T_0}{3n_0}\int\left(\frac{v^2}{v_m^2} - \frac{3}{2}\right)F_{eq}\phi d\boldsymbol{v}, \quad \boldsymbol{q} = k_B T_0\int\left(\frac{v^2}{v_m^2} - \frac{5}{2}\right)\boldsymbol{v}F_{eq}\phi d\boldsymbol{v}. \tag{2.48}$$

Also, the eigenvalues λ_{02} and λ_{11} are, respectively, related to the relaxation rates of stress and heat flux in spatial-homogeneous problems (and hence the viscosity and thermal conductivity; details can be found in Appendix B):

$$\frac{\partial\sigma_{ij}}{\partial t} = n\lambda_{02}\sigma_{ij} = -\frac{p}{\mu}\sigma_{ij},$$
$$\frac{\partial q_i}{\partial t} = n\lambda_{11}q_i = -\frac{2}{3}\frac{p}{\mu}q_i. \tag{2.49}$$

That the eigenvalues are not positive can be understood below. If we define

$$[\psi,\phi] = -\frac{m\pi^{3/2}}{\rho}\int \exp(-\xi^2)\psi\mathcal{J}(\phi)d\boldsymbol{\xi}$$
$$= \frac{m\pi^{3/2}}{4\rho}\int \exp(-\xi^2)[\psi(\boldsymbol{v}') + \psi(\boldsymbol{v}'_*) - \psi(\boldsymbol{v}) + \psi(\boldsymbol{v}_*)]\mathcal{J}(\phi)d\boldsymbol{\xi}, \tag{2.50}$$

then we have $[\phi,\phi] = -\lambda\frac{m\pi^{3/2}}{\rho}\int\exp(-\xi^2)\phi^2 d\boldsymbol{\xi} \geq 0$, where the equality holds only when ϕ is the linear combination of the collisional invariants. Therefore, we have $\lambda \leq 0$. These non-positive value of eigenvalues guarantee the system relax to equilibrium.

2.4 Wang-Chang and Uhlenbeck Equation

The Boltzmann equation can be heuristically extended to molecular gas, dense gas, granular gas, and quantum gas. We refer to Kremer's book for details [13]. In this section, we introduce the Wang-Chang and Uhlenbeck equation for molecular (diatomic and polyatomic) gas [12].

The molecular gas possesses internal (rotational, vibrational, or electronic) energy. Wang-Chang and Uhlenbeck treated the internal motion quantum mechanically and assigned each energy level a VDF. Suppose $f_i(t, \boldsymbol{x}, \boldsymbol{v})$ is the VDF for the internal energy level i with internal energy e_i, then its evolution is given by

$$\frac{\partial f_i}{\partial t} + \boldsymbol{v} \cdot \frac{\partial f_i}{\partial \boldsymbol{x}} + \boldsymbol{a} \cdot \frac{\partial f_i}{\partial \boldsymbol{v}} = Q_i, \tag{2.51}$$

where

$$Q_i = \sum_{jkl} \iint P_{ij}^{kl} B(\cos\theta, v_r) \left[f_l(\boldsymbol{v}_*') f_k(\boldsymbol{v}') - f_j(\boldsymbol{v}_*) f_i(\boldsymbol{v}) \right] d\Omega d\boldsymbol{v}_*. \tag{2.52}$$

This collision integral describes a collision between two "particles". Before the collision, the first particle is in the ith state and has a velocity $\boldsymbol{v}$, while the second particle is in the jth state and has a velocity $\boldsymbol{v}_*$. After the collision, the first particle jumps to the kth state, while the second particle jumps to the lth state. The transfer probability from the levels (i, j) to (k, l) is P_{ij}^{kl}. Since the total energy is conserved during collision, such a transition occurs only when

$$v_r'^2 = v_r^2 + \frac{4}{m}(e_i + e_j - e_k - e_l) > 0. \tag{2.53}$$

When the collision is admissible, the molecular velocity after collision is

$$\boldsymbol{v}' = \frac{\boldsymbol{v} + \boldsymbol{v}_*}{2} + \frac{v_r'}{2}\boldsymbol{\Omega}, \quad \boldsymbol{v}_*' = \frac{\boldsymbol{v} + \boldsymbol{v}_*}{2} - \frac{v_r'}{2}\boldsymbol{\Omega}. \tag{2.54}$$

An example is introduced to show the complexity of the Wang-Chang and Uhlenbeck equation even if only the rotational modes are excited. For molecular nitrogen interacting with the (6–12) Lennard-Jones potential with a potential depth of 91 K, the ith rotational energy level has an energy of

$$e_i = 2.9i(i+1)\ \text{K}, \quad i = 0, 1, 2, \ldots, \tag{2.55}$$

and a degeneracy of

$$g_i = 2i + 1. \tag{2.56}$$

By grouping the degenerate internal energy levels, the collision operator is modified into

$$Q_i = \sum_{jkl} \iint P_{ij}^{kl} B \left[\omega_{ij}^{kl} f_l(\boldsymbol{v}'_*) f_k(\boldsymbol{v}') - f_j(\boldsymbol{v}_*) f_i(\boldsymbol{v})\right] d\Omega d\boldsymbol{v}_*, \tag{2.57}$$

where $\omega_{ij}^{kl} = g_k g_l / g_i g_j$. The transfer probability is found by fitting the MD simulation data [14, 15]

$$P_{ij}^{kl} = P_0 \omega_{ij}^{kl} \left[\frac{2e_{tot}}{5e_{tr0}} \exp(-\Delta_1 - \Delta_2 - \Delta_3 - \Delta_4) + \frac{5e_{tr0}}{2e_{tot}} \exp(-\Delta_3 - \Delta_4)\right],$$

where e_{tr0} is the relative translational energy of the two molecules, e_{tot} is the total energy of them, and

$$\Delta_1 = \frac{|e_i + e_j - e_k - e_l|}{e_{tr0}}, \quad \Delta_2 = 2\frac{|e_i + e_l - e_k - e_j|}{e_{tot}},$$

$$\Delta_3 = 4\frac{|e_i - e_k|}{e_{tr0} + e_i}, \quad \Delta_4 = 2\frac{|e_j - e_l|}{e_{tr0} + e_j},$$

$$e_{tr0} = \frac{m}{4} v_r^2, \quad e_{tot} = \frac{m}{4} v_r^2 + e_i + e_j.$$

If the total number of energy levels is N, the computational cost of Eq. (2.57) is about N^4 times of that for the BCO; the cost is huge, as for nitrogen gas N can be as high as 70 in strong shock waves [15]. Therefore, a kinetic model to simplify Eq. (2.57) is highly desired. This will be discussed in Chap. 7.

Finally, it is noted that the equilibrium distribution function can be given as $f_i = o_i F_{eq}$, where (the Einstein summation is not used here)

$$o_i = \frac{g_i \exp(-e_i / k_B T)}{\sum_j g_j \exp(-e_j / k_B T)} \tag{2.58}$$

is the occupation fraction of molecules with the internal energy e_i.

2.5 Enskog Equation

Since the MFP of gas molecules is much larger than the molecular dimension in most situations (e.g., the nominal diameter of methane is about 0.4 nm, while the MFP of methane at standard temperature and pressure is about 76 nm), the BCO is localized. However, the nonlocal collision emerges when the ratio of the molecular dimension to the MFP becomes appreciable, e.g., when the pressure is increased to 100 and 300 atmospheric pressure, the MFP is reduced to 0.9 nm and 0.35 nm, respectively; the rarefaction effect may be still important when the characteristic flow size is small, say, in the gas flow in ultra-tight shale strata [16, 17].

The kinetic theory for moderately dense gas of HS molecules is developed by Enskog, who extended the BCO to the following form:

$$Q = \sigma^2 \iint d\boldsymbol{v}_* d\boldsymbol{k}(\boldsymbol{k} \cdot \boldsymbol{v}_r) \times \left[g(\boldsymbol{x}, \boldsymbol{k}, \boldsymbol{v}'_*) f(\boldsymbol{x}, \boldsymbol{v}') - g(\boldsymbol{x}, -\boldsymbol{k}, \boldsymbol{v}_*) f(\boldsymbol{x}, \boldsymbol{v}) \right]. \tag{2.59}$$

The binary collision occurs at a distance of one molecular diameter, hence is nonlocal. Due to the excluded volume effect, the collision frequency is increased comparing to that of dilute gas, and this effect is characterized by the product of distribution function and pair correlation function at different locations:

$$g(\boldsymbol{x}, \boldsymbol{k}, \boldsymbol{v}) = \chi\left(\boldsymbol{x} + \frac{\sigma}{2}\boldsymbol{k}\right) f(\boldsymbol{x} + \sigma\boldsymbol{k}, \boldsymbol{v}), \tag{2.60}$$

where the pair correlation function χ is independent of the molecular velocity [18]:

$$\chi(n) = \frac{1}{2}\frac{2-\eta}{(1-\eta)^3}, \tag{2.61}$$

but is related to the solid fraction $\eta = \pi n \sigma^3/6$.

Due to the nonlocal collision, the equation of state does not follow the ideal gas law anymore, but is modified to

$$p = nk_B T \frac{1+\eta+\eta^2-\eta^3}{(1-\eta)^3}. \tag{2.62}$$

Note that the shear viscosity of HS gas is proportional to the square root of temperature. To have a general temperature-dependent viscosity, the collision kernel $\boldsymbol{k} \cdot \boldsymbol{v}_r$ is modified to be a function of the relative collision velocity [19]. Note that the mass is conserved in the Enskog collision operator, but the momentum and energy (even in the case of elastic collision) are not conserved.

2.5.1 Liquid-Vapor Flow

The Enskog equation can be further extended to describe the liquid-vapor two phase flows. The rarefaction effects may appear when the system is tightly confined [16, 17], or at the liquid-vapor interface during evaporation and condensation [20, 21].

To model the liquid-vapor flow, the repulsive and attractive potentials are treated separately. For former is modeled by the Enskog collision operator (2.59), while the latter is incorporated into the mean-filed force in the streaming operator, resulting in the Enskog-Vlasov equation:

$$\frac{\partial f}{\partial t} + \boldsymbol{v} \cdot \frac{\partial f}{\partial \boldsymbol{x}} + \left(\boldsymbol{a} + \frac{\boldsymbol{F}}{m} \right) \cdot \frac{\partial f}{\partial \boldsymbol{v}} = Q, \tag{2.63}$$

where the mean-field force from the surrounding gas molecules is

$$\boldsymbol{F}(\boldsymbol{x}) = \int_{r>\sigma} \frac{d\phi}{dr} \frac{\boldsymbol{x}_1 - \boldsymbol{x}}{|\boldsymbol{x}_1 - \boldsymbol{x}|} n(\boldsymbol{x}_1) d\boldsymbol{x}_1. \tag{2.64}$$

Usually, the attractive potential $\phi(r)$ is assumed of Sutherland's type:

$$\phi(r) = \begin{cases} +\infty, & r < \sigma, \\ -\phi_a \left(\frac{r}{\sigma}\right)^{-\gamma}, & r \geq \sigma, \end{cases} \tag{2.65}$$

where $r = |\boldsymbol{x}_1 - \boldsymbol{x}|$ is the intermolecular distance, ϕ_a controls the depth of potential well, and γ controls the range of attractive interaction.

Due to the attractive force, the equation of state takes the generalized van der Waals form [20]:

$$p = nk_B T \frac{1 + \eta + \eta^2 - \eta^3}{(1-\eta)^3} - \frac{2\pi\sigma^3}{3} \frac{\gamma}{\gamma - 3} \phi_a n^2. \tag{2.66}$$

At a critical packing density $\eta_c = \pi\sigma^3 n_c = 0.130443$, the critical temperature is reached:

$$T_c = 0.094329 \frac{4\gamma}{\gamma - 3} \frac{\phi_a}{k_B}. \tag{2.67}$$

2.5.2 Granular Gas

The granular gas is composed of many small discrete grains in rapid motion, so that the binary collision approximately holds. The kinetic energy is, however, not conserved during collision. When the rotation of grains is neglected, the post-collision velocities are given by

$$\boldsymbol{v}' = \boldsymbol{v} - \frac{1+\alpha}{2}(\boldsymbol{v}_r \cdot \boldsymbol{k})\boldsymbol{k}, \quad \boldsymbol{v}'_* = \boldsymbol{v}_* + \frac{1+\alpha}{2}(\boldsymbol{v}_r \cdot \boldsymbol{k})\boldsymbol{k}, \tag{2.68}$$

where $0 \leq \alpha \leq 1$ is the restitution coefficient. The Enskog collision operator for dense granular gases in the d-dimensional space reads

$$Q = \sigma^{d-1} \int_{\mathbb{R}^d} \int_{\mathbb{S}^{d-1}} d\boldsymbol{v}_* d\boldsymbol{k} (\boldsymbol{k} \cdot \boldsymbol{v}_r) \times \left[\frac{J}{\alpha} g(\boldsymbol{x}, \boldsymbol{k}, \tilde{\boldsymbol{v}}_*) f(\boldsymbol{x}, \tilde{\boldsymbol{v}}) - g(\boldsymbol{x}, -\boldsymbol{k}, \boldsymbol{v}_*) f(\boldsymbol{x}, \boldsymbol{v}) \right]. \tag{2.69}$$

It should be noted that here $\tilde{\boldsymbol{v}}$ and $\tilde{\boldsymbol{v}}_*$ are the pre-collision velocities, which yield post-collision velocities $\boldsymbol{v}$ and $\boldsymbol{v}_*$ after the binary collision, and J is the Jacobian of the transformation from $(\tilde{\boldsymbol{v}}, \tilde{\boldsymbol{v}}_*)$ to $(\boldsymbol{v}, \boldsymbol{v}_*)$. When the restitution coefficient α is a constant, we have

$$\tilde{\boldsymbol{v}} = \boldsymbol{v} - \frac{1+\alpha}{2\alpha}(\boldsymbol{v}_r \cdot \boldsymbol{k})\boldsymbol{k}, \quad \tilde{\boldsymbol{v}}_* = \boldsymbol{v}_* + \frac{1+\alpha}{2\alpha}(\boldsymbol{v}_r \cdot \boldsymbol{k})\boldsymbol{k}, \tag{2.70}$$

and the Jacobian is $J = 1/\alpha$. When α is a function of the normal relative velocity $|\boldsymbol{v}_r \cdot \boldsymbol{k}|$, it is hard to calculate the Jacobian; alternatively, Eq. (2.69) is written in the weak form as

$$\begin{aligned} \int Q\Psi(\boldsymbol{v})d\boldsymbol{v} = \sigma^{d-1} \iiint d\boldsymbol{v}_* d\boldsymbol{k} d\boldsymbol{v} (\boldsymbol{k} \cdot \boldsymbol{v}_r) \\ \times g(\boldsymbol{x}, -\boldsymbol{k}, \boldsymbol{v}_*) f(\boldsymbol{x}, \boldsymbol{v}) \left[\Psi(\boldsymbol{v}') - \Psi(\boldsymbol{v})\right], \end{aligned} \tag{2.71}$$

where $\Psi(\boldsymbol{v})$ is any smooth function of $\boldsymbol{v}$. Note that the mass is conserved in the Enskog collision operator, but the momentum and energy (even in the case of elastic collision) are not conserved.

The Enskog equation for dense granular flows will be solved in Chap. 14.

2.6 Gas-Surface Boundary Condition

To completely determine the RGD in the presence of solid surface, the VDF for the molecules reflected from the solid surface should be specified; this is dictated by the gas-surface interaction, which plays an important role in rarefied gas flows, especially in microelectromechanical system where the ratio between surface area and volume is large.

Physically, when the wall structure and the gas-wall potential are known, the gas-surface interaction can be captured by the MD simulation [22, 23]. However, the time step in microscopic MD simulation is far smaller than that used in mesoscopic Boltzmann equation, which greatly limit its application. On the other hand, although attempts have been made to model the gas-surface interaction at the mesoscopic level, e.g., the Enskog-Vlasov collision operator [24, 25] and the Boltzmann-type gas-phonon collision operator [26], the accuracy has not been fully assessed.

Therefore, an empirical gas kinetic boundary condition (BC), which determines the VDF of reflected molecules from the solid wall in terms of the VDF of incident molecules, is highly needed. In RGD, the kinetic BC takes the general form [27]:

$$v_n f(\boldsymbol{v}) = \int_{v_n'<0} |v_n'| R(\boldsymbol{v}' \to \boldsymbol{v}) f(\boldsymbol{v}') d\boldsymbol{v}', \quad v_n > 0, \tag{2.72}$$

where $\boldsymbol{v}$ and $\boldsymbol{v}'$ are the velocities of reflected and incident molecules, respectively, and v_n is the normal component of the molecular velocity $\boldsymbol{v}$ pointing into the gas.

The scattering kernel $R(\boldsymbol{v}' \to \boldsymbol{v})$ in Eq. (2.72) is non-negative. Without considering adsorption, desorption, or chemical reactions, it obeys the normalization condition $\int_{v_n>0} R(\boldsymbol{v}' \to \boldsymbol{v})d\boldsymbol{v} = 1$, and the reciprocity relation

$$|v_n'|f_0(T_w, \boldsymbol{v}')R(\boldsymbol{v}' \to \boldsymbol{v}) = |v_n|f_0(T_w, \boldsymbol{v})R(-\boldsymbol{v} \to -\boldsymbol{v}'), \tag{2.73}$$

with

$$f_0(T_w, \boldsymbol{v}) = \exp\left(-\frac{m|\boldsymbol{v} - \boldsymbol{u}_w|^2}{2k_B T_w}\right) \tag{2.74}$$

being the equilibrium VDF and $\boldsymbol{u}_w$ the wall velocity. That is, in the thermodynamic equilibrium, both the incident and reflected molecules obey the Maxwellian distribution.

Typical kinetic BCs are introduced below.

2.6.1 *Maxwell Boundary Condition*

It was proposed by Maxwell in 1879, and is known as the diffuse-specular BC. The scattering kernel reads [28]

$$R_M(\boldsymbol{v}' \to \boldsymbol{v}) = \alpha_M \frac{m^2 v_n}{2\pi(k_B T_w)^2} \exp\left(-\frac{mv^2}{2k_B T_w}\right) + (1 - \alpha_M)\delta(\boldsymbol{v}' - \boldsymbol{v} + 2\boldsymbol{n}v_n), \tag{2.75}$$

where δ is the delta function, and $0 \le \alpha_M \le 1$ is the accommodation coefficient. This BC states that, after the collision with a surface, a molecule is diffusely scattered with the probability α_M, otherwise it is reflected specularly. Purely specular and diffuse reflections occurs when $\alpha_M = 0$ and $\alpha_M = 1$, respectively.

2.6.2 *Epstein Boundary Condition*

The accommodation coefficient in the Maxwell BC is independent of the impinging velocities of molecules, which does not agree with both theoretical and experimental investigations. To remedy this, Epstein proposed the following scattering kernel in 1967 [29] :

$$R_E(\boldsymbol{v}' \to \boldsymbol{v}) = \frac{v_n \Theta(\boldsymbol{v})\Theta(\boldsymbol{v}')}{\int_{v_n>0} v_n f_0(T_w, \boldsymbol{v})\Theta(\boldsymbol{v})d\boldsymbol{v}} f_0(T_w, \boldsymbol{v}) + [1 - \Theta(\boldsymbol{v}')]\delta(\boldsymbol{v}' - \boldsymbol{v} + 2\boldsymbol{n}v_n), \tag{2.76}$$

where the probability of diffuse reflection is a function of molecular velocity, $\Theta(\boldsymbol{v})$. When $\Theta(\boldsymbol{v}) = \alpha_M$, the Epstein BC reduces to the Maxwell BC.

Following the arguments that

- "for incident molecules with very low energies, most of the molecules are trapped by the attractive part of the gas-surface interaction and hence are almost completely accommodated. That is, $\Theta(\boldsymbol{v}) \approx 1$";
- "at high energies the degree of accommodation decreases because an increasing fraction of the molecules have sufficient energy to overcome the trapping effect";
- "at sufficiently high energies it is found that the accommodation coefficient increases again toward a high energy asymptote".

Epstein designed the following form of $\Theta(\boldsymbol{v})$:

$$\Theta(\boldsymbol{v}) = \Theta_0 \exp\left(-\alpha \frac{mv^2}{2k_B T_w}\right) + \Theta_1 \left[1 - \exp\left(-\beta \frac{mv^2}{2k_B T_w}\right)\right], \tag{2.77}$$

where Θ_0, α, β, and Θ_1 are constants. This simple expression predicts the thermal accommodation coefficient well, for various kinds of gases interacting with tungsten [29].

2.6.3 Cercignani-Lampis Boundary Condition

The scattering kernel reads

$$\begin{aligned} R_{CL}(\boldsymbol{v}' \to \boldsymbol{v}) =& \frac{m^2 v_n}{2\pi \alpha_n \alpha_t (2-\alpha_t)(k_B T_w)^2} I_0\left(\frac{\sqrt{1-\alpha_n} m v_n v_n'}{\alpha_n k_B T_w}\right) \\ & \times \exp\left\{-\frac{m[v_n^2 + (1-\alpha_n)v_n'^2]}{2k_B T_w \alpha_n} - \frac{m|\boldsymbol{v}_t - (1-\alpha_t)\boldsymbol{v}_t'|^2}{2k_B T_w \alpha_t (2-\alpha_t)}\right\}, \end{aligned} \tag{2.78}$$

where $\boldsymbol{v}_t$ is the tangential velocity and

$$I_0(x) = \frac{1}{2\pi} \int_0^{2\pi} \exp(x \cos\phi) d\phi. \tag{2.79}$$

The tangential momentum accommodation coefficient (TMAC) α_t and energy accommodation coefficient (EAC) α_n are limited to [0, 2] and [0, 1], respectively. Diffuse reflection occurs when $\alpha_t, \alpha_n = 1$, while specular reflection occurs when $\alpha_t, \alpha_n = 0$. The "backwards" scattering is captured when $\alpha_t = 2$ and $\alpha_n = 0$.

The Cercignani-Lampis BC has been widely used because of the following two major factors:

- In the experiment of thermal beam scattering, the Cercignani-Lampis BC recovers the exact shape of reflection plume [30].
- In the thermal transpiration, the exponent of thermomolecular pressure difference, an important indicator of the performance of Knudsen pump, can be less than 0.5 in

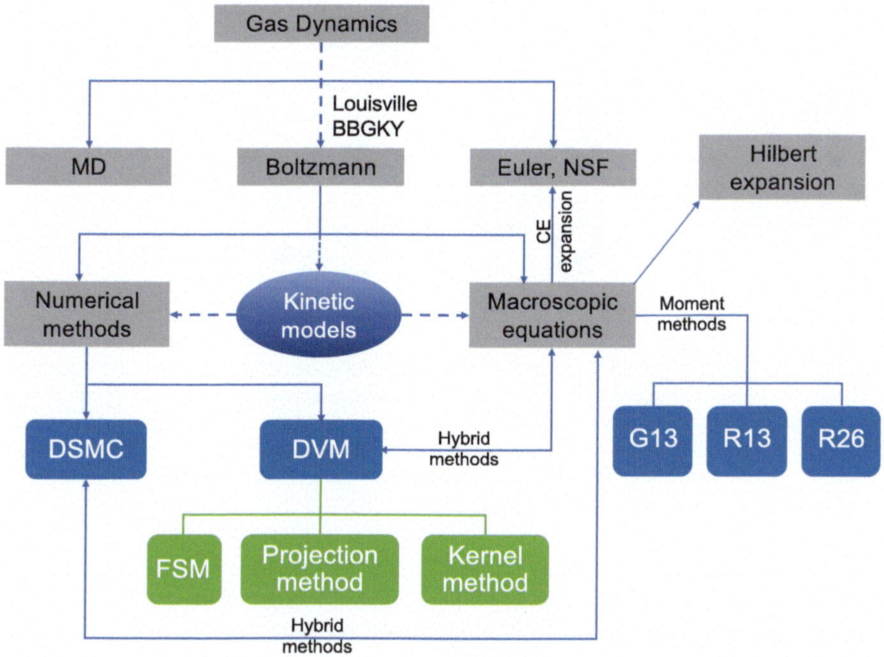

Fig. 2.3 Hierarchy of analytical and numerical methods in gas dynamics. The Boltzmann equation, which bridges the macroscopic and microscopic description of the gas dynamics, is a fundamental equation in RGD. It can be solved directly by the stochastic or deterministic numerical methods, or it produces the macroscopic equations, through the Hilbert expansion, Chapman-Enskog expansion, or the Grad moment method, which are then solved numerically. Due to the complexity of the Boltzmann equation, sometimes it is replaced by simpler kinetic models. Note that while the Boltzmann equation uses the one-particle VDF, it can be derived from the Bogoliubov-Born-Green-Kirkwood-Yvon (BBGKY) hierarchy in statistical physics

the free-molecular flow, which agrees well with some experimental measurements [31], while Maxwell's BC model always predicts an exponent of 0.5 at any value of α_M.

The performance of these kinetic BCs will be assessed in Chap. 11.

2.7 Numerical Methods

The hierarchy of analytical and numerical methods in gas dynamics is depicted in Fig. 2.3, where the Boltzmann equation plays a critical role. The analytical methods (e.g., Hilbert expansion, Chapman-Enskog expansion, and Grad's moment methods) try to derive equations of macroscopic quantities to describe the RGD, but they are only valid up to a certain value of the Knudsen number. Therefore, it is inevitable to solve the Boltzmann equation numerically, by using either deterministic

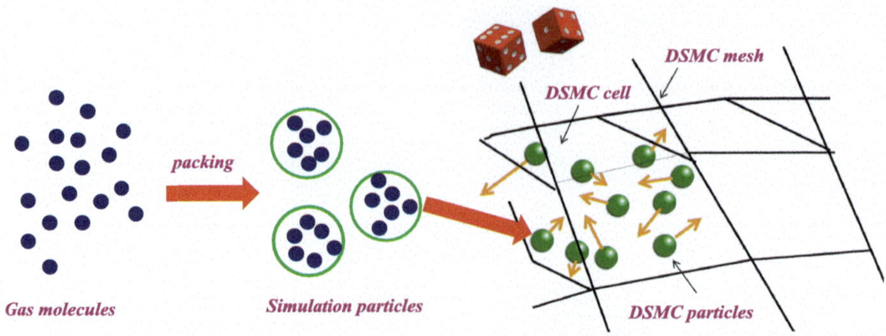

Fig. 2.4 Schematic of the DSMC method. Real gas molecules are packed into simulation particles to reduce the computational cost. The mass of each simulated particle is equal to the molecular mass, but the collision diameter (or collision probability) is carefully chosen so that the simulated shear viscosity in DSMC is the same as that of the real gas [35]; since the collision dynamics is the same as that in the Boltzmann equation, the thermal conductivity is automatically recovered

or stochastic numerical methods [32–35, 38]. Compared to the NSF equations, the Boltzmann equation and its simplified kinetic models are expensive to solve in terms of computation time and memory. This is mainly due to the following facts:

- Additional dimensions of molecular velocity space are discretized in deterministic methods, while simulation particles are generated in spatial cells in stochastic methods. Therefore, the computer memory usage and computational time are much higher.
- The gas dynamics is simulated on finite length and time scales. In order to reduce the numerical dissipation, the spatial cell size and time step must be smaller than the kinetic scale, i.e., the MFP and mean collision time of gas molecules, respectively. Therefore, the computational cost is much higher than the NSF equations where the hydrodynamic scale is used.

Typical numerical methods in RGD are introduced below.

2.7.1 Direct Simulation Monte Carlo

The direct simulation Monte Carlo (DSMC) is a stochastic method that uses simulation particles to mimic the binary collision of real gas molecules. As sketched in Fig. 2.4, the computational domain is divided into spatial cells, and in each cell there are several simulation particles. Each simulation particle represents F_N real gas molecules, so that the computational memory and time are much reduced when compared to MD simulation. In DSMC, the simulation particles follow the streaming and collision processes:

$$\begin{aligned}\text{Streaming:}\quad & \frac{\partial f}{\partial t} = -\boldsymbol{v}\cdot\frac{\partial f}{\partial \boldsymbol{x}} - \boldsymbol{a}\cdot\frac{\partial f}{\partial \boldsymbol{v}},\\ \text{Collision:}\quad & \frac{\partial f}{\partial t} = Q(f, f_*).\end{aligned} \tag{2.80}$$

The streaming is solved exactly by the method of characteristics, while the collision in each spatial cell is solved by selecting the collision pair with certain probability.

Consider the collision of $N = nV_c/F_N$ simulate particles inside a DSMC cell of volume V_c. The collision probability P between two simulated particles over the time interval Δt is calculated from Eq. (2.20) as $P = (F_N/V_c)\sigma_T v_r \Delta t$. Since the total cross section obtained from classical physics is divergent in the grazing collision limit, in DSMC the collision models, such as the variable-hard-sphere model and variable-soft-sphere model [35], are carefully chosen so that it would produce a required value of shear viscosity, thermal conductivity, or even diffusion coefficient. In the no-time-counter method,

$$\frac{1}{2}N^2\frac{F_N}{V_c}(\sigma_T v_r)_{\max}\Delta t \tag{2.81}$$

collision pairs are selected from the cell at the time step, and the collision occurs with the probability

$$\frac{\sigma_T v_r}{(\sigma_T v_r)_{\max}}. \tag{2.82}$$

When the collision pair is chosen, a uniformly distributed variable is generated, and the collision is allowed to happen when the random variable is less than the value in Eq. (2.82). The post-collision velocities are updated according to the conservation of momentum and energy in Eq. (2.4). For example, in the variable-hard-sphere model where the collision kernel is independent of the deflection angle, the deflection angle is chosen randomly and uniformly.

Due to the stochastic nature, macroscopic quantities have to be sampled and averaged in DSMC. It has been proved that DSMC solutions converge to those of the Boltzmann equation for monatomic gases in the limit of vanishing discretization and stochastic errors [36]. DSMC is also extended to deal with molecular gas, dense gas, and even chemical reactions. However, in contrast to monatomic gas, there is no rigorous proof. Some of the problems will be discussed in Chap. 7.

Main advantages of DSMC are as follows:

- It is unconditionally stable.
- It does not need artificial boundaries in velocity space. Also, simulation particles automatically concentrate in regions where the VDF is not small so that the computer memory is not wasted.
- Simulation particles represent a large number of real molecules so that the number of operations is greatly reduced.
- The computing time is linear with N, since if N is doubled through halving F_N, the number of pair selections in Eq. (2.81) also doubles.

These properties make the DSMC efficient in simulating high-speed rarefied gas flows. However, it is time-consuming when simulates the near-continuum flow, low-speed flow, or transient flow. Hybrid continuum/particle approaches [37, 39, 40] and the low-noise DSMC [41–43] have partially easen these difficulties.

2.7.2 *Discrete Velocity Methods*

Contrasting to the DSMC, the discrete velocity method (DVM) solves the Boltzmann equation deterministically, where the molecular velocity space is truncated and discretized. In addition to the time-splitting scheme (2.80), the Boltzmann equation can be solved implicitly, by the conventional iterative scheme (CIS).

In this book, the CIS will be used to find the steady-state solution of the Boltzmann equation. This is because in rarefied gas flows the Reynolds number (which is proportional to the Mach number but inversely proportional to the Knudsen number) is small, so that steady-state solutions are of particular interest. In CIS, given the VDF f^k at the kth step, its value at the next iteration step is calculated as

$$\overline{\nu} f^{k+1} + \boldsymbol{v} \cdot \frac{\partial f^{k+1}}{\partial \boldsymbol{x}} = \overline{\nu} f^k - \boldsymbol{a} \cdot \frac{\partial f^k}{\partial \boldsymbol{v}} + Q(f^k, f^k_*), \tag{2.83}$$

where $\overline{\nu}$ is the mean collision frequency or its inverse is the effective time step when the backward Euler method is adopted for the time discretization. The steady-state solution is found when the relative error in macroscopic quantities between two consecutive steps is small.

Note that $\overline{\nu}$ determines the numerical efficiency and stability. If it is large, it may need large number of iterations to find the steady-state solution, but if it is too small, the iteration will blow up. Normally, it is safe to replace $\overline{\nu}$ by the local collision frequency, resulting in the following iterative scheme with gain term Q^+ of BCO:

$$\nu(f^k) f^{k+1} + \boldsymbol{v} \cdot \frac{\partial f^{k+1}}{\partial \boldsymbol{x}} = -\boldsymbol{a} \cdot \frac{\partial f^k}{\partial \boldsymbol{v}} + Q^+(f^k, f^k_*). \tag{2.84}$$

In the deterministic numerical methods, the BCO is the most difficult and time-consuming part to solve, due to its multidimensional and nonlinear structure. A brief introduction on this topic is given below.

2.7.2.1 Conservative Projection Method

In the original DVM, a fixed set of discrete velocities (e.g., the cross points of dashed lines in Fig. 2.5) is chosen to approximate the molecular velocity space, and the binary collision occurs only when the post-collision velocities lie on the velocity nodes [44]. To improve the accuracy and efficiency of DVM, Tcheremissine constructed the

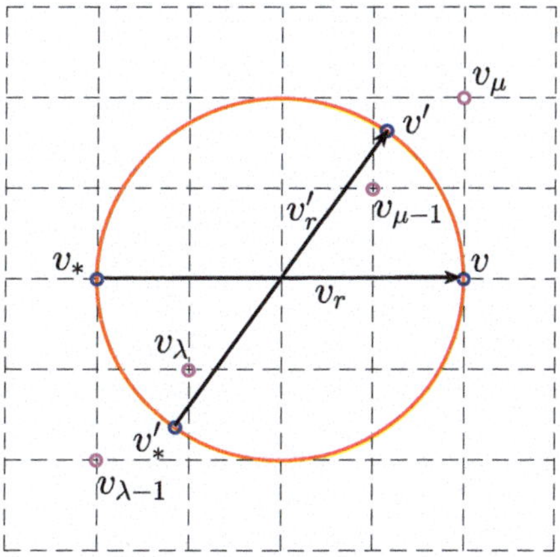

Fig. 2.5 Schematic of the conservative projection method for solving the Boltzmann equation, where the post-collision velocities fall off the pre-designed velocity grids; therefore, the VDFs are interpolated based on those of the neighboring discrete velocity grids

conservative projection method, where the post-collision velocity is projected to nearby velocity grids, see Fig. 2.5. That is, when the pre-collision velocities $\boldsymbol{v}$ and $\boldsymbol{v}_*$ are chosen in a collision, and the post-collision velocities $\boldsymbol{v}'$ and $\boldsymbol{v}_*'$ fall off the pre-designed velocity grids, the VDFs of the post-collisions are reconstructed, based on the VDF in neighboring discrete velocity nodes:

$$f(\boldsymbol{v}_*')f(\boldsymbol{v}') = [f(\boldsymbol{v}_\lambda)f(\boldsymbol{v}_\mu)]^{1-r}[f(\boldsymbol{v}_{\lambda-1})f(\boldsymbol{v}_{\mu-1})]^r, \tag{2.85}$$

where the parameter r is chosen to fulfill the conservation of energy and momentum. In Tcheremissine's method, the BCO is turned to an eight-dimensional integral, which is approximated by a numerical quadrature based on the Korobov sequence in number theory [45, 46]. Together with the time-splitting scheme, this method has been used to solve a number of problems in RGD [15, 47–49].

2.7.2.2 Numerical Kernel Method

The numerical kernel method is firstly developed for the linearized BCO [50], and then to the full BCO [51, 52]. In 2001, it is extended to the Boltzmann equation for a binary mixture [53]. Together with the CIS, where the streaming operator is approximated by the finite-difference method, their solutions obtained some 30 years ago are still very accurate by today's standard. However, this method seems to be restricted to HS molecules, and usually only 1D problems such as normal shock flow and planar Fourier flow are considered, where the VDF has a cylindrical symmetry.

2.7.2.3 Spectral Methods

After the breakthrough in the analytical treatment of the BCO of Maxwell gas using the Fourier transform [54], many spectral methods are proposed to solve the BCO numerically. The fast spectral method (FSM) is the most powerful one originally proposed by Mouhot and Pareschi in 2006, where the BCO is solved with a spectral accuracy, and a computational complexity of $O(M^2 N_v^3 \log N_v)$; M^2 is the number of grid points in the discretization of solid angle, and N_v^3 is the number of total discrete velocities. In practical calculations, $M \ll N_v$ [55, 56]. Nowadays, the FSM has been widely applied to solve rarefied gas flows [9, 10, 16, 57–64].

2.7.3 *Multi-scale Simulation*

The time-splitting scheme or CIS is used to find steady-state solutions of the Boltzmann equation. The former has a time step limited by the mean collision time and the Courant-Friedrichs-Lewy (CFL) condition. The latter is time-implicit, which is efficient when the Knudsen number is high, where converged solution can be found in dozens of iterations [65]. However, it converges extreme slowly for flows with low Knudsen numbers [63]. Worse still, the solution is subject to large numerical dissipation when coarse spatial grids are used. The simulation time in DSMC also increases significantly in near-continuum flows due to the frequent binary collision. Therefore, it is of great significance to develop multi-scale numerical schemes that work efficiently and accurately in the whole range of Knudsen number.

2.7.3.1 Hybrid Methods

Hybrid numerical methods combine models working at different regions. Normally, the computational fluid dynamics, which solves the NSF equations, are used in the continuum flow regime, while the kinetic models are applied in regions where rarefaction effects dominate [37, 39, 66–70]. However, intrinsic difficulties arise in the coupling. First, the criterion to determine where the continuum region is relies on empirical parameters that vary in different flow conditions [39]. Second, the continuum-kinetic coupling is required to lie in the region that can be accurately modeled by NSF equations, so that the kinetic equation is still employed in regions with low Knudsen number. In this case, CIS still needs lots of iterations to converge to the steady state; similarly, DSMC needs small cell size and time step, hence large number of evolution steps to find steady-state solutions.

Recently higher-order macroscopic equations derived from the Grad moment method, which are more accurate than the NSF equations, are coupled with kinetic equations to simulate multi-scale flows [71]; the benefit is that more spatial regions will be covered by macroscopic equations, so that the computational memory and cost are reduced when compared to the hybrid methods of NSF equations and gas kinetic model.

2.7.3.2 Unified Gas Kinetic Scheme

The unified gas kinetic scheme (UGKS) developed by Prof. Kun Xu offers a novel paradigm to deal with multi-scale problems by the sole use of gas kinetic equation [72]. Its most successful contribution is that the streaming and collision are handled simultaneously. For example, when the gas dynamics is described by the Bhatnagar, Gross, and Krook (BGK) model [73]

$$\frac{\partial f}{\partial t} + v_x \frac{\partial f}{\partial x} = \frac{F_{eq} - f}{\tau},$$

the VDF in the finite-volume framework is updated as

$$\begin{aligned} f_j^{n+1} &= f_j^n + \frac{1}{\Delta x} \int_{t^n}^{t^{n+1}} [v_x \hat{f}_{j-1/2}(t) - v_x \hat{f}_{j+1/2}(t)] dt \\ &+ \frac{1}{\Delta x} \int_{t^n}^{t^{n+1}} \int_{x_{j-1/2}}^{x_{j+1/2}} \frac{F_{eq} - f}{\tau} dx dt, \end{aligned}$$

with Δx is the length of the jth cell and $\hat{f}$ is the VDF at the cell interface. Instead of using the upwind scheme for the evaluation of VDF at the cell interface, the solution $\hat{f}_{j+1/2}$ is constructed from the integral solution of the BGK model:

$$\begin{aligned} \hat{f}_{j+1/2} = f(x_{j+1/2}, t, v_x) &= \frac{1}{\tau} \int_{t^n}^{t} F_{eq}(x', t', v_x) \exp\left(-\frac{t - t'}{\tau}\right) dt' \\ &+ \exp\left(-\frac{t - t^n}{\tau}\right) f_0^n(x_{j+1/2} - v_x(t - t^n), t^n, v_x), \end{aligned}$$

where $x' = x_{j+1/2} - v_x(t - t')$ is the molecular trajectory and f_0^n is the initial gas distribution function of f at time t^n around the cell interface $x_{j+1/2}$.

Due to this natural coupling, UGKS and its variants [74–78] remove the restrictions on spatial cell size and time step (but it is still limited by the CFL condition), and its accuracy and efficiency have been demonstrated in many rarefied gas flow simulations. However, as information exchanging relies on the time evolution of VDF, UGKS needs a large number of evolution steps to obtain steady-state solutions in near-continuum flows [79, 80]. Also, UGKS is difficult to deal with the Boltzmann equation, as the complicated form of the BCO prohibits the analytical treatment.

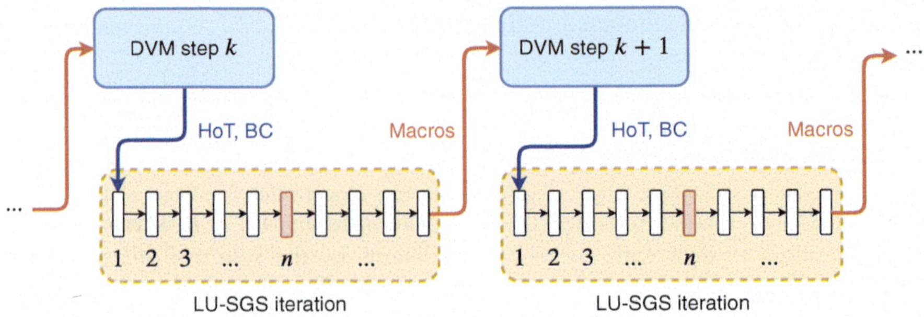

Fig. 2.6 Computational procedures in GSIS. HoT stands for the high-order terms extracted from the DVM solution of the Boltzmann equation. LU-SGS is the implicit time-stepping procedure solving the macroscopic synthetic equations which are exactly derived from the Boltzmann equation [81]. The Boltzmann equation and synthetic equations can be solved by different numerical methods with different order of accuracy, thus sophisticated methods in computational fluid dynamics can be used

2.7.3.3 General Synthetic Iterative Scheme

To develop a multi-scale numerical method for the Boltzmann equation, the general synthetic iterative scheme (GSIS) is proposed [61, 62, 81–83]. Its key ingredient is that macroscopic synthetic equations are simultaneously solved with the Boltzmann equation. The synthetic equations are exactly derived from the Boltzmann equation, where the stress and heat flux not only contain the Newton law of stress and the Fourier law of heat conduction, but also contain higher order constitutive relations extracted from the VDF. At each iteration of GSIS, the synthetic equations are solved to the steady state, and their solutions are used to modify the VDF in the Boltzmann equation, see Fig. 2.6. This two-way coupling enables the GSIS to find the steady-state solution within dozens of iterations at any values of Knudsen number and guarantees the solution's accuracy even at the hydrodynamic scale, rather than the tiny kinetic scale [83]; details will be presented in Chap. 8.

References

1. Sharipov F, Bertoldo G (2009) Numerical solution of the linearized Boltzmann equation for an arbitrary intermolecular potential. J Comput Phys 228:3345–3357
2. Sharipov F, Benites VJ (2017) Transport coefficients of helium-neon mixtures at low density computed from ab initio potentials. J Chem Phys 147:224302
3. Zeng JN, Li Q, Wu L (2022) Kinetic modeling of rarefied molecular gas dynamics. Acta Aerodyn Sinica 40:1–30 (in Chinese)
4. Chapman S, Cowling TG (1970) The mathematical theory of non-uniform gases. Cambridge University Press, Cambridge
5. Sharipov F (2018) Influence of quantum intermolecular interaction on internal flows of rarefied gases. Vacuum 156:146–153

6. Zhu LH, Wu L, Zhang YH, Sharipov F (2019) Ab initio calculation of rarefied flows of helium-neon mixture: classical vs. quantum scatterings. Int J Heat Mass Trans 145:118765
7. Cercignani C (1990) Mathematical methods in kinetic theory. Plenum Publishing Inc., 223 Spring Street, New York, N.Y. 10013
8. Struchtrup H (2005) Macroscopic transport equations for rarefied gas fows: approximation methods in kinetic theory. Springer, Heidelberg
9. Wu L, White C, Scanlon TJ, Reese JM, Zhang YH (2013) Deterministic numerical solutions of the Boltzmann equation using the fast spectral method. J Comput Phys 250:27–52
10. Wu L, Reese JM, Zhang YH (2014) Solving the Boltzmann equation by the fast spectral method: application to microflows. J Fluid Mech 746:53–84
11. Sugawara A, Yip S, Sirovich L (1968) Spectrum of density fluctuations in gases. Phys Fluids 11:925–932
12. Wang-Chang CS, Uhlenbeck GE (1951) Transport phenomena in polyatomic gases. University of Michigan Engineering Research Rept. No. CM-681
13. Kremer GM (2009) An introduction to the Boltzmann equation and transport processes in gases. Springer, Berlin
14. Beylich AA (2000) An interlaced system for Nitrogen gas. In: Proceedings of CECAM workshop
15. Tcheremissine FG, Agarwal RK (2008) Computation of hypersonic shock waves in diatomic gases using the generalized Boltzmann equation. AIP Conf Proc 1084:427–433
16. Wu L, Zhang YH, Reese JM (2015) Fast spectral solution of the generalized Enskog equation for dense gases. J Comput Phys 303:66–79
17. Wu L, Liu HH, Reese JM, Zhang YH (2016) Non-equilibrium dynamics of dense gas under tight confinement. J Fluid Mech 794:252–266
18. Sanchez IC (1994) Virial coefficients and close-packing of hard spheres and discs. J Chem Phys 101:7003–7006
19. Kurochkin VI (2002) Enskog kinetic equation modified for a dense soft-sphere gas. Tech Phys 47:19–23
20. Frezzotti A, Gibelli L, Lorenzani S (2005) Mean field kinetic theory description of evaporation of a fluid into vacuum. Phys Fluids 17:012102
21. Busuioc S, Gibelli L (2020) Mean-field kinetic theory approach to langmuir evapration of polyatomic liquids. Phys Fluids 32:093314
22. Barisik M, Beskok A (2014) Scale effects in gas nano flows. Phys Fluids 26:052003
23. Barisik M, Beskok A (2016) "Law of the nano-wall" in nano-channel gas flows. Microfluid Nanofluid 20:46
24. Mech J, Frezzotti A, Gibelli L (2008) A kinetic model for fluid wall interaction. Proc IMechE, Part C Eng Sci 222:787–795
25. Barbante P, Frezzotti A, Gibelli L (2015) A kinetic theory description of liquid menisci at the microscale. Kinet Relat Models 8:235–254
26. Brull S, Charrier P, Mieussens L (2016) Nanoscale roughness effect on Maxwell-like boundary conditions for the Boltzmann equation. Phys Fluids 28:082004
27. Cercignani C (1988) The Boltzmann equation and its applications. Springer, New York
28. Maxwell JC (1879) On stresses in rarefied gases arising from inequalities of temperature. Philosoph Trans R Soc Part 1(170):231–256
29. Epstein M (1967) A model of the wall boundary condition in kinetic theory. AIAA J 5:1797–1800
30. Cercignani C (1971) Model for gas surface interaction: comparison between theory and experiments. In: 7th international symposium on rarefied gas dynamics, vol 1, pp 75–79
31. Sharipov F (2003) Application of the Cercignani–Lampis scattering kernel to calculations of rarefied gas flows. III. Poiseuille flow and thermal creep through a long tube. Eur J Mech B/Fluids 22:145–154
32. Chu CK (1965) Kinetic-theoretic description of the formation of a shock wave. Phys Fluids 8:12

33. Aristov VV (2001) Direct methods for solving the Boltzmann equation and study of nonequilibrium flows. Springer, Netherlands
34. Bird GA (1970) Direct simulation and the Boltzmann equation. Phys Fluids 13:2676–2681
35. Bird GA (1994) Molecular gas dynamics and the direct simulation of gas flows. Oxford Science Publications, Oxford University Press Inc, New York
36. Wagner W (1992) A convergence proof for Bird's direct simulation Monte Carlo method for the Boltzmann equation. J Stat Phys 66:1011–1044
37. Sun QH, Boyd ID, Boyd GV (2003) A hybrid continuum/particle approach for micro-scale gas flows. Rarefied Gas Dyn 663
38. Nanbu K (1983) Direct simulation scheme derived from the Boltzmann equation. i. Monocomponent gases. J Phys Soc Jpn 52:2042–2049
39. Schwartzentruber TE, Scalabrin LC, Boyd ID (2007) A modular particle-continuum numerical method for hypersonic non-equilibrium gas flows. J Comput Phys 225:1159–1174
40. Burt JM, Boyd ID (2009) A hybrid particle approach for continuum and rarefied flow simulation. J Comput Phys 228(2):460–475
41. Baker LL, Hadjiconstantinou NG (2005) Variance reduction for Monte Carlo solutions of the Boltzmann equation. Phys Fluids 17(5):051703
42. Homolle TMM, Hadjiconstantinou NG (2007) A low-variance deviational simulation Monte Carlo for the Boltzmann equation. J Comput Phys 226(2):2341–2358
43. Radtke GA, Hadjiconstantinou NG, Wagner W (2011) Low-noise Monte Carlo simulation of the variable hard sphere gas. Phys Fluids 23(3):030606
44. Goldstein D, Sturtevant B, Boardwell JE (1989) Investigation of the motion of discrete-velocity gases. Rarefied Gas Dynamics. Progress in Aeronautics and Astronautics, 118
45. Tcheremissine FG (2005) Direct numerical solution of the Boltzmann equation. Rar Gas Dynam 677–685
46. Tcheremissine FG (2006) Solution to the Boltzmann kinetic equation for high-speed flows. Comput Math Math Phys 46:315
47. Kloss YY, Shuvalov PV, Tcheremissine FG (2010) Solving Boltzmann equation on GPU. Procedia Comput Sci 1(1):1083–1091
48. Dodulad OI, Tcheremissine FG (2013) Computation of a shock wave structure in monatomic gas with accuracy control. Comput Math Math Phys 53:827–844
49. Dodulad OI, Kloss YY, Savichkin DO, Tcheremissine FG (2014) Knudsen pumps modeling with Lennard-Jones and ab initio intermolecular potentials. Vacuum 109:360–367
50. Sone Y, Ohwada T, Aoki K (1989) Temperature jump and Knudsen layer in a rarefied-gas over a plane wall: numerical-analysis of the linearized boltzmann-equation for hard-sphere molecules. Phys Fluids A-Fluid Dyn 1(2):363–370
51. Ohwada T (1993) Structure of normal shock-waves: direct numerical analysis of the Boltzmann equation for hard-sphere molecules. Phys Fluids A-Fluid Dyn 5(1):217–234
52. Ohwada T (1996) Heat flow and temperature and density distributions in a rarefied gas between parallel plates with different temperatures. Finite-difference analysis of the nonlinear Boltzmann equation for hard-sphere molecules. Phys Fluids 8(8):2153–2160
53. Kosuge S, Aoki K, Takata S (2001) Shock-wave structure for a binary gas mixture: finite-difference analysis of the Boltzmann equation for hard-sphere molecules. Eur J Mech B/Fluids 20(1):87–126
54. Bobylev AV (1988) The theory of the nonlinear spatially uniform Boltzmann equation for Maxwell molecules. Math Phys Rev 7:111–233
55. Mouhot C, Pareschi L (2006) Fast algorithms for computing the Boltzmann collision operator. Math Comput 75(256):1833–1852
56. Filbet F, Mouhot C, Pareschi L (2006) Solving the Boltzmann equation in NlogN. SIAM J Sci Comput 28(3):1029–1053
57. Wu L, Reese JM, Zhang YH (2014) Oscillatory rarefied gas flow inside rectangular cavities. J Fluid Mech 748:350–367
58. Wu L, White C, Scanlon TJ, Reese JM, Zhang YH (2015) A kinetic model of the Boltzmann equation for non-vibrating polyatomic gases. J Fluid Mech 763:24–50

59. Wu L, Zhang J, Reese JM, Zhang YH (2015) A fast spectral method for the Boltzmann equation for monatomic gas mixtures. J Comput Phys 298:602–621
60. Wu L, Liu HH, Zhang YH, Reese JM (2015) Influence of intermolecular potentials on rarefied gas flows: fast spectral solutions of the Boltzmann equation. Phys Fluids 27:082002
61. Su W, Zhu LH, Wang P, Zhang YH, Wu L (2020) Can we find steady-state solutions to multiscale rarefied gas flows within dozens of iterations? J Comput Phys 407:109245
62. Su W, Zhang YH, Wu L (2021) Multiscale simulation of molecular gas flows by the general synthetic iterative scheme. Comput Methods Appl Mech Engrg 373:113548
63. Wu L, Zhang J, Liu HH, Zhang YH, Reese JM (2017) A fast iterative scheme for the linearized Boltzmann equation. J Comput Phys 338:431–451
64. Wu L (2019) A fast spectral method for the Uehling-Uhlenbeck equation for quantum gas mixtures: homogeneous relaxation and transport coefficients. J Comput Phys 399:108924
65. Ho MT, Zhu LH, Wu L, Wang P, Guo ZL, Li ZH, Zhang YH (2019) A multi-level parallel solver for rarefied gas flows in porous media. Comput Phys Commun 234:14–25
66. Roveda R, Goldstein DB, Varghese PL (1998) Hybrid Euler/particle approach for continuum/rarefied flows. J Spacecr Rocket 35(3):258–265
67. Wang WL, Boyd I (2003) Hybrid DSMC-CFD simulations of hypersonic flow over sharp and blunted bodies. In: 36th AIAA thermophysics conference, pp 1–13
68. Wijesinghe HS, Hornung RD, Garcia AL, Hadjiconstantinou NG (2004) Three-dimensional hybrid continuum-atomistic simulations for multiscale hydrodynamics. J Fluids Eng 126(5):768–777
69. Kolobov VI, Arslanbekov RR, Aristov VV, Frolova AA, Zabelok SA (2007) Unified solver for rarefied and continuum flows with adaptive mesh and algorithm refinement. J Comput Phys 223(2):589–608
70. Tang ZY, He BJ, Cai GB (2014) Investigation on a coupled Navier-Stokes-Direct simulation Monte Carlo method for the simulation of plume flowfield of a conical nozzle. Int J Numer Meth Fluids 76(2):95–108
71. Yang WQ, Gu XJ, Wu L, Emerson DR, Zhang YH, Tang S (2020) A hybrid approach to couple the discrete velocity method and method of moments for rarefied gas flows. J Comput Phys 410:109397
72. Xu K, Huang JC (2010) A unified gas-kinetic scheme for continuum and rarefied flows. J Comput Phys 229:7747–7764
73. Bhatnagar PL, Gross EP, Krook M (1954) A model for collision processes in gases. I. Small amplitude processes in charged and neutral one-component systems. Phys Rev 94:511–525
74. Guo ZL, Xu K, Wang RJ (2013) Discrete unified gas kinetic scheme for all Knudsen number flows: low-speed isothermal case. Phys Rev E 88:033305
75. Guo ZL, Wang RJ, Xu K (2015) Discrete unified gas kinetic scheme for all Knudsen number flows. II. Thermal compressible case. Phys Rev E 91:033313
76. Zhu YJ, Zhong CW, Xu K (2016) Implicit unified gas-kinetic scheme for steady state solutions in all flow regimes. J Comput Phys 315:16–38
77. Yang LM, Shu C, Yang WM, Chen Z, Dong H (2018) An improved discrete velocity method (DVM) for efficient simulation of flows in all flow regimes. Phys Fluids 30:062005
78. Yang LM, Chen Z, Shu C, Yang WM, Wu J, Zhang LQ (2019) Improved fully implicit discrete-velocity method for efficient simulation of flows in all flow regimes. Phys Rev E 98:063313
79. Chen SZ, Xu K (2015) A comparative study of an asymptotic preserving scheme and unified gas-kinetic scheme in continuum flow limit. J Comput Phys 288:52–65
80. Wang P, Ho MT, Wu L, Guo ZL, Zhang YH (2018) A comparative study of discrete velocity methods for low-speed rarefied gas flows. Comput Fluids 161:33–46
81. Zhu LH, Pi XC, Su W, Li ZH, Zhang YH, Wu L (2021) General synthetic iterative scheme for nonlinear gas kinetic simulation of multi-scale rarefied gas flows. J Comput Phys 430:110091
82. Su W, Ho MT, Zhang YH, Wu L (2020) GSIS: an efficient and accurate numerical method to obtain the apparent gas permeability of porous media. Comput Fluids 206:104576
83. Su W, Zhu LH, Wu L (2020) Fast convergence and asymptotic preserving of the general synthetic iterative scheme. SIAM J Sci Comput 42:B1517–B1540

Chapter 3
Fluid-Dynamic Equation

Since the Boltzmann equation is defined in high-dimensional phase space, it is useful to derive fluid-dynamic equations using only a few macroscopic quantities. This is one of the major challenges in Hilbert's problems: "Thus Boltzmann's work on the principles of mechanics suggests the problem of developing mathematically the limiting processes, which lead from the atomistic view to the laws of motion of continua" [1], which is not fully solved even today. In this chapter we first introduce how macroscopic equations are derived, and then assess their accuracy.

3.1 Hilbert Expansion

Macroscopic equations like Eq. (1.1) can be obtained by multiplying the Boltzmann equation (2.19) with the collision invariants and integrating with respect to the molecular velocity $\boldsymbol{v}$. However, they are not closed because expressions for the stress and heat flux are not known. One may continue this process to find the evolution equations for the stress, heat flux, and other high order moments, but due to the streaming operator, higher order unclosed moments always appear.

Hilbert proposed to solve the Boltzmann equation via a formal asymptotic expansion of the VDF as [2]:

$$f(t, \boldsymbol{x}, \boldsymbol{v}) = \sum_{n=0}^{\infty} \epsilon^n f^{(n)}(t, \boldsymbol{x}, \boldsymbol{v}), \tag{3.1}$$

where ϵ can be regarded as a measure of the spatial Knudsen number; it is assumed to be a small parameter for ordering terms and quantities appearing in the expansion, which will be replaced by unity when the solutions are obtained. Meanwhile, the BCO is written as (we shall see in Sect. 4.2 that after normalization the Knudsen number appears in the denominator):

L. Wu, *Rarefied Gas Dynamics*,
https://doi.org/10.1007/978-981-19-2872-7_3

$$\frac{Q}{\epsilon} = \sum_{n=0}^{\infty} \epsilon^{n-1} Q^{(n)}, \tag{3.2}$$

where $Q^{(0)} = Q(f^{(0)}, f^{(0)})$ and $Q^{(1)} = f^{(0)}\mathcal{J}(\phi)$ if we write $f^{(1)} = f^{(0)}\phi$. Macroscopic variables (2.2) are also expanded:

$$\left[\rho, \rho\boldsymbol{u}, \frac{3}{2}k_B\rho T\right] = \sum_{n=0}^{\infty} \epsilon^n \int \left[1, \boldsymbol{v}, \frac{1}{2}mc^2\right] mf^{(n)} d\boldsymbol{v}, \tag{3.3}$$

and

$$p_{ij} = \sum_{n=0}^{\infty} \epsilon^n p_{ij}^{(n)}, \quad \boldsymbol{q} = \sum_{n=0}^{\infty} \epsilon^n \boldsymbol{q}^{(n)}, \tag{3.4}$$

with $p_{ij}^{(n)} = \int mc_ic_j f^{(n)} d\boldsymbol{v}$ and $\boldsymbol{q}^{(n)} = \int \frac{1}{2}mc^2\boldsymbol{c} f^{(n)} d\boldsymbol{v}$.

Substituting Eq. (3.1) into the Boltzmann equation and collecting powers of ϵ yields an infinite system of integro-differential equations. At the order $O(\epsilon^{-1})$ we have $Q^{(0)} = 0$. Therefore, $f^{(0)}$ is the equilibrium distribution function (2.31):

$$f^{(0)} = \frac{\rho^{(0)}}{m}\left(\frac{m}{2\pi k_B T^{(0)}}\right)^{3/2} \exp\left(-\frac{m(\boldsymbol{v} - \boldsymbol{u}^{(0)})^2}{2k_B T^{(0)}}\right), \tag{3.5}$$

but the density, velocity and temperature are evaluated at the zeroth-order expansion, e.g., $\rho^{(0)} = \int mf^{(0)} d\boldsymbol{v}$. If only this zeroth-order expansion is adopted, we have $\rho^{(0)} = \rho$, $\boldsymbol{u}^{(0)} = \boldsymbol{u}$, and $T^{(0)} = T$. Therefore, the corresponding set of macroscopic equations are the Euler equations, i.e., Eq. (1.1) with $p_{ij} = nk_BT\delta_{ij}$ and $\boldsymbol{q} = 0$.

When the first-order expansion in VDF is considered: $f = f^{(0)}1 + \phi$, we have

$$f^{(0)}\mathcal{J}(\phi) = \frac{\partial f^{(0)}}{\partial t} + \boldsymbol{v} \cdot \frac{\partial f^{(0)}}{\partial \boldsymbol{x}} + \boldsymbol{a} \cdot \frac{\partial f^{(0)}}{\partial \boldsymbol{v}}. \tag{3.6}$$

By solving Eq. (3.6), one can obtain ϕ and hence $f^{(1)}$; and the process can be carried on to obtain high-order approximations; we refer to the monograph of Sone for the complicated derivation of macroscopic equations and their applications [3]. However, it should be noted that the NSF equations never emerge according to the Hilbert expansion.

3.2 Chapman-Enskog Expansion

Chapman and Enskog independently proposed a new technique [4, 5], which has become the standard procedure to derive macroscopic equations from the kinetic equation. The most successful part is that the NSF equations are derived, where the transport coefficients agree with experimental data. The unsuccessful parts are that macroscopic equations obtained from higher order expansion are (i) not necessary more accurate than the NSF equations [3, 6], and (ii) unstable to perturbations [7, 8].

In the Chapman-Enskog expansion, the VDF, stress and heat flux remain the same as the Hilbert expansion. However, the five conservative variables $C_M = \{\rho, \boldsymbol{u}, T\}$ are defined only by the zeroth-order expansion. That is,

$$\left[\rho, \rho\boldsymbol{u}, \frac{3}{2}k_B\rho T\right] = m\int\left[1, \boldsymbol{v}, \frac{1}{2}mc^2\right]f^{(0)}d\boldsymbol{v}, \tag{3.7}$$

with the compatibility conditions

$$\int f^{(n)}d\boldsymbol{v} = \int \boldsymbol{v}f^{(n)}d\boldsymbol{v} = \int c^2 f^{(n)}d\boldsymbol{v} = 0, \quad \text{for} \quad n \geq 1. \tag{3.8}$$

On substituting the above equations to Eq. (1.1), one finds that the time derivatives should be formally written as a series in ϵ [9]:

$$\frac{\partial}{\partial t} = \sum_{n=0}^{\infty}\epsilon^n\frac{\partial}{\partial t_n}. \tag{3.9}$$

Therefore, at the zeroth-order approximation, we have

$$\begin{aligned}\frac{\partial\rho}{\partial t_0} + \frac{\partial(\rho u_j)}{\partial x_j} &= 0,\\ \frac{\partial(\rho u_i)}{\partial t_0} + \frac{\partial(\rho u_i u_j + nk_BT\delta_{ij})}{\partial x_j} &= \rho a_i,\\ \frac{\partial(\rho E)}{\partial t_0} + \frac{\partial\left(\rho E u_j + u_i nk_BT\delta_{ij}\right)}{\partial x_j} &= \rho a_j u_j,\end{aligned} \tag{3.10}$$

while at other orders ($n > 0$) we have

$$\frac{\partial\rho}{\partial t_n} = 0, \quad \frac{\partial(\rho u_i)}{\partial t_n} + \frac{\partial p_{ij}^{(n)}}{\partial x_j} = 0, \quad \frac{\partial(\rho E)}{\partial t_n} + \frac{\partial\left(u_i p_{ij}^{(n)} + q_j^{(n)}\right)}{\partial x_j} = 0. \tag{3.11}$$

At the order $O(\epsilon^{-1})$, like the Hilbert expansion, we have $Q(f^{(0)}, f^{(0)}) = 0$, but here $f^{(0)}$ is just the equilibrium distribution function (2.31), because the conservative variables C_M are determined only according to the zeroth-order expansion. At the order $O(1)$, the VDF $f^{(1)} = F_{eq}\phi$ satisfies

$$
\begin{aligned}
F_{eq}\mathcal{J}(\phi) =& \frac{\partial F_{eq}}{\partial t_0} + \boldsymbol{v}\cdot\frac{\partial F_{eq}}{\partial \boldsymbol{x}} + \boldsymbol{a}\cdot\frac{\partial F_{eq}}{\partial \boldsymbol{v}} \\
=& \left[2\xi_{\langle i}\xi_{j\rangle}\frac{\partial u_{\langle i}}{\partial x_{j\rangle}} + \left(\xi^2 - \frac{5}{2}\right)\xi_i\sqrt{\frac{2k_B T}{m}}\frac{\partial \ln T}{\partial x_i} \right] F_{eq}.
\end{aligned} \tag{3.12}
$$

Note that the final equation is obtained with the help of $\frac{\partial F_{eq}}{\partial t_0} = \frac{\partial F_{eq}}{\partial C_M}\frac{\partial C_M}{\partial t_0}$, $\frac{\partial F_{eq}}{\partial \boldsymbol{x}} = \frac{\partial F_{eq}}{\partial C_M}\frac{\partial C_M}{\partial \boldsymbol{x}}$, and Eq. (3.10).

The solution of the integral equation (3.12) is decomposed into homogeneous and inhomogeneous parts. The homogeneous part of solution satisfied $\mathcal{J}(\phi) = 0$, thus ϕ is a linear combination of the five fundamental collisional invariants. However, the compatibility condition (3.8) requires that all the combination coefficients are zero. The inhomogeneous solution must be read

$$
\phi = -A_\mu(\xi)\xi_{\langle i}\xi_{j\rangle}\frac{\partial u_{\langle i}}{\partial x_{j\rangle}} - A_\kappa(\xi)\xi_i\sqrt{\frac{2k_B T}{m}}\frac{\partial \ln T}{\partial x_i}, \tag{3.13}
$$

where $\int \xi^2 A_\kappa(\xi) F_{eq} d\boldsymbol{v} = 0$ for the compatibility condition (3.8), and the solutions of $A_\kappa(\xi)$ and $A_\mu(\xi)$ satisfy

$$
\mathcal{J}(A_\mu\xi_{\langle i}\xi_{j\rangle}) = -2\xi_{\langle i}\xi_{j\rangle}, \text{ and } \mathcal{J}(A_\kappa\xi_i) = -\left(\xi^2 - \frac{5}{2}\right)\xi_i. \tag{3.14}
$$

Once $A_\mu(\xi)$ and $A_\kappa(\xi)$ are known, Newton's law for stress and Fourier's law for heat conduction can be obtained as

$$
\sigma_{ij}^{(1)} = -2\mu\frac{\partial u_{\langle i}}{\partial x_{j\rangle}}, \text{ and } q_i^{(1)} = -\kappa\frac{\partial T}{\partial x_i}, \tag{3.15}
$$

where the shear viscosity and heat conductivity are

$$
\begin{aligned}
\mu =& \frac{2p}{15\pi^{3/2}}\int \exp(-\xi^2)A_\mu(\xi)\xi^4 d\boldsymbol{\xi}, \\
\kappa =& \frac{2p}{3\pi^{3/2}}\frac{k_B}{m}\int \exp(-\xi^2)A_\kappa(\xi)\xi^4 d\boldsymbol{\xi}.
\end{aligned} \tag{3.16}
$$

The NSF equations are obtained by combining Eqs. (3.9), (3.10) and (3.11) with $n = 1$, which is accurate to $O(\text{Kn})$. The Burnett and super-Burnett equations can be derived from the second- and third-order Chapman-Enskog expansions [10].

3.2.1 Expansion in Sonine Polynomials

Although Eq. (3.14) can be solved exactly by the fast spectral method so that the transport coefficients in Eq. (3.16) can be calculated exactly (see Sect. 5.5.2), an analytical derivation will be beneficial.

With the Sonine polynomials in Appendix A, the integral equation (3.14) can be written as $\mathcal{J}(A_\kappa\xi_i) = S_{\frac{3}{2}}^{(1)}(\xi^2)\xi_i$ and $\mathcal{J}(A_\mu\xi_{\langle i}\xi_{j\rangle}) = -2S_{\frac{5}{2}}^{(0)}(\xi^2)\xi_{\langle i}\xi_{j\rangle}$. For their solution, $A_\kappa(\xi)$ and $A_\mu(\xi)$ are expanded as

$$A_\kappa(\xi) = -\sum_{r=1}^{n_a} a_r S_{\frac{3}{2}}^{(r)}(\xi^2), \text{ and } A_\mu(\xi) = \sum_{r=0}^{n_b} b_r S_{\frac{5}{2}}^{(r)}(\xi^2). \tag{3.17}$$

Multiplying Eq. (3.14) with $\exp(-\xi^2)S_{\frac{3}{2}}^{(s)}(\xi^2)\xi_i$ and $\exp(-\xi^2)S_{\frac{5}{2}}^{(s)}(\xi^2)\xi_{\langle i}\xi_{j\rangle}$, respectively, and taking advantage of the orthogonality of Sonine's polynomials, we have

$$\sum_{r=1}^{n_a} \alpha_{sr}a_r = \frac{15}{4}\pi^3\frac{m}{\rho}\delta_{s1}, \qquad \sum_{r=1}^{n_a} \beta_{sr}b_r = 5\pi^3\frac{m}{\rho}\delta_{s0}, \tag{3.18}$$

where the matrices α_{sr} and β_{sr} are

$$\alpha_{sr} = \left[S_{\frac{3}{2}}^{(s)}(\xi^2)\xi_i, S_{\frac{3}{2}}^{(r)}(\xi^2)\xi_i\right], \quad \beta_{sr} = \left[S_{\frac{5}{2}}^{(s)}(\xi^2)\xi_{\langle i}\xi_{j\rangle}, S_{\frac{5}{2}}^{(r)}(\xi^2)\xi_{\langle i}\xi_{j\rangle}\right], \tag{3.19}$$

with the bracket operator defined in Eq. (2.50).

Note that with the compatibility equation, the last equation in Eq. (3.16) can be rewritten as $\kappa = \frac{2p}{3\pi^{3/2}}\frac{k_B}{m}\int \exp(-\xi^2)A_\kappa(\xi)\xi_i\left(\xi^2 - \frac{5}{2}\right)\xi_i d\boldsymbol{\xi}$. Therefore, together with Eq. (3.17), we have

$$\mu = \frac{p}{2}b_0, \quad \kappa = \frac{p}{2}a_1. \tag{3.20}$$

In order to obtain the analytical expressions for shear viscosity and thermal conductivity, we need to compute α_{sr} and β_{sr}, and their inverses. Their values depend on the values of n_a and n_b in Eq. (3.17). However, for Maxwellian molecules, according to Eq. (2.42), α_{sr} and β_{sr} are diagonal, so

$$\mu = \frac{p}{2}b_0 = \frac{5\pi^3}{2}k_BT\beta_{00}^{-1}, \quad \kappa = \frac{p}{2}a_1 = \frac{75\pi^3}{8}\frac{k_B}{m}k_BT\alpha_{11}^{-1}. \tag{3.21}$$

For other intermolecular potentials, this approximation already gives a very accurate estimation of the transport coefficients, as we will see immediately in the following subsections.

3.2.2 Expansion to the First Order

If only one element in each expansion in Eq. (3.17) is used, we get

$$\begin{aligned}\alpha_{11} =& \frac{1}{4}\int \exp(-\xi^2 - \xi_*^2)\mathcal{D}^2(\xi^2\xi_i)B(\cos\theta, v_r)d\Omega d\boldsymbol{\xi}_* d\boldsymbol{\xi},\\ \beta_{00} =& \frac{1}{4}\int \exp(-\xi^2 - \xi_*^2)\mathcal{D}^2(\xi_{\langle i}\xi_{j\rangle})B(\cos\theta, v_r)d\Omega d\boldsymbol{\xi}_* d\boldsymbol{\xi},\end{aligned} \tag{3.22}$$

where $\mathcal{D}(\psi) = \psi + \psi_* - \psi' - \psi'_*$.

If we define $\eta_i = \xi_i - \xi_{*i}$, $\zeta_i = \frac{1}{2}(\xi_i + \xi_{*i}) = \frac{1}{2}(\xi'_i + \xi'_{*i})$, $\eta'_i = \xi'_i - \xi'_{*i}$, and $v_r = v_m\eta$, we have $\xi^2 + \xi_*^2 = 2\zeta^2 + \frac{1}{2}\eta^2$, $d\boldsymbol{\xi}_* d\boldsymbol{\xi} = d\boldsymbol{\eta} d\boldsymbol{\zeta}$, $D(\xi^2\xi_i) = \zeta_i(\eta_i\eta_j - \eta'_i\eta'_j)$, $\mathcal{D}(\xi_{\langle i}\xi_{j\rangle}) = \frac{1}{2}(\eta_i\eta_j - \eta'_i\eta'_j)$. Thus,

$$\mathcal{D}^2(\xi_{\langle i}\xi_{j\rangle}) = \frac{1}{4}(\eta_i\eta_j - \eta'_i\eta'_j)(\eta_i\eta_j - \eta'_i\eta'_j) = \frac{1}{2}(\eta^4 - \eta_i\eta'_i\eta_j\eta'_j) = \frac{1}{2}\eta^4\sin^2\theta, \tag{3.23}$$

where the last equation is obtained by the fact that the relative velocity is deflected by an angle of θ after the binary collision, while its magnitude remains unchanged, see Fig. 2.1.

In the calculation of $D^2(\xi^2\xi_i)$, if we ignore the terms which have zero contribution to the integral in Eq. (3.22), we have

$$\begin{aligned}D^2(\zeta^2\zeta_i) =& \zeta_1^2(\eta_1\eta_j - \eta'_1\eta'_j)(\eta_1\eta_j - \eta'_1\eta'_j) + \zeta_2^2(\eta_2\eta_j - \eta'_2\eta'_j)(\eta_2\eta_j - \eta'_2\eta'_j)\\ &+ \zeta_3^2(\eta_3\eta_j - \eta'_3\eta'_j)(\eta_3\eta_j - \eta'_3\eta'_j)\\ =& \frac{\zeta^2}{3}(\eta_1\eta_j - \eta'_1\eta'_j)(\eta_1\eta_j - \eta'_1\eta'_j) + \frac{\zeta^2}{3}(\eta_2\eta_j - \eta'_2\eta'_j)(\eta_2\eta_j - \eta'_2\eta'_j)\\ &+ \frac{\zeta^2}{3}(\eta_3\eta_j - \eta'_3\eta'_j)(\eta_3\eta_j - \eta'_3\eta'_j)\\ =& \frac{2}{3}\zeta^2(\eta^4 - \eta_i\eta'_i\eta_j\eta'_j) = \frac{2}{3}\zeta^2\eta^4\sin^2\theta.\end{aligned} \tag{3.24}$$

After integration with respect to ζ it can be shown that

$$\alpha_{11} = \beta_{00} = \frac{1}{8}\sqrt{\frac{\pi}{2}}^{-3} 4\pi v_m \int_0^\infty \exp\left(-\frac{\eta^2}{2}\right)\eta^7\sigma_\mu d\eta, \tag{3.25}$$

where

$$\sigma_\mu = \int \frac{B(\cos\theta, v_r)}{v_r}\sin^2\theta d\Omega = 2\pi\int_0^\pi \frac{B(\cos\theta, v_m\eta)}{v_m\eta}\sin^3\theta d\theta \tag{3.26}$$

is the viscosity cross-section. Therefore, the shear viscosity, which is obtained from Eq. (3.21) for the first approximation, is given by

$$\mu^{[1]} = \frac{5\sqrt{\pi m k_B T}}{8D}, \tag{3.27}$$

where

$$D = \left(\frac{m}{4k_B T}\right)^4 \int_0^\infty v_r^7 \sigma_\mu \exp\left(-\frac{m v_r^2}{4k_B T}\right) dv_r. \tag{3.28}$$

The corresponding thermal conductivity is given by

$$\kappa^{[1]} = \frac{15}{4}\frac{k_B}{m}\mu^{[1]}, \tag{3.29}$$

which results in a Prandtl number

$$\mathrm{Pr} = \frac{\mu c_p}{\kappa} = \frac{5k_B}{2m}\frac{\mu}{\kappa} = \frac{2}{3}, \tag{3.30}$$

where c_p is the specific heat at constant pressure.

3.2.3 Expansion to Higher Orders

It turns out that the expansion to first order is accurate enough for the shear viscosity and thermal conductivity: the results do not change much if we consider larger values of n_a and n_b in Eq. (3.17). For example, for the power-law potential (2.9), expansion to the second order leads to

$$\begin{aligned} \mu^{[2]} &= \left[1 + \frac{3(\eta-5)^2}{2(\eta-1)(101\eta-113)}\right]\mu^{[1]}, \\ \kappa^{[2]} &= \left[1 + \frac{(\eta-5)^2}{4(\eta-1)(11\eta-13)}\right]\kappa^{[1]}. \end{aligned} \tag{3.31}$$

For Maxwell molecules, Eqs. (3.28) and (3.29) are exact, while for HS molecules, if $n_a = n_b = 4$, we have

$$\mu^{[4]} = 1.016\mu^{[1]}, \text{ and } \kappa^{[4]} = 1.025\kappa^{[1]}. \tag{3.32}$$

For power-law potentials with exponents $5 \le \eta < \infty$, the correction factors fall between these for Maxwellian and HS molecules.

3.3 Moment Methods

Grad proposed another way to obtain the closed macroscopic equations [11]. In addition to the five equations from the conservation laws, he obtained the evolution equations of shear stress and heat flux by respectively multiplying the Boltzmann equation with $c_i c_j$ and $c_i c^2$ and integrating over molecular velocity space:

$$\begin{aligned}
&\frac{D\sigma_{ij}}{Dt} + \frac{\partial u^0_{ijk}}{\partial x_k} + \frac{4}{5}\frac{\partial q_{\langle i}}{\partial x_{j\rangle}} + 2\sigma_{k\langle i}\frac{\partial u_{j\rangle}}{\partial x_k} + \sigma_{ij}\frac{\partial u_k}{\partial x_k} + 2p\frac{\partial u_{\langle i}}{\partial x_{j\rangle}} = -\frac{p}{\mu}\sigma_{ij}, \\
&\frac{Dq_i}{Dt} + \frac{5}{2}p\frac{\partial \theta}{\partial x_i} - \sigma_{ik}\frac{\partial \theta}{\partial x_k} - \sigma_{ik}\theta\frac{\partial \ln\rho}{\partial x_k} - \frac{\sigma_{ik}}{\rho}\frac{\partial \sigma_{kl}}{\partial x_l} - \frac{5}{2}\theta\frac{\partial \sigma_{ik}}{\partial x_k} + \frac{1}{2}\frac{\partial u^1_{ik}}{\partial x_k} \\
&\qquad + \frac{1}{6}\frac{\partial w}{\partial x_i} + u^0_{ijk}\frac{\partial u_k}{\partial x_l} + \frac{7}{5}q_i\frac{\partial u_k}{\partial x_k} + \frac{7}{5}q_k\frac{\partial u_i}{\partial x_k} + \frac{2}{5}q_k\frac{\partial u_k}{\partial x_i} = -\frac{p}{\mu}q_i,
\end{aligned} \tag{3.33}$$

where $\theta = k_B T/m$, the terms in the right-hand side of this equation are obtained from the Boltzmann equation for Maxwellian molecules, and

$$\frac{D}{Dt} = \frac{\partial}{\partial t} + u_k\frac{\partial}{\partial x_k} \tag{3.34}$$

is the material derivative.

The system is not closed because of the following unknown quantities:

$$[u^0_{ijk}, u^1_{ij}] = m\int [c_{\langle i}c_j c_{k\rangle}, c^2 c_{\langle i}c_{j\rangle}] f d\boldsymbol{v}, \quad w = m\int c^4(f - F_{eq})d\boldsymbol{v}. \tag{3.35}$$

They are determined to be $u^0_{ijk} = w = 0$ and $u^1_{ij} = 7\theta\sigma_{ij}$, when the VDF is expanded into the Hermite polynomials of the peculiar velocity:

$$f = F_{eq}\left[1 + \frac{\sigma_{ij}}{2\rho(k_B T/m)^2}c_{\langle i}c_{j\rangle} + \frac{2}{5}\frac{q_i}{\rho(k_B T/m)^2}c_i\left(\frac{c^2}{2(k_B T/m)} - \frac{5}{2}\right)\right]. \tag{3.36}$$

Higher order equations such as the Grad 26 moments equations are constructed by adding more moments and their evolution equations. Generally speaking, when the rarefaction effects become significant, the number of moments should be increased. For example, for normal shock waves with Mach number up to 1.65 and 1.887, at least 13 and 21 moments are needed [12].

The moment systems are linearly stable, and hence research into this direction remains active and significant progress has been made. To remove the problem of hyperbolicity that results in discontinuities in the simulation of strong shock waves, Struchtrup and Torrihon derived the regularized 13 (R13) equations by combining Grad's method and Chapman-Enskog expansion [13], while Gu and Emerson developed the R26 equations [14]. R13 and R26 are both linearly stable and are accurate to the order of Kn^3 and Kn^5, respectively.

3.4 Accuracy of Macroscopic Equations

We assess the fluid-dynamic equations by considering the Rayleigh-Brillouin scattering (RBS), where the light is scattered by the spontaneous density fluctuation of gas molecules; more details are presented in Chap. 15. This problem is ideal, because the gas-wall interactions are absent so only the macroscopic equations are assessed. Moreover, a solution can always be found in this 1D and linearizable problem.

Introducing the dimensionless quantities:

$$\tilde{x}=\frac{x_1}{L_0},\quad \tilde{t}=\frac{t}{L/v_0},\quad \tilde{\rho}=\frac{\rho}{\rho_0}-1,\quad \tilde{u}=\frac{u_1}{v_0},$$
$$\tilde{T}=\frac{T}{T_0}-1,\quad \tilde{\sigma}=\frac{\sigma_{11}}{p_0},\quad \tilde{q}=\frac{q_1}{p_0v_0},\tag{3.37}$$

where $v_0=\sqrt{k_BT_0/m}$, Eq. (1.1) is linearized as (tildes are omitted for clarity):

$$\frac{\partial\rho}{\partial t}+\frac{\partial u}{\partial x}=0,$$
$$\frac{\partial u}{\partial t}+\frac{\partial\rho}{\partial x}+\frac{\partial T}{\partial x}+\frac{\partial\sigma}{\partial x}=0,$$
$$\frac{3}{2}\frac{\partial T}{\partial t}+\frac{\partial u}{\partial x}+\frac{\partial q}{\partial x}=0.\tag{3.38}$$

3.4.1 *Equations from Chapman-Enskog Expansion*

The stress and heat flux in the NSF equations are given by[1]

$$\sigma^{(NS)}=-\frac{4}{3}\text{Kn}\frac{\partial u}{\partial x},\quad q^{(NS)}=-\frac{15}{4}\text{Kn}\frac{\partial T}{\partial x}.\tag{3.39}$$

The stress and heat flux in the Burnett equations are (for Maxwellian molecules)

$$\sigma^{(B)}=\sigma^{(NS)}-\text{Kn}^2\left(\frac{4}{3}\frac{\partial^2\rho}{\partial x^2}-\frac{2}{3}\frac{\partial^2T}{\partial x^2}\right),$$
$$q^{(B)}=q^{(NS)}-\frac{7}{4}\text{Kn}^2\frac{\partial^2u}{\partial x^2},\tag{3.40}$$

while those in the super-Burnett equations are [15]

[1] Note that in this chapter Kn is $\sqrt{2/\pi}$ times of the Knudsen number defined in Eq. (1.10).

$$\sigma^{(SB)} = \sigma^{(B)} + \frac{2}{9}\mathrm{Kn}^3 \frac{\partial^3 u}{\partial x^3},$$
$$q^{(SB)} = q^{(B)} + \mathrm{Kn}^3 \left(-\frac{157}{16}\frac{\partial^3 T}{\partial x^3} - \frac{5}{8}\frac{\partial^3 \rho}{\partial x^3} \right). \tag{3.41}$$

Applying the Laplace transform for t and the Fourier transform for x in the spontaneous RBS, Eqs. (3.38) and (3.39) are turned into the following matrix form [16]:

$$\begin{bmatrix} -i\varpi & 2i\pi & 0 \\ 2i\pi & -i\varpi + \frac{16}{3}\pi^2\mathrm{Kn} & 2\,i\pi \\ 0 & 2\,i\pi & -\frac{3}{2}i\varpi + 15\pi^2\mathrm{Kn} \end{bmatrix} \begin{bmatrix} \hat{\rho} \\ \hat{u} \\ \hat{T} \end{bmatrix} = \begin{bmatrix} 1 \\ 0 \\ 0 \end{bmatrix}, \tag{3.42}$$

where $\hat{\rho}$, $\hat{u}$, and $\hat{T}$ are the spectra of the corresponding quantities, and ϖ is the scattering frequency normalized by v_0/L (and hence $\varpi = 2\sqrt{2}\pi f_s$). Note that the value of 1 in the right-hand-side of the equation comes from the Laplace transform of the initial density impulse. And for this reason the Hilbert expansion is not considered because we do not know how to distribute this initial density impulse to the different order of $\rho^{(n)}$. The spectrum of spontaneous RBS is the real part density disturbance $\hat{\rho}$. By solving Eq. (3.42) we get

$$S_s^{NS} = \Re \left(\frac{160\,i\mathrm{Kn}^2\pi^4 + 46\,\mathrm{Kn}\,\varpi\,\pi^2 - 3\,i\varpi^2 + 8\,i\pi^2}{160\,\mathrm{Kn}^2\varpi\,\pi^4 - 46\,i\mathrm{Kn}\,\varpi^2\pi^2 + 120\,i\mathrm{Kn}\,\pi^4 - 3\,\varpi^3 + 20\,\varpi\,\pi^2} \right), \tag{3.43}$$

and those of the Burnett and super-Burnett equations are obtained in a same manner.

The spectra obtained from the NSF, Burnett, super-Burnett equations are compared with those from the linearized Boltzmann equation in Fig. 3.1. The NSF equations work up to $\mathrm{Kn} \approx 0.02$. The Burnett equations do not improve the accuracy, although it is obtained using one more order of expansion than that of NSF equations. And the super-Burnett equations are even more inaccurate, especially when $\mathrm{Kn} = 0.1$. This conforms that one does not know what step of the approximation is required or sufficient, in the Chapman-Enskog expansion, to obtain a solution that is correct up to certain order [3, 6].

3.4.2 *Moment Equations*

In the spontaneous RBS, the stress and heat flux in the Grad 13 (G13) equations are described by the following equations:

$$\frac{\partial \sigma}{\partial t} + \frac{4}{3}\frac{\partial u}{\partial x} + \frac{8}{15}\frac{\partial q}{\partial x} = -\frac{\sigma}{\mathrm{Kn}},$$
$$\frac{\partial q}{\partial t} + \frac{\partial \sigma}{\partial x} + \frac{5}{2}\frac{\partial T}{\partial x} = -\frac{2}{3}\frac{q}{\mathrm{Kn}}. \tag{3.44}$$

The spectrum of spontaneous RBS can be obtained by solving the following matrix:

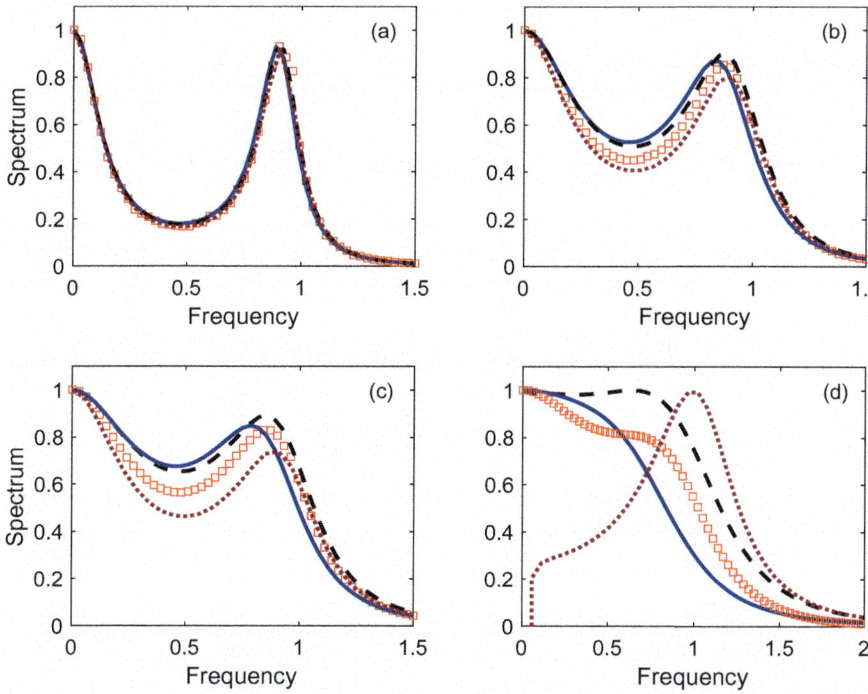

Fig. 3.1 Spontaneous RBS spectra obtained from the Chapman-Enskog expansion [17]. The Knudsen number is **a** 0.02, **b** 0.04, **c** 0.05, and **d** 0.1. Solid, dashed and dotted lines are results from the NS, Burnett and super-Burnett equations, respectively. For figures in this section, squares are the LBE solutions of Maxwellian gas

$$\begin{bmatrix} -i\varpi & 2i\pi & 0 & 0 & 0 \\ 2i\pi & -i\varpi & 2i\pi & 2i\pi & 0 \\ 0 & 2i\pi & -\frac{3}{2}i\varpi & 0 & 2i\pi \\ 0 & \frac{8}{3}i\pi & 0 & -i\varpi+\frac{1}{\text{Kn}} & \frac{16}{15}i\pi \\ 0 & 0 & 5i\varpi & 2i\varpi & -i\varpi+\frac{2}{3\text{Kn}} \end{bmatrix} \begin{bmatrix} \hat{\rho} \\ \hat{u} \\ \hat{T} \\ \hat{\sigma} \\ \hat{q} \end{bmatrix} = \begin{bmatrix} 1 \\ 0 \\ 0 \\ 0 \\ 0 \end{bmatrix}, \tag{3.45}$$

where $\hat{\sigma}$ and $\hat{q}$ are, respectively, the Laplace-Fourier transform of σ and q in the temporal-spatial domains.

The governing equations for the stress and heat flux in the R13 moment system are

$$\begin{aligned} \frac{\partial \sigma}{\partial t} + \frac{4}{3}\frac{\partial u}{\partial x} + \frac{8}{15}\frac{\partial q}{\partial x} - \frac{6}{5}\text{Kn}\frac{\partial^2 \sigma}{\partial x^2} &= -\frac{\sigma}{\text{Kn}}, \\ \frac{\partial q}{\partial t} + \frac{\partial \sigma}{\partial x} + \frac{5}{2}\frac{\partial T}{\partial x} - \frac{18}{5}\text{Kn}\frac{\partial^2 q}{\partial x^2} &= -\frac{2}{3}\frac{q}{\text{Kn}}. \end{aligned} \tag{3.46}$$

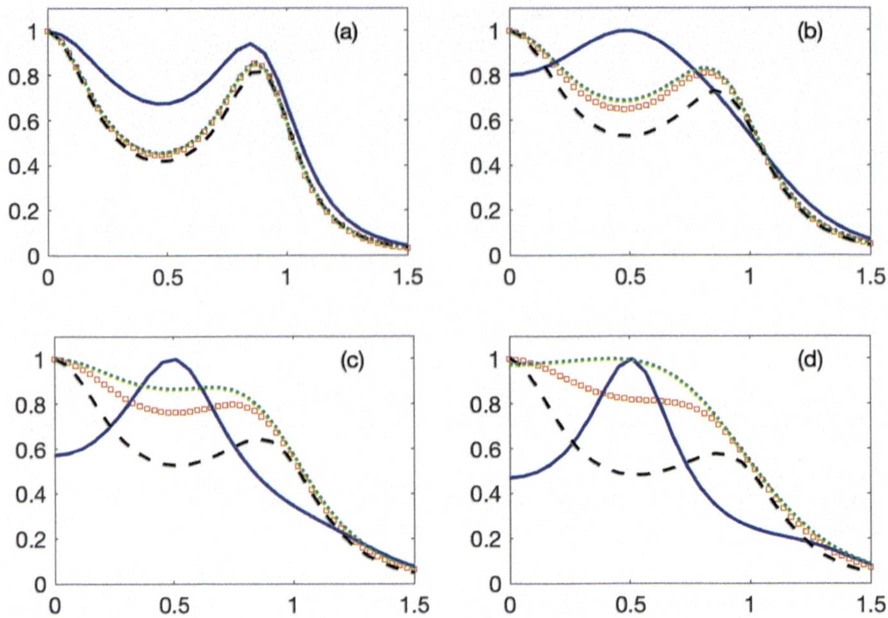

Fig. 3.2 Spontaneous RBS spectra obtained from the moment equations [17]. The Knudsen number is **a** =0.04, **b** 0.06, **c** 0.08, and **d** 0.1. Horizontal and vertical axis are the normalized frequency and spectrum, respectively. Solid, dashed, and dotted lines are results from the G13, R13, and R26 equations, respectively

Similarly, the evolution of stress and heat flux in the linearized R26 equations are governed by [14]:

$$\begin{aligned}\frac{\partial\sigma}{\partial t}+\frac{4}{3}\frac{\partial u}{\partial x}+\frac{8}{15}\frac{\partial q}{\partial x}+\frac{\partial\bar{m}}{\partial x}&=-\frac{\sigma}{\mathrm{Kn}},\\ \frac{\partial q}{\partial t}+\frac{5}{2}\frac{\partial T}{\partial x}+\frac{\partial\sigma}{\partial x}+\frac{1}{6}\frac{\partial\bar{\Delta}}{\partial x}+\frac{1}{2}\frac{\partial\bar{R}}{\partial x}&=-\frac{2}{3}\frac{q}{\mathrm{Kn}},\end{aligned}\tag{3.47}$$

where the higher order moments $\bar{m}$, $\bar{\Delta}$, and $\bar{R}$ are governed by

$$\begin{aligned}\frac{\partial\bar{m}}{\partial t}+\frac{9}{5}\frac{\partial\sigma}{\partial x}+\frac{9}{35}\frac{\partial\bar{R}}{\partial x}-\frac{16}{7}\frac{\mathrm{Kn}}{A_{\phi1}}\frac{\partial^2\bar{m}}{\partial x^2}&=-\frac{3}{2}\frac{\bar{m}}{\mathrm{Kn}},\\ \frac{\partial\bar{\Delta}}{\partial t}+8\frac{\partial q}{\partial x}-\frac{7\mathrm{Kn}}{3}\frac{\partial^2\bar{\Delta}}{\partial x^2}-4\mathrm{Kn}\frac{\partial^2\bar{R}}{\partial x^2}&=-\frac{2}{3}\frac{\bar{\Delta}}{\mathrm{Kn}},\\ \frac{\partial\bar{R}}{\partial t}+\frac{56}{15}\frac{\partial q}{\partial x}+2\frac{\partial\bar{m}}{\partial x}-\frac{\mathrm{Kn}}{5}\left(\frac{54}{7A_{\psi1}}+\frac{16}{3}\right)\frac{\partial^2\bar{R}}{\partial x^2}-\frac{28\mathrm{Kn}}{45}\frac{\partial^2\bar{\Delta}}{\partial x^2}&=-\frac{7}{6}\frac{\bar{R}}{\mathrm{Kn}},\end{aligned}\tag{3.48}$$

with $A_{\phi1}=2.097$ and $A_{\psi1}=1.698$.

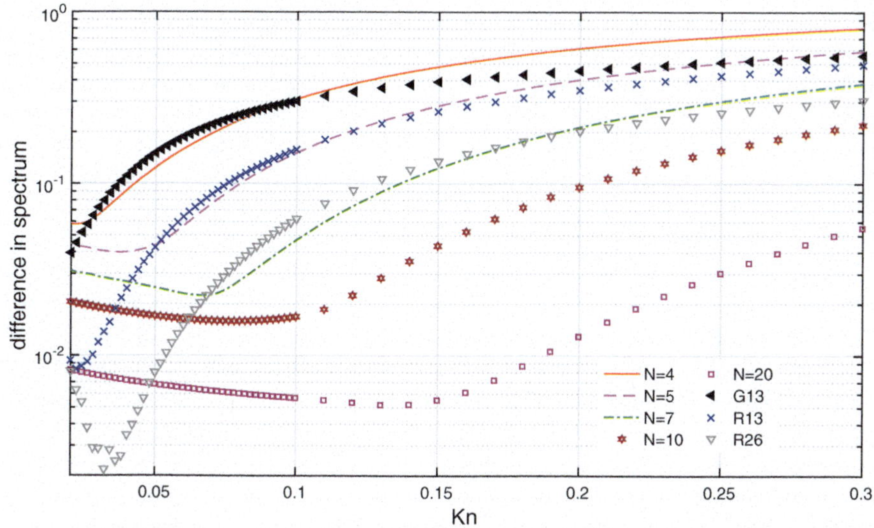

Fig. 3.3 Convergence test of equivalent moment systems [17]. Difference $\int_{\infty}^{\infty} |S^{LBE} - S_{GH}^{N}| df_s$ in the spectrum of spontaneous RBS, where S_{GH}^{N} is the spectrum obtained by solving the Shakhov kinetic model in Sect. 6.2.3 with the Gauss-Hermite quadrature of order N (these solutions are equivalent to those of Grad moment equations having $(N+1)(N+2)(N+3)/6$ moments [18] that are accurate up to the order of Kn^{N-2}). Note that areas of RBS spectra are normalized to unity. The solution may be viewed accurate when the difference is less than 0.05

Figure 3.2 compares the RBS spectra obtained from the moment equations with those from the linearized Boltzmann equation. It can be seen that the accuracy increases with the number of moments.

3.5 Convergence of Moment Equations

Results in the previous section show that higher order Chapman-Enskog expansion is not necessarily more accurate than lower order expansion, while the accuracy of moment method increases with the number of moments included. Now we consider the speed of convergence of the moment systems, i.e., how many moments should be used to predict the RGD.

3.5.1 Rayleigh-Brillouin Scattering

The derivation of high-order moment equations is cumbersome. Thus, the linearized Shakhov equation [19] is solved numerically using the DVM, and the molecular velocity space is discretized by the Gauss-Hermite quadrature of different orders; this is equivalent to the moment method with different order of approximations [20]. To be specific, if the Nth order Gauss-Hermite quadrature is used, the moment up to the order $N-1$ is captured accurately. Considering the fact that the G13 equations, where the VDF is expanded up to third-order Hermite polynomials, have accuracy of $O(\text{Kn}^2)$, it can be inferred that the numerical results of the Shakhov equation based on the Gauss-Hermite quadrature of order N has an accuracy of $O(\text{Kn}^{N-2})$.

This is confirmed in Fig. 3.3: when $N = 4, 5$, and 7, the Gauss-Hermite quadrature yields an equivalent moment system accurate to $O(\text{Kn}^2)$, $O(\text{Kn}^3)$, and $O(\text{Kn}^5)$, respectively. As a result, the difference in spectrum almost overlaps that from the G13, R13, and R26 moment equations, respectively, in a wide range of Kn.

Figure 3.3 further shows the accuracy increases monotonically with the order of the Gauss-Hermite quadrature. This is equivalent to say that higher order moment system has higher accuracy. However, the convergence of moment equations is rather slow at large values of Kn, e.g., when $\text{Kn} = 0.3$, the error is only reduced by about one order of magnitude when N is increased from 5 to 20, where the number of moment equations (proportional to N^3) is increased by two orders of magnitude.

3.5.2 *Sound Propagation*

We further assess the convergence of moment systems in the sound propagation in rarefied gas [17]. The numerical method will be presented in Sect. 9.4. Part of the results is shown in Fig. 1.2. Note that the rarefaction effects become stronger when $\varpi\tau_c$ increases, because $\varpi\tau_c$ is proportional to the temporal Knudsen number Kn_t as $\varpi\tau_c = 2\sqrt{2\pi}\,\text{Kn}_t$. The NSF equations are inaccurate when $\varpi\tau_c = 0.3$, while R13 (R26) equations give good results up to $\varpi\tau_c = 0.3$ ($\varpi\tau_c = 0.67$). This finding again confirms that the accuracy of moment equations increases when more moments are included in macroscopic equations.

Figure 3.4 shows that, when $\varpi\tau_c = 1.09$, the Gauss-Hermite quadrature of order 8 is more accurate than the R26 equations, since the former is equivalent to the Grad moment equations of accuracy $O(\text{Kn}^6)$, one order more accurate than the R26 equations. To obtain reliable results, the VDF should be approximated by Hermite polynomials to the 20th order. When $\varpi\tau_c$ is increased to 2.72, Gauss-Hermite quadrature of order higher than 50 is needed.

3.5.2.1 Reason of Slow Convergence

The slow convergence of the moment systems is demonstrated at the mesoscopic level. To this end, the marginal VDF $h_m = \iint h dv_2 v_3$ at the resting receiver is plotted in Fig. 3.5. When $\varpi\tau_c = 1.09$, the temporal and spatial Knudsen numbers are, respectively, $\text{Kn}_t = 0.123$ and $\text{Kn} = 0.069$. The VDF is smooth, except it has a discontinuity around $v_1 = 0$. The Gauss-Hermite quadrature with $N = 8$ cannot predict

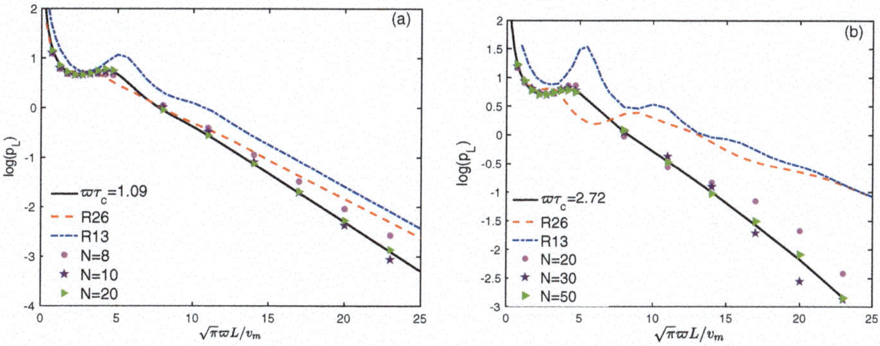

Fig. 3.4 Convergence test of equivalent moment systems [17]: sound amplitude at the receiver versus the dimensionless length $\sqrt{\pi}\varpi L/v_m$. Solid lines: accurate results from the Shakhov model. Symbols: solutions of the Shakhov model when the velocity space is discretized by the Gauss-Hermite quadrature of order N

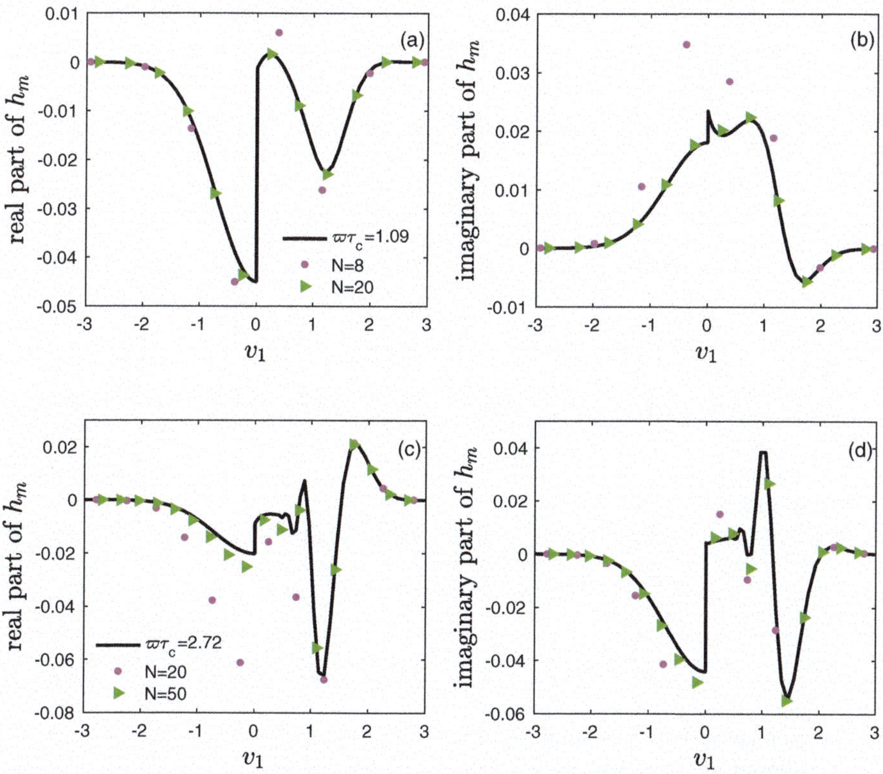

Fig. 3.5 Convergence test of equivalent moment systems [17]. Marginal VDF at the receiver when $\sqrt{\pi}\varpi L/v_m = 20$. **a**, **b** $\varpi\tau_c = 1.09$ and **c**, **d** $\varpi\tau_c = 2.72$. Solid lines: accurate results from the Shakhov model. Symbols: solutions of the Shakhov model when the velocity space is discretized by the Gauss-Hermite quadrature of order N

the gas pressure, because the VDF in Fig. 3.5a, b cannot be fitted by the Hermite polynomials up to the 8th order. Only the Gauss-Hermite quadrature of order higher than $N = 20$ can give reasonable good prediction.

When $\varpi \tau_c = 2.72$, the rarefaction effects are even stronger as the temporal and spatial Knudsen numbers are respectively $\text{Kn}_t = 0.306$ and $\text{Kn} = 0.171$. The VDF is rather irregular: it not only has discontinuity around $v_1 = 0$, but also has rapid variations. Figures 3.4b and 3.5c, d show that, although the Gauss-Hermite quadrature of order $N = 50$ predicts the gas pressure, it is not accurate enough in resolving the VDF.

Thus, the underlying reason for the slow convergence of Grad moment systems is the large discontinuities and rapid variations in VDF. Adding more discrete velocity grids in the framework of Gauss-Hermite quadrature is not economic; which means that adding more moments to the moment system is not economic as long as the VDF is expanded into Hermite polynomials. In highly rarefied gas flows, it is more advantageous to solve the kinetic equation by using non-uniform discrete velocity grids [21–23].

References

1. Hilbert D (1902) Mathematical problems. Bull Am Math Soc 8:437–479
2. Hilbert D (1912) Begrundung der kinetische gastheorie. Math Ann 72:562–577
3. Sone Y (2002) Kinetic theory and fluid dynamics. Birkhauser Boston
4. Chapman S (1916) On the law of distribution of molecular velocities, and the theory of viscosity and thermal conduction, in a non-uniform simple monatomic gas. Phil Trans R Soc A 216:538–548
5. Enskog D (1917) Kinetische Theorie der Vorgänge in mässig verdünnten Gasen. Uppsala University, Sweden
6. Cercignani C (1988) The Boltzmann equation and its applications. Springer, New York
7. Bobylev AV (2006) Instabilities in the Chapman-Enskog expansion and hyperbolic Burnett equations. J Stat Phys 124:371–399
8. Garcia-Colin LS, Velasco RM, Uribe FJ (2008) Beyond the Navier-Stokes equations: Burnett hydrodynamics. Phys Rep 465:149–189
9. Struchtrup H (2005) Macroscopic Transport Equations for Rarefied gas fows: approximation methods in kinetic theory. Springer, Heidelberg
10. Chapman S, Cowling TG (1970) The mathematical theory of non-uniform gases. Cambridge University Press, Cambridge
11. Grad H (1949) On the kinetic theory of rarefied gases. Comm Pure Appl Math 2:331–407
12. Weiss W (1995) Continuous shock structure in extended thermodynamics. Phys Rev E 52:5760
13. Struchtrup H, Torrilhon M (2003) Regularization of Grad's 13 moment equations: derivation and linear analysis. Phys Fluids 15:2668–2690
14. Gu XJ, Emerson DR (2009) A high-order moment approach for capturing non-equilibrium phenomena in the transition regime. J Fluid Mech 636:177–216
15. Shavaliyev MSh (1993) Super-Burnett corrections to the stress tensor and the heat flux in a gas of Maxwellian molecules. J Appl Math Mech 57:573–576
16. Marques W Jr, Kremer GM (1993) Spectral distribution of scattered light in polyatomic gases. Phys A 197:352–363
17. Wu L, Gu XJ (2020) On the accuracy of macroscopic equations for linearized rarefied gas flows. Advances in Aerodynamics 2:2

18. Fan YW (2016) Development and application of moment method in gas kinetic theory. PhD Thesis, Peking University
19. Shakhov EM (1968) Generalization of the Krook kinetic relaxation equation. Fluid Dyn 3(5):95–96
20. Shan XW, Yuan XF, Chen HD (2006) Kinetic theory representation of hydrodynamics: a way beyond the Navier-Stokes equation. J Fluid Mech 550:413–441
21. Su W, Lindsay S, Liu HH, Wu L (2017) Comparative study of the discrete velocity and lattice Boltzmann methods for rarefied gas flows through irregular channels. Phys Rev E 96:023309
22. Naris S, Valougeorgis D (2005) The driven cavity flow over the whole range of the Knudsen number. Phys Fluids 17:097106
23. Ambruş VE, Sofonea V (2012) High-order thermal lattice Boltzmann models derived by means of Gauss quadrature in the spherical coordinate system. Phys Rev E 86:016708

Chapter 4
Fast Spectral Method for Monatomic Gas Flow

The Boltzmann collision operator is a generalized convolution; therefore, it is better solved by the Fourier spectral method. First, we show how to design the collision kernel to recover the transport coefficients. Second, we present a detailed spectral approximation of the Boltzmann collision operator and assess its accuracy by comparing it with exact solutions for Maxwellian molecules. Third, we validate the accuracy of fast spectral method by the numerical kernel method, molecular dynamics simulation, and direct simulation Monte Carlo method.

4.1 Inverse Design of Collision Kernel

To fully harness the power of the fast spectral method, the collision kernel $B(\theta, v_r)$ in the BCO should be carefully designed, otherwise the numerical complexity will be increased by one order of magnitude. Here we show how to design the collision kernel to recover the temperature dependence of shear viscosity; when the shear viscosity is fixed, according to the structure of BCO, the thermal conductivity is automatically recovered.

4.1.1 Power-Law Potential

From Eqs. (2.13) and (2.14) it is seen that the collision kernel for the power-law potential (2.9) can be rewritten as

$$B = \frac{b|db|}{\sin\theta|d\theta|}v_r \equiv c_\alpha(\theta)v_r^\alpha, \quad \alpha = \frac{\eta-5}{\eta-1}, \tag{4.1}$$

and it is shown in Eq. (2.17) that $c_\alpha(\theta)$ approaches $\theta^{(\alpha-5)/2}$ when $\theta \to 0$, and results in an infinite total cross-section. Usually, a finite cutoff in the aiming distance b

L. Wu, *Rarefied Gas Dynamics*,
https://doi.org/10.1007/978-981-19-2872-7_4

is introduced so that the total cross-section is given by πb_{max}^2, where b_{max} is the maximum allowed aiming distance. Alternatively, $c_\alpha(\theta)$ is replaced by a constant C_α, yielding the variable-hard-sphere (VHS) model in DSMC [1]:

$$B = \frac{3}{4}\left(\frac{2K}{m}\right)^{\frac{2}{\eta-1}} A_2(\eta)v_r^\alpha \equiv C_\alpha v_r^\alpha, \tag{4.2}$$

where $A_2(\eta) = \int_0^\infty \sin^2\theta W_0 dW_0$ and $W_0 = b(mv_r^2/2K)^{1/(\eta-1)}$ [1] so that DSMC recovers the shear viscosity

$$\mu = \mu(T_0)\left(\frac{T}{T_0}\right)^\omega, \quad \omega = \frac{\eta+3}{2(\eta-1)}, \tag{4.3}$$

with ω being the viscosity index.

To recover both the shear viscosity and self-diffusion coefficient in DSMC, the variable-soft-sphere model is used, where the deflection angle satisfies $b = \sigma\cos^{\alpha'}(\theta/2)$, and hence the collision kernel is

$$B = \frac{\alpha b\sigma}{4}\cos^{\alpha'-2}\left(\frac{\theta}{2}\right)v_r^\alpha. \tag{4.4}$$

Likewise, to achieve maximum efficiency in the numerical approximation of BCO, special forms of collision kernel are needed. For example, it is suggested to use the following collision kernel

$$B = \frac{(\alpha+3)(\alpha+5)C_\alpha}{24}\sin^{\alpha-1}\left(\frac{\theta}{2}\right)v_r^\alpha, \tag{4.5}$$

so that the FSM is not limited to HS gas [2]. The collision kernel can be generalized to

$$B = \frac{\Gamma[(7+\alpha)/2]C_\alpha}{6\Gamma[(3+\alpha+\gamma)/2]\Gamma(2-\gamma/2)}\sin^{\alpha+\gamma-1}\left(\frac{\theta}{2}\right)\cos^{-\gamma}\left(\frac{\theta}{2}\right)v_r^\alpha. \tag{4.6}$$

The additional parameter γ has two benefits. First, the FSM can be extended to solve all inverse power-law potentials except the Coulomb potential. Second, the ratio between shear viscosity and self-diffusion coefficient can be recovered, which is useful in the simulation of gas mixtures.

4.1.2 Lennard-Jones Potential

For monatomic gas, the Lennard-Jones potential (2.10) is more accurate than the phenomenological power-law potential (2.9). Only when the temperature does not

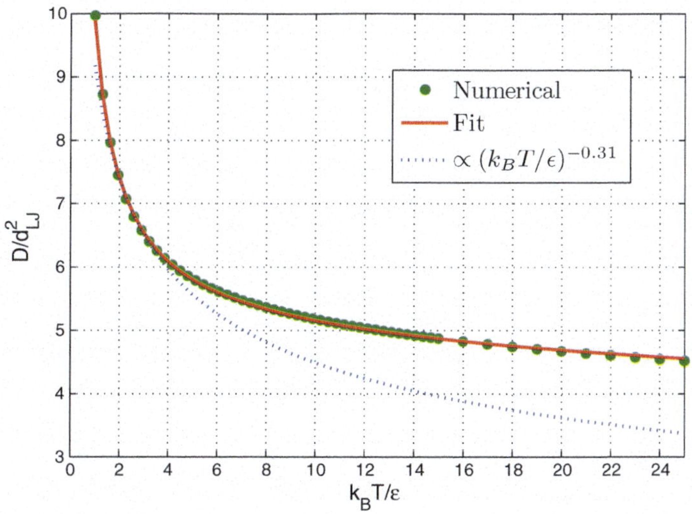

Fig. 4.1 The parameter D (3.28) as a function of $k_B T/\epsilon$ in the Lennard-Jones potential (2.10), which is fitted as $D/d_{LJ}^2 = b_1(k_B T/\epsilon)^{-0.4} + b_2(k_B T/\epsilon)^{-0.45} + b_3(k_B T/\epsilon)^{-0.5}$, with $b_1 = 407.4$, $b_2 = -811.9$, and $b_3 = 414.4$ [3]

vary too much could the shear viscosity be approximated by a single power-law function of temperature. For instance, when $k_B T/\epsilon$ is large, the repulsive intermolecular force dominants, and $\mu \propto T^{2/3}$; when $k_B T/\epsilon$ is small, the attractive intermolecular force dominants, and $\mu \propto T^{5/6}$. Also, when $2 < k_B T/\epsilon < 3$, $\mu \propto T^{0.81}$. In these regions, the VHS model is successfully implemented in DSMC, and the collision kernel (4.6) can be safely used.

To design a collision kernel valid over a wider temperature range, from Eqs. (3.27) and (3.28) we know that D should be fitted by a linear combination of power functions of temperature. For example, for argon with the potential depth $\epsilon = 119.18k_B$, the fitted D in Fig. 4.1 is valid when 120 K$< T <$ 3000 K, while the single power-law fit with $\mu \propto T^{0.81}$ works only when 240 K$< T <$ 360 K. Thus, the collision kernel is designed as

$$B = \frac{d_{LJ}^2}{8\pi} \sum_{j=1}^{3} \frac{(m/4\epsilon)^{(\alpha_j-1)/2}}{\Gamma(\frac{3+\alpha_j}{2})} b_j \sin^{\alpha_j-1}\left(\frac{\theta}{2}\right) v_r^{\alpha_j}, \tag{4.7}$$

where $\alpha_1 = 0.2$, $\alpha_2 = 0.1$, and $\alpha_3 = 0$. This form is chosen not only to recover D but also make the computational time be the same as that for Eq. (4.6). For wider temperature range, more terms with different values of α_j and b_j may be added into Eq. (4.7).

4.2 Normalization

It is useful to introduce the dimensionless variables:

$$\widetilde{f} = \frac{v_m^3}{n_0} f, \quad \widetilde{x} = \frac{x}{L}, \quad (\widetilde{v}, \widetilde{u}, \widetilde{c}) = \frac{(v, u, c)}{v_m}, \quad \widetilde{t} = \frac{v_m}{L} t,$$
$$\widetilde{a} = \frac{L}{v_m^2} a, \quad \widetilde{n} = \frac{n}{n_0}, \quad \widetilde{T} = \frac{T}{T_0}, \quad \widetilde{p} = \frac{p}{n_0 k_B T_0}, \quad \widetilde{q} = \frac{q}{n_0 k_B T_0 v_m}, \tag{4.8}$$

where n_0 is the average molecular number density, L is the characteristic flow length, and v_m is the most probable speed (2.45) at the reference temperature T_0. For simplicity the tildes will be omitted. The normalized macroscopic quantities are calculated as

$$[n, u, T, p_{ij}, q] = \int \left[1, v, \frac{2}{3n} c^2, 2c_i c_j, c^2 c\right] f dv. \tag{4.9}$$

After normalization, the Boltzmann equation (2.19) with the collision kernel (4.6) takes the following form:

$$\frac{\partial f}{\partial t} + v \cdot \frac{\partial f}{\partial x} + a \cdot \frac{\partial f}{\partial v} = Q^+ - \nu f. \tag{4.10}$$

Here, the gain term of the collision operator is

$$Q^+ = \frac{1}{\mathrm{Kn}'} \iint \sin^{\alpha+\gamma-1}\left(\frac{\theta}{2}\right) \cos^{-\gamma}\left(\frac{\theta}{2}\right) v_r^\alpha f(v_*') f(v') d\Omega dv_*, \tag{4.11}$$

while the loss term is νf, where the collision frequency is

$$\nu = \frac{1}{\mathrm{Kn}'} \iint \sin^{\alpha+\gamma-1}\left(\frac{\theta}{2}\right) \cos^{-\gamma}\left(\frac{\theta}{2}\right) v_r^\alpha f(v_*) d\Omega dv_*, \tag{4.12}$$

and

$$\mathrm{Kn}' = \frac{64\sqrt{2}^\alpha}{5} \Gamma\left(\frac{\alpha+\gamma+3}{2}\right) \Gamma\left(2 - \frac{\gamma}{2}\right) \mathrm{Kn}. \tag{4.13}$$

Note that in DSMC, the MFP of VHS model (i.e., λ_{vhs} in Eq. (4.52) in Ref. [1]) is frequently used. In order to compare the numerical results between the FSM and DSMC, the following relation should be considered:

$$\mathrm{Kn} = \frac{15\pi}{2(7-2\omega)(5-2\omega)} \mathrm{Kn}_{vhs}. \tag{4.14}$$

For the Lennard-Jones potential, when the collision kernel is approximated by Eq. (4.7), the term $\sin^{\alpha+\gamma-1}(\theta/2)\cos^{-\gamma}(\theta/2)v_r^{\alpha}/\text{Kn}'$ in Eqs. (4.11) and (4.12) is replaced by

$$\frac{5\sum_{j=1}^{3} b_j(k_B T_0/2\epsilon)^{(\alpha_j-1)/2}\sin^{\alpha_j-1}(\theta/2)v_r^{\alpha_j}/\Gamma(\frac{\alpha_j+3}{2})}{64\sqrt{2}\text{Kn}\sum_{j=1}^{3} b_j(k_B T_0/\epsilon)^{(\alpha_j-1)/2}}. \tag{4.15}$$

4.3 Fast Spectral Method

The numerical approximation of the BCO by the FSM is elaborated [3]. We refer to the paper by Mouhot and Pareschi [2] for its main properties.

4.3.1 Carleman Representation

With the basic identity

$$\begin{aligned}
2\int_{\mathbb{R}^3}\delta(2\boldsymbol{y}\cdot\boldsymbol{v}_r+y^2)f(\boldsymbol{y})d\boldsymbol{y} &= 2\int_{\mathbb{R}^3}\delta(|\boldsymbol{y}+\boldsymbol{v}_r|^2-v_r^2)f(\boldsymbol{y})d\boldsymbol{y}\\
&= 2\int_{\mathbb{R}^3}\delta(y^2-v_r^2)f(\boldsymbol{y}-\boldsymbol{v}_r)d\boldsymbol{y} = v_r\int_{\mathbb{S}^2}f(v_r\Omega-\boldsymbol{v}_r)d\Omega,
\end{aligned} \tag{4.16}$$

the gain part of the BCO is rewritten as

$$\begin{aligned}
Q^+ =&\frac{1}{\text{Kn}'}\iint\Theta v_r f\left(\boldsymbol{v}_*-\frac{v_r\Omega-\boldsymbol{v}_r}{2}\right)f\left(\boldsymbol{v}+\frac{v_r\Omega-\boldsymbol{v}_r}{2}\right)d\Omega d\boldsymbol{v}_*\\
=&\frac{2}{\text{Kn}'}\iint\Theta\delta(2\boldsymbol{y}\cdot\boldsymbol{v}_r+y^2)f\left(\boldsymbol{v}_*-\frac{\boldsymbol{y}}{2}\right)f\left(\boldsymbol{v}+\frac{\boldsymbol{y}}{2}\right)d\boldsymbol{y}d\boldsymbol{v}_*\\
=&\frac{4}{\text{Kn}'}\iint\Theta\delta(\boldsymbol{y}\cdot\boldsymbol{v}_r+y^2)f(\boldsymbol{v}_*-\boldsymbol{y})f(\boldsymbol{v}+\boldsymbol{y})d\boldsymbol{y}d\boldsymbol{v}_*\\
=&\frac{4}{\text{Kn}'}\iint\Theta\delta(\boldsymbol{y}\cdot\boldsymbol{z})f(\boldsymbol{v}+\boldsymbol{z})f(\boldsymbol{v}+\boldsymbol{y})d\boldsymbol{y}d\boldsymbol{z},
\end{aligned} \tag{4.17}$$

where

$$\Theta = \sin^{\alpha+\gamma-1}\left(\frac{\theta}{2}\right)\cos^{-\gamma}\left(\frac{\theta}{2}\right)v_r^{\alpha-1} \tag{4.18}$$

when the collision kernel (4.6) is used. Other collision kernels can be handled in the same way.

Notice that in the step-by-step calculations of Eq. (4.17) the transformations $\boldsymbol{y}=(v_r\Omega-\boldsymbol{v}_r)/2$ and $\boldsymbol{z}=\boldsymbol{v}_*-\boldsymbol{v}-\boldsymbol{y}=-\boldsymbol{v}_r-\boldsymbol{y}$ are used; also the delta function requires $\boldsymbol{y}$ to be perpendicular to $\boldsymbol{z}$. Therefore,

$$\cos\theta = \frac{\boldsymbol{\Omega}\cdot\boldsymbol{v}_r}{v_r} = \frac{-(\boldsymbol{y}-\boldsymbol{z})\cdot(\boldsymbol{y}+\boldsymbol{z})}{|\boldsymbol{y}+\boldsymbol{z}|^2} \stackrel{\boldsymbol{y}\perp\boldsymbol{z}}{=} \frac{z^2-y^2}{y^2+z^2},$$
$$\sin\left(\frac{\theta}{2}\right) = \frac{|\boldsymbol{y}|}{\sqrt{y^2+z^2}}, \quad \cos\left(\frac{\theta}{2}\right) = \frac{|\boldsymbol{z}|}{\sqrt{y^2+z^2}}. \tag{4.19}$$

Hence $\Theta = |\boldsymbol{y}|^{\alpha+\gamma-1}|\boldsymbol{z}|^{-\gamma}$ and the BCO becomes

$$Q = \frac{4}{\mathrm{Kn}'}\iint d\boldsymbol{y}d\boldsymbol{z}\delta(\boldsymbol{y}\cdot\boldsymbol{z})|\boldsymbol{y}|^{\alpha+\gamma-1}|\boldsymbol{z}|^{-\gamma}$$
$$\times\left[f(\boldsymbol{v}+\boldsymbol{z})f(\boldsymbol{v}+\boldsymbol{y}) - f(\boldsymbol{v}+\boldsymbol{y}+\boldsymbol{z})f(\boldsymbol{v})\right]. \tag{4.20}$$

4.3.2 Fourier-Galerkin Spectral Method

The VDF is periodized on the velocity domain $\mathcal{D}_L = [-L_v, L_v]^3$, which is discretized by N_k equally spaced points in the kth direction:

$$v_k(j_k) = 2\frac{L_v}{N_k}j_k, \tag{4.21}$$

where $k = 1, 2, 3,$ and $j_k \in [-N_k/2, -N_k/2+1, \ldots, N_k/2-1]$. The VDF is approximated by a truncated Fourier series:

$$f(\boldsymbol{v}) = \sum_{\boldsymbol{j}=-(N_1,N_2,N_3)/2}^{(N_1,N_2,N_3)/2-1} \hat{f}_{\boldsymbol{j}}\exp(i\boldsymbol{\xi}_{\boldsymbol{j}}\cdot\boldsymbol{v}),$$
$$\hat{f}_{\boldsymbol{j}} = \frac{1}{(2L_v)^3}\int f(\boldsymbol{v})\exp(-i\boldsymbol{\xi}_{\boldsymbol{j}}\cdot\boldsymbol{v})d\boldsymbol{v}, \tag{4.22}$$

where $\boldsymbol{\xi}_{\boldsymbol{j}} = \boldsymbol{j}\pi/L$ are the frequency components with $\boldsymbol{j} = (j_1, j_2, j_3)$.

The BCO (4.20) is also truncated, with the infinite velocity space replaced by the finite one in $\mathcal{B}_R$, a sphere of radius R centered at the origin:

$$Q = \frac{4}{\mathrm{Kn}'}\int_{\mathcal{B}_R}\int_{\mathcal{B}_R} d\boldsymbol{y}d\boldsymbol{z}\delta(\boldsymbol{y}\cdot\boldsymbol{z})|\boldsymbol{y}|^{\alpha+\gamma-1}|\boldsymbol{z}|^{-\gamma}$$
$$\times\left[f(\boldsymbol{v}+\boldsymbol{z})f(\boldsymbol{v}+\boldsymbol{y}) - f(\boldsymbol{v}+\boldsymbol{y}+\boldsymbol{z})f(\boldsymbol{v})\right], \tag{4.23}$$

where the choice of R will be discussed in the numerical simulation later. Expanding the truncated BCO by Fourier series, it is found that its $\boldsymbol{j}$th spectrum is related to the spectrum $\hat{f}$ as

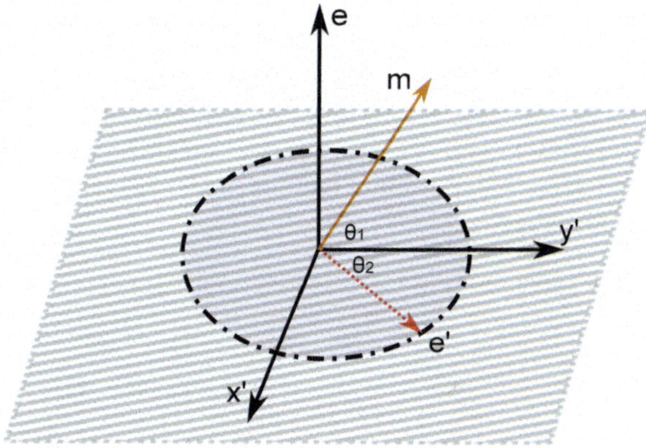

Fig. 4.2 Demonstration of the integration with respect to e' in the kernel mode [3]. The 3D Cartesian coordinate system $x'y'e$ is formed when the y'-axis is chosen as the projection of m into the plane perpendicular to e (the angle between m and y'-axis is θ_1). e' is in the $x'y'$ plane, and its angle with y'-axis is θ_2

$$\widehat{Q}_j = \sum_{\substack{l+m=j \\ l,m=-(N_1,N_2,N_3)/2}}^{(N_1,N_2,N_3)/2-1} \hat{f}_l \hat{f}_m [\beta(\boldsymbol{l}, \boldsymbol{m}) - \beta(\boldsymbol{m}, \boldsymbol{m})], \tag{4.24}$$

where $\boldsymbol{l} = (l_1, l_2, l_3)$, $\boldsymbol{m} = (m_1, m_2, m_3)$, and the kernel mode $\beta(\boldsymbol{l}, \boldsymbol{m})$ is

$$\begin{aligned}
\beta(\boldsymbol{l}, \boldsymbol{m}) =& \frac{4}{\text{Kn}'} \iint \delta(\boldsymbol{y} \cdot \boldsymbol{z}) |\boldsymbol{y}|^{\alpha+\gamma-1} |\boldsymbol{z}|^{-\gamma} \exp(i\xi_l \cdot \boldsymbol{y} + i\xi_m \cdot \boldsymbol{z}) d\boldsymbol{y} d\boldsymbol{z} \\
=& \frac{1}{\text{Kn}'} \int \int \delta(\boldsymbol{e} \cdot \boldsymbol{e}') \left[\int_{-R}^{R} |\rho|^{\alpha+\gamma} \exp(i\rho\xi_l \cdot \boldsymbol{e}) d\rho \right] \\
& \times \left[\int_{-R}^{R} |\rho'|^{1-\gamma} \exp(i\rho'\xi_m \cdot \boldsymbol{e}') d\rho' \right] d\boldsymbol{e}' d\boldsymbol{e} \\
=& \frac{1}{\text{Kn}'} \int_{\mathbb{S}^2} \phi_{\alpha+\gamma}(\xi_l \cdot \boldsymbol{e}) \left[\int_{\mathbb{S}^2} \delta(\boldsymbol{e} \cdot \boldsymbol{e}') \phi_{1-\gamma}(\xi_m \cdot \boldsymbol{e}') d\boldsymbol{e}' \right] d\boldsymbol{e},
\end{aligned} \tag{4.25}$$

with $\boldsymbol{e}$, $\boldsymbol{e}'$ being the unit vectors in the sphere $\mathbb{S}^2$, and

$$\phi_\delta(s) = 2 \int_0^R \rho^\delta \cos(\rho s) \, d\rho. \tag{4.26}$$

A new Cartesian coordinate system is constructed in Fig. 4.2 to simplify the kernel mode as

$$\begin{aligned}
\beta(\mathbf{l}, \mathbf{m}) &= \frac{1}{\text{Kn}'} \int_{\mathcal{S}^2} \phi_{\alpha+\gamma}(\xi_\mathbf{l} \cdot \mathbf{e}) \left[\int_0^{2\pi} \phi_{1-\gamma}(|\xi_\mathbf{m}| \cos\theta_1 \cos\theta_2) d\theta_2 \right] d\mathbf{e} \\
&= \frac{2}{\text{Kn}'} \int_{\mathcal{S}^2} \phi_{\alpha+\gamma}(\xi_\mathbf{l} \cdot \mathbf{e}) \cdot \psi_\gamma(|\xi_\mathbf{m}| \cos\theta_1) d\mathbf{e},
\end{aligned} \tag{4.27}$$

where

$$\psi_\gamma(s) = \int_0^\pi \phi_{1-\gamma}(s\cos\theta_2)d\theta_2 = 2\pi\int_0^R \rho^{1-\gamma}J_0(\rho s)d\rho, \tag{4.28}$$

with J_0 being the zeroth-order Bessel function.

In order to apply the Fast Fourier Transform (FFT) based convolution in Eq. (4.24), $\boldsymbol{l}$ and $\boldsymbol{m}$ are separated into two functions. To achieve this, two numerical quadratures are considered:

- The integration in Eq. (4.27) is approximated by the trapezoidal rule:

$$\beta(\boldsymbol{l},\boldsymbol{m}) \simeq \frac{4\pi^2}{\mathrm{Kn}'M^2}\sum_{p,q=1}^{M-1,M}\phi_{\alpha+\gamma}(\xi_{\boldsymbol{l}}\cdot\boldsymbol{e}_{\theta_p,\varphi_q})\psi_\gamma(\xi_{\boldsymbol{m}}^\perp)\cdot\sin\theta_p, \tag{4.29}$$

where the unit vector is $\boldsymbol{e}_{\theta_p,\varphi_q} = (\sin\theta_p\cos\varphi_q, \sin\theta_p\sin\varphi_q, \cos\theta_p)$, and

$$\xi_{\boldsymbol{m}}^\perp = \sqrt{|\xi_{\boldsymbol{m}}|^2 - (\xi_{\boldsymbol{m}}\cdot\boldsymbol{e}_{\theta_p,\varphi_q})^2}, \tag{4.30}$$

with $\theta_p = p\pi/M$ the discrete polar angle, $\varphi_q = q\pi/M$ the discrete azimuthal angle, and $p, q = 1, 2, \ldots, M$.
- The integration in Eq. (4.27) is approximated by a Gauss-Legendre quadrature of order M (see the MATLAB code in Appendix C):

$$\beta(\boldsymbol{l},\boldsymbol{m}) \simeq \frac{4}{\mathrm{Kn}'}\sum_{p,q=1}^{M}\omega_p\omega_q\phi_{\alpha+\gamma}(\xi_{\boldsymbol{l}}\cdot\boldsymbol{e}_{\theta_p,\varphi_q})\psi_\gamma(\xi_{\boldsymbol{m}}^\perp)\cdot\sin\theta_p, \tag{4.31}$$

where θ_p (φ_q) and ω_p (ω_q) are the p (q)th abscissas and weights in the Gauss-Legendre quadrature for the polar angle θ and azimuthal angle φ, both in the region of $[0, \pi]$.

The analytical form of $\phi_{\alpha+\gamma}(s)$ can be obtained when $\alpha+\gamma$ is an integer. For instance, when $\gamma = 0$, for Maxwellian molecules ($\eta = 5, \alpha = 0$) and HS ($\eta = \infty$, $\alpha = 1$) molecules, we have

$$\begin{aligned}\phi_0(s) &= \frac{2\sin(Rs)}{s},\\ \phi_1(s) &= \frac{2R\sin(Rs)}{s} - \frac{4\sin^2(Rs/2)}{s^2},\end{aligned} \tag{4.32}$$

while for other cases $\phi_{\alpha+\gamma}(s)$ and $\psi_\gamma(s)$ can be accurately calculated by numerical quadrature. Typical profiles of $\phi_{\alpha+\gamma}$ and ψ_γ are shown in Fig. 4.3.

For the power-law potential, $2 \le \eta < \infty$, so $-3 \le \alpha \le 1$. From Eq. (4.26) we see that δ must be larger than -1; this requires $\alpha+\gamma, 1-\gamma > -1$. If $\gamma = 0$, α must be larger than -1; this means that the collision kernel (4.5) cannot deal with the soft potential with $\eta \le 3$. In the modified collision kernel (4.6), if $\gamma \to 2$, the

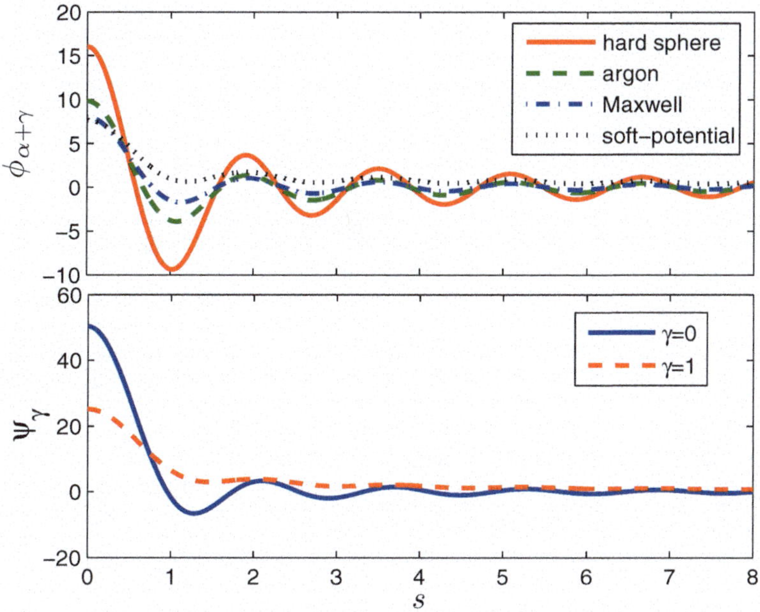

Fig. 4.3 Profiles of $\phi_{\alpha+\gamma}$ and ψ_γ according to Eqs. (4.26) and (4.28) when $R = 4$ and $\gamma = 0$. Because of symmetry, the region $s < 0$ is not plotted. For the soft potential, we use $\alpha = -0.4$ and the shear viscosity is proportional to $T^{1.2}$ [3]

FSM can be applied to all inverse power-law potentials except the Coulomb potential with $\eta = 2$. Note that in reality the charged particles interact through the shielded Coulomb potential (2.11), therefore the FSM can be applied to any form of realistic intermolecular potential [4].

4.3.3 Detailed Implementation

The kernel modes are pre-computed and stored. The storage of $\phi_{\alpha+\gamma}$ and ψ_γ requires $2M^2N_1N_2N_3$ units of computational memory. We also need $N_1N_2N_3$ units of storage for

$$\phi_{loss} = \sum_{p,q=1}^{M-1,M} \phi_{\alpha+\gamma}(\xi_{\boldsymbol{m}}, \theta_p, \varphi_q)\psi_\gamma(\xi_{\boldsymbol{m}}, \theta_p, \varphi_q) \sin\theta_p, \tag{4.33}$$

which is used to calculate the collision frequency in Eq. (4.12). For space-homogeneous problems, such storage is relatively large when compared to the storage of VDF. However, for space-inhomogeneous problems, the storage will be relatively small because different spatial grids share the same kernel modes.

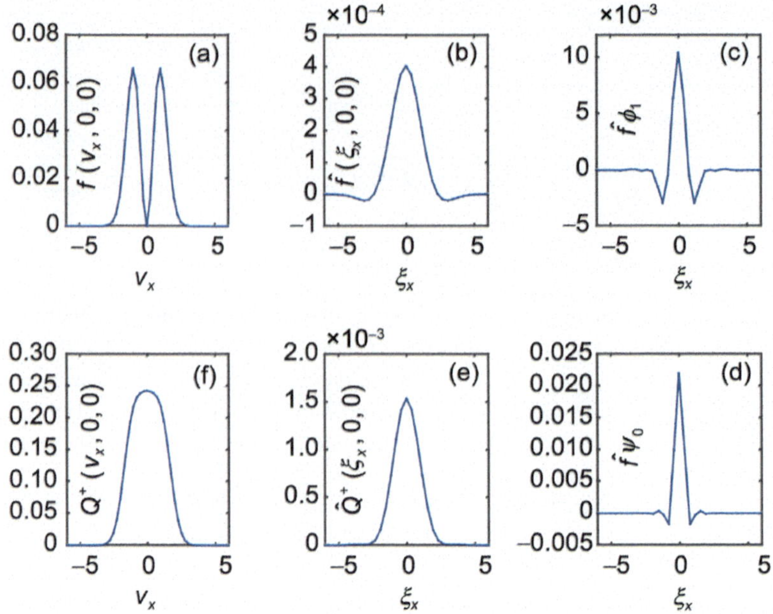

Fig. 4.4 Numerical implementation of the FSM for the gain part of BCO [5]. Given the VDF in (**a**), its spectrum is obtained in (**b**) by inverse FFT. Then the spectrum is respectively multiplied with the kernel modes ϕ and ψ (the results are in (**c**) and (**d**), respectively), and their convolution is shown in (**e**). The FFT of (**e**) gives the gain part of the BCO for specific combination of p and q in (**f**). Summering Q^+ for all combinations of p and q gives the gain part of the BCO. Note that the whole process is done in 3D velocity space and spectrum space, but for clarity only the 1D results are shown. The pseudo-code is given in Appendix D

The detailed implementation of FSM is visualized in Fig. 4.4 and described in the caption. One can find the MATLAB code in Appendix E.

One of the drawbacks of FSM, as with any spectral methods in the approximation of BCO, is that the momentum and energy are not exactly conserved. To ensure the conservation, the method of Lagrangian multipliers is used [6]: Q^{new} is reconstructed by minimizing the function $\sum_j (Q_j - Q_j^{new})^2$ under the constraints

$$\sum_j Q_j^{new} = \sum_j \boldsymbol{v} Q_j^{new} = \sum_j v^2 Q_j^{new} = 0.$$

That is, $Q^{new} = Q - (\lambda_n + \boldsymbol{\lambda}_v \cdot \boldsymbol{v} + \lambda_e v^2)$, where the five Lagrangian multipliers satisfy

$$\sum_j Q = \sum_j (\lambda_n + \lambda_v \cdot v + \lambda_e |v|^2),$$

$$\sum_j vQ = \sum_j v(\lambda_n + \lambda_v \cdot v + \lambda_e v^2),$$

$$\sum_j v^2 Q = \sum_j v^2(\lambda_n + \lambda_v \cdot v + \lambda_e v^2).$$

Note that the Lagrangian multipliers are very small due to the spectral accuracy of FSM. In practice, we normally do not enforce conservation if the steady-state solution of the Boltzmann equation can be quickly found, e.g., by using the general synthetic iterative scheme in Chap. 8.

4.3.4 Non-uniform Discretization of Velocity Space

In the simulation of rarefied gas flows with large Knudsen numbers, the "over concentration" phenomenon are frequently encountered [7], where the VDF concentrates or has sharp variations around $v \sim 0$. Accordingly, the molecular velocity space should be discretized non-uniformly, with more points placed near $v = 0$. In microflows, we usually use the following non-uniform discretization:

$$v_i = \frac{L_v}{(N_v - 1)^\iota}(-N_v + 1, -N_v + 3, \ldots, N_v - 1)^\iota, \tag{4.34}$$

where ι is a positive odd integer that usually takes the value of 3.

However, the frequency space must be uniformly discretized, otherwise the FFT-based convolution cannot be applied. This means that Steps 2 and 3 in the algorithm of Appendix D remain unchanged, but the Fourier transform in Steps 1 and 4 is calculated by direct summation. Fortunately, the computational cost can be reduced by considering the following two factors:

- The number of discretized frequency components N_ξ can be different from the corresponding velocity components N_v. In fact, N_ξ can be smaller than N_v due to the spectral accuracy of FSM; hence the computational cost of the convolution is $O(M^2 N_\xi^3 \ln N_\xi)$.
- In the rectangular Cartesian coordinates, the Fourier transform in Steps 1 and 4 can be done by direct summation in each direction sequentially, thus the cost will be at the order of $N_v^3 N_\xi \ln N_v$. This cost is certainly higher than the FFT on uniform grids but is only comparable to the FFT-based convolution in Steps 2 and 3.

Therefore, compared to the uniform discretization with a large number of velocity grids, the use of non-uniform velocity grids will not only reduce computational memory but also computational time. The sample MATLAB code can be found in Appendix F.

4.4 Homogeneous Relaxation

The relax-to-equilibrium process of Maxwellian molecules is used to assess the performance of FSM. This is a spatial-homogeneous problem, where the Boltzmann equation becomes

$$\frac{\partial f}{\partial t}=\frac{5}{32\pi}\iint \sin^{-1}\left(\frac{\theta}{2}\right)\left[f(\boldsymbol{v}'_*)f(\boldsymbol{v}')-f(\boldsymbol{v}_*)f(\boldsymbol{v})\right]d\Omega d\boldsymbol{v}_*. \tag{4.35}$$

4.4.1 *Bobylev-Krook-Wu Solution*

The Bobylev-Krook-Wu (BKW) solution for Eq. (4.35) reads [8]:

$$f(\boldsymbol{v},t)=\frac{1}{2(2\pi K)^{3/2}}\exp\left(-\frac{v^2}{2K}\right)\left(\frac{5K-3}{K}+\frac{1-K}{K^2}v^2\right), \tag{4.36}$$

where $K=1-0.4\exp(-t/6)$. In order to assess how accurately the FSM can approximate the BCO, we compare Q^{nu}, the numerical approximation of Q, to the analytical solution $Q^{an}=\partial f/\partial t$.

We first analyze the influence of the truncation radius R used in Eq. (4.23). Suppose the VDF has a support of S, then the BCO has a support $\sqrt{2}S$ [2, 9]. To avoid the aliasing error, the velocity space is normally truncated at $L_v=(3+\sqrt{2})S/2$ [10].

We vary the value of R to see its influence on the accuracy; large values of N and M are chosen so that they will not introduce additional errors. Figure 4.5 shows that the FSM reaches minimum error when

$$L_v\geq R\geq\frac{2\sqrt{2}}{3+\sqrt{2}}L_v. \tag{4.37}$$

This is because some collisions are ignored in truncating the BCO if R is too small, while the aliasing error due to the periodicity of VDF emerges if R is too large.

We then analyze the influence of solid angle discretization and the number of frequency components based on the numerical results in Table 4.1, where $R=6$ so that the truncation of BCO has a minimum error. The FSM has large error when $N=16$, because the range of frequency domain is not wide enough to cover the spectrum $\hat{f}$. When $N=24$, this problem is solved and the error is reduced by one order of magnitude.

When M is fixed, the error does not decrease when $N\geq 32$. This means that the discretization in solid angle has large error, which is explained as follows. From Eq. (4.32) and Fig. 4.3 we know $\phi_{\alpha+\gamma}$ has a quasi-period $2\pi/R$. Then, for a fixed value of ξ_l, the integral kernel in Eq. (4.27) oscillates $R|\xi_l|/\pi$ times as the solid angle varies. In the worst case of $\xi_l\rightarrow N\pi/2L$, $\phi_{\alpha+\gamma}$ oscillates $NR/4L_v$ times,

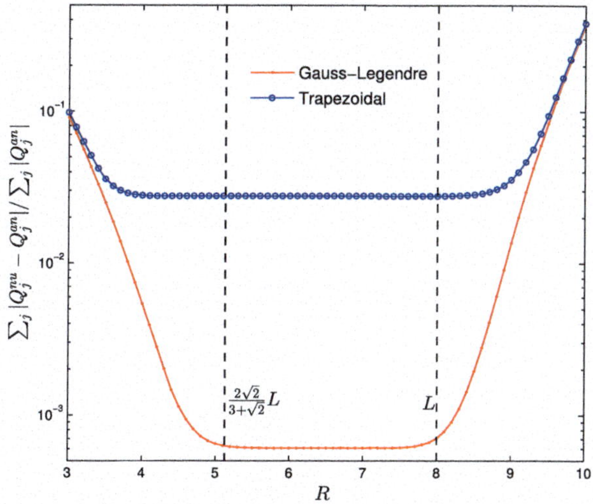

Fig. 4.5 The relative error versus the truncation radius R. Parameters are $L_\upsilon = 8$, $N = 48$, and $M = 7$. The Gauss-Legendre quadrature is used in the approximation of kernel mode [3]

Table 4.1 Relative error $\sum_j |Q_j^{nu} - Q_j^{an}| / \sum_j |Q_j^{an}|$ in the approximation of BCO. T(G) stands for the trapezoidal (Gauss-Legendre) quadrature. Parameters are $L_\upsilon = 8$ and $R = 6$

N		$M = 5$	6	7	8	12	16
16	T	4.58E-1	4.73E-1	4.55E-1	4.52E-1	4.78E-1	4.83E-1
	G	2.10E-1	3.35E-1	2.48E-1	2.77E-1	2.74E-1	2.69E-1
24	T	7.94E-2	5.20E-2	4.73E-2	3.93E-2	2.92E-2	2.59E-2
	G	4.61E-2	2.09E-2	9.16E-3	2.10E-2	1.72E-2	1.37E-2
32	T	5.54E-2	3.51E-2	2.57E-2	1.93E-2	8.39E-3	4.75E-3
	G	4.26E-2	6.18E-3	6.49E-4	2.11E-4	1.86E-4	1.57E-4
48	T	4.26E-2	3.88E-2	2.77E-2	2.08E-2	8.99E-3	5.01E-3
	G	4.31E-2	6.17E-3	6.09E-4	4.56E-5	4.94E-6	3.85E-6
64	T	5.90E-2	3.87E-2	2.77E-2	2.08E-2	8.99E-3	5.02E-3
	G	4.30E-2	6.16E-3	6.10E-4	4.70E-5	3.87E-6	4.31E-6

which means that

$$M \sim \frac{N}{4}\frac{R}{L_\upsilon}. \tag{4.38}$$

Thus, if M and N are doubled, the spectral accuracy of FSM is maintained.

From Fig. 4.5 and Table 4.1 it is seen that the Gauss-Legendre quadrature is better than the trapezoidal rule.

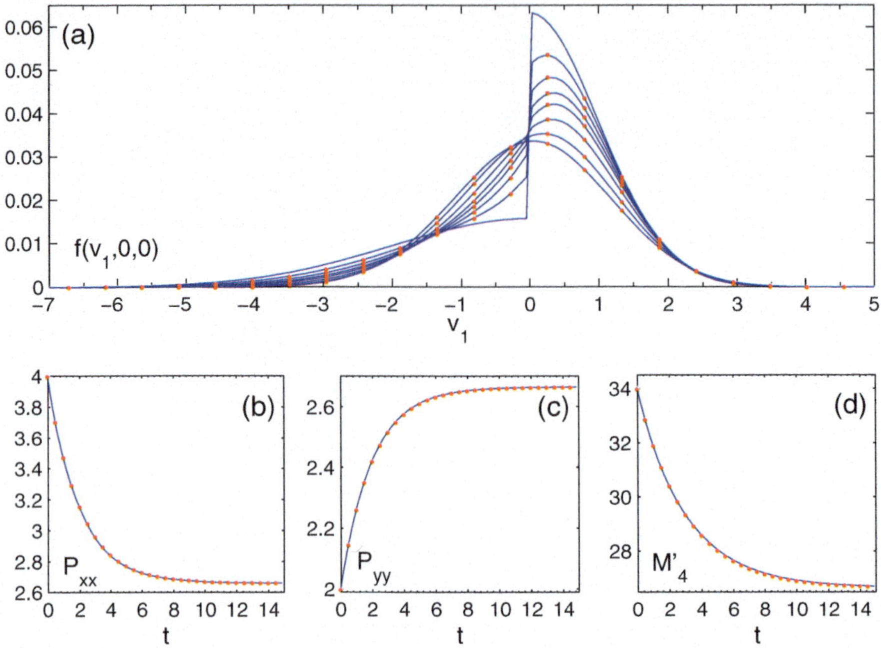

Fig. 4.6 **a** Relaxation of an initial discontinuous VDF [3]. From top to bottom (at $v_1 > 0$), the time for each line is $t = 0, 0.5, 1, 1.5, 2, 3, 5$, and 9, respectively. **b–d** Evolution of the second- and fourth-order moments. Dots: FSM results with $N_1, N_2, N_3 = 42$. Solid lines in **a** are the FSM results with $N_1 = 256$, $N_2, N_3 = 42$, in **b–d** are analytical solutions. Other parameters are $L_v = 11$, $R = 2\sqrt{2}L_v/(3+\sqrt{2})$, and $M = 5$ with Eq. (4.29)

4.4.2 Discontinuous Velocity Distribution

The initial VDF used in the preceding case is smooth, where the spectral accuracy of FSM is proved analytically [2]. Here we consider the relaxation of an initial discontinuous VDF

$$f(\boldsymbol{v}, t=0) = \frac{1}{3(2\pi)^{3/2}} \begin{cases} 4\exp\left(-\frac{v^2}{2}\right), & v_1 \ge 0, \\ \exp\left(-\frac{v_1^2}{8} - \frac{v_2^2+v_3^2}{2}\right), & v_1 < 0, \end{cases} \tag{4.39}$$

toward the Maxwell equilibrium distribution. This kind of VDF is ubiquitous in wall-bounded rarefied gas flows, so it is important that the FSM can capture the evolution of both macroscopic and mesoscopic quantities.

Although the analytical solution for VDF cannot be obtained, these for the second- and fourth-order moments exist:

$$
\begin{aligned}
P_{xx} &= \int f v_x^2 d\boldsymbol{v} = \frac{4}{3}\exp\left(-\frac{t}{2}\right) + \frac{8}{3}, \\
P_{yy} &= \int f v_y^2 d\boldsymbol{v} = -\frac{2}{3}\exp\left(-\frac{t}{2}\right) + \frac{8}{3}, \\
M_4' &= \int f v^4 d\boldsymbol{v} = \frac{22}{3}\exp\left(-\frac{t}{3}\right) + \frac{80}{3}.
\end{aligned}
\tag{4.40}
$$

Figure 4.6 demonstrates that the FSM can accurately capture the evolution of these moments, even when the initial VDF has a large jump. The maximum relative error in the second-order moments is less than 10^{-3}, while that in the fourth-order moment is less than 3×10^{-3} [3].

4.5 Accuracy in Inhomogeneous Problems

The Boltzmann equation is solved by the CIS (2.84), where the spatial derivative $\partial f/\partial \boldsymbol{x}$ is approximated by the second-order upwind scheme.

4.5.1 *Normal Shock Waves*

The normal shock wave problem is ideal for assessing the capability of FSM in capturing strong rarefaction effects, since it is a 1D problem without the influence of gas-wall interaction. The number density and temperature are normalized by n_0 and T_0 in the upstream part; the spatial coordinate is normalized by the upstream MFP

$$
\lambda = \frac{16}{5\pi}\sqrt{\frac{\pi}{2mk_B T_0}}\frac{\mu(T_0)}{n_0}
\tag{4.41}
$$

so that $\text{Kn} = 5\pi/16$.

Suppose the shock wave varies in the x_2 direction. Then the normalized VDF at the upstream end is

$$
f = \frac{1}{\pi^{3/2}}\exp\left[-v_1^2 - \left(v_2 - \sqrt{\frac{5}{6}}\text{Ma}\right)^2 - v_3^2\right],
\tag{4.42}
$$

and that at the downstream end is

$$
f = \frac{n_d}{(\pi T_d)^{3/2}}\exp\left[-\frac{v_1^2 + (v_2 - u_d)^2 + v_3^2}{T_d}\right],
\tag{4.43}
$$

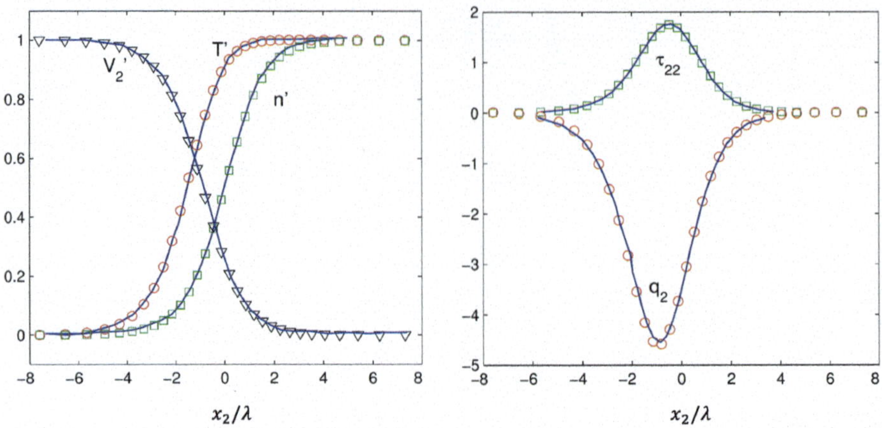

Fig. 4.7 The structure of normal shock wave when Ma = 3, where the reduced macroscopic quantity $\mathcal{M} = \rho, u, T$ in the left figure is normalized as $(\mathcal{M} - \mathcal{M}_u)/(\mathcal{M}_d - \mathcal{M}_u)$, with the subscripts u and d represent upstream and downstream, respectively. Lines: numerical kernel method [11]. Symbols: FSM [3]

where n_d, u_d, and T_d are the downstream density, velocity, and temperature of the normal shock wave, respectively, whose values are determined by the Rankine-Hugoniot relation:

$$
\begin{aligned}
n_d &= \frac{4\mathrm{Ma}^2}{\mathrm{Ma}^2 + 3}, \\
u_d &= \sqrt{\frac{5}{96}}\frac{\mathrm{Ma}^2 + 3}{\mathrm{Ma}}, \\
T_d &= \frac{(5\mathrm{Ma}^2 - 1)(\mathrm{Ma}^2 + 3)}{16\mathrm{Ma}^2},
\end{aligned} \tag{4.44}
$$

with Ma being the Mach number in the upstream part.

4.5.1.1 Hard-Sphere Potential

Consider a normal shock wave in HS gas with Ma = 3, where the downstream temperature is about four times of upstream temperature. Therefore, the velocity domain is chosen as $[-10, 10]^3$ to cover the velocity space in the whole shock wave. The truncated velocity domain is approximated by $42 \times 42 \times 42$ uniformly distributed grid points. We take $M = 5$ in the discretization of solid angle. It can be seen from Fig. 4.7 that the FSM and numerical kernel method [11] give identical shock wave structure.

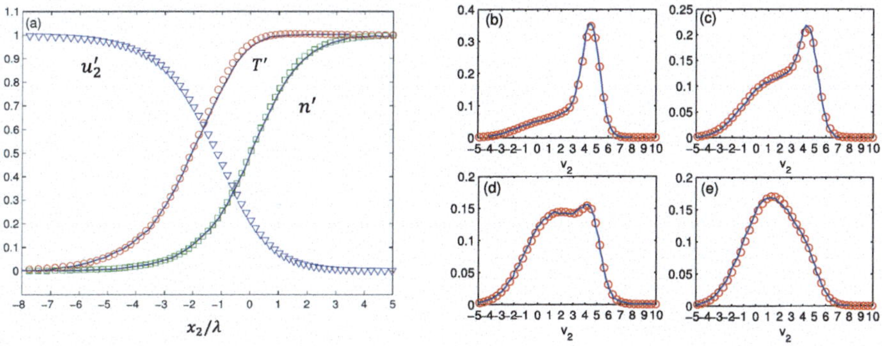

Fig. 4.8 **a** Shock structure in the argon gas with Ma = 5. The marginal VDF $\int\int f dv_1 dv_3/n$ is shown at a reduced density of **b** 0.151, **c** 0.350, **d** 0.511, and **e** 0.759. Lines: MD results [12]. Symbols: FSM. The velocity domain $[-18, 18]^3$ is divided into $42 \times 84 \times 42$ grid points [3]

4.5.1.2 Lennard-Jones Potential

We simulate the normal shock wave using the collision kernel (4.15) for the Lennard-Jones potential (2.10). For argon gas with Ma = 5, we find in Fig. 4.8 that the FSM predicts both the macroscopic quantities and mesoscopic VDFs in good agreement with the MD simulation [12]. Note that the upstream and downstream temperatures are about 300 K and 2600 K, respectively; the excellent agreement with MD data validates the accuracy of the collision kernel (4.15) for the Lennard-Jones potential.

One finds that the VDF significantly deviates from the Maxwell equilibrium distribution (2.31), and this is the reason why rarefied gas flow is also called non-equilibrium gas flow.

4.5.2 Force-Driven Poiseuille Flows

Consider a rarefied gas between two parallel plates located at $x_2 = L/2$ and $x_2 = -L/2$. The plates are stationary, with the temperature T_0. A uniform normalized acceleration $a_1 = 0.11$ is applied to the gas in the x_1 direction, where the acceleration term $\boldsymbol{a} \cdot \partial f/\partial \boldsymbol{v}$ in the Boltzmann equation is calculated by FFT. The HS gas is simulated at Kn = 0.1 and 0.5.

Due to symmetry, only the region $-0.5 \le x_2 \le 0$ is considered, which is divided into 50 unequally spaced points. At the $(k+1)$th iteration step, the VDF at the wall is determined by the diffuse BC as

$$\begin{aligned} f^{(k+1)}(x_2 = -0.5, v_2 \ge 0) =& \frac{|n|}{\pi^{3/2}} \exp(-v^2), \\ n =& 2\sqrt{\pi} \int_{v_2<0} v_2 f^{(k)}(x_2 = -0.5, v) d\boldsymbol{v}. \end{aligned} \tag{4.45}$$

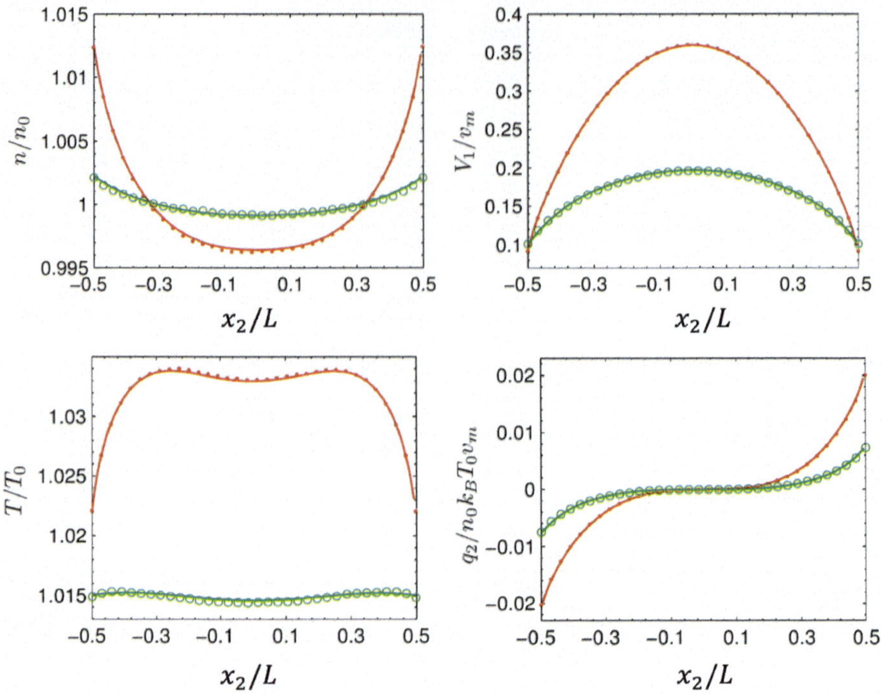

Fig. 4.9 Force-driven Poiseuille flow of HS molecules when Kn = 0.1 (dots) and Kn = 0.5 (circles) [3]. Lines: FSM. Symbols: DSMC [13]

To ensure the conservation of total mass, the VDF is re-scaled at the end of each iteration so that the average number density in the computational domain is 1 [14].

We use $32 \times 32 \times 32$ uniform velocity grids in the region $[-6, 6]^3$. Numerical results in Fig. 4.9 show that FSM agrees well with DSMC. Both methods capture the double peaks in the temperature profile, while the NSF equations always produce parabolic temperature profiles [15].

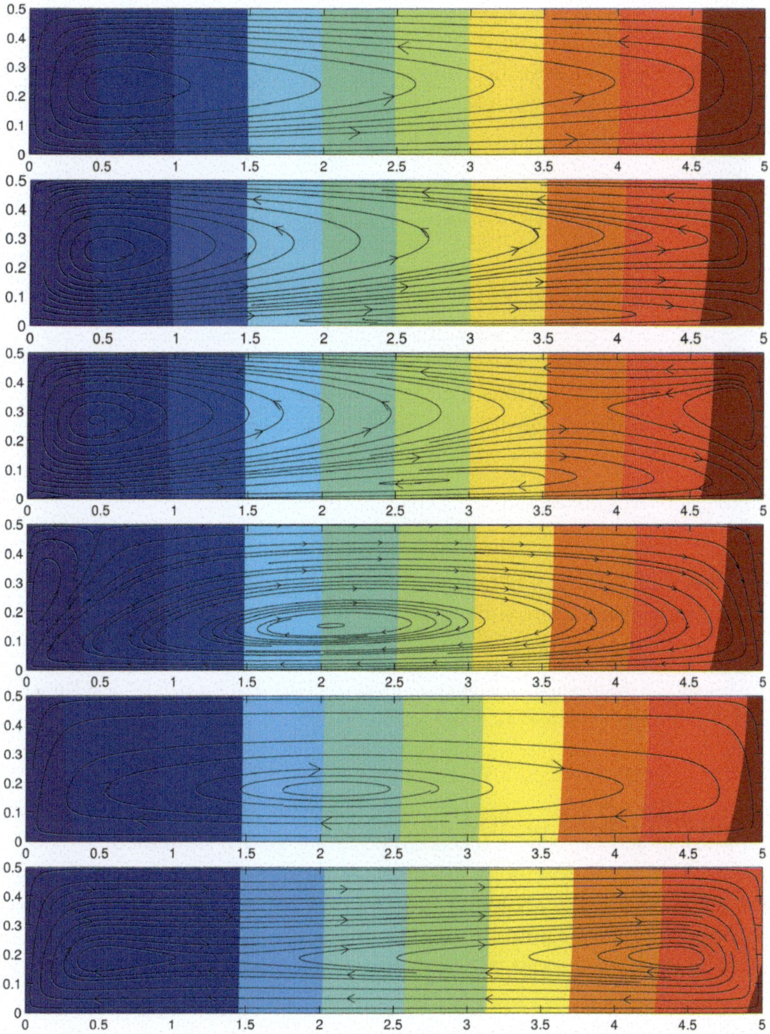

Fig. 4.10 Thermal transpiration of argon gas in a cavity [16]. In each figure, from the left to right, the dimensionless temperature of each contour is $1 + 0.1i$, where $i = 1, 2, \ldots, 9$. From the top to bottom, Kn = 0.08, 0.2, 0.25, 0.6, 2, and 10, respectively. Due to symmetry, the top half spatial region is not shown

4.5.3 Thermal Transpiration in a Cavity

Consider the thermal transpiration in a cavity with the aspect ratio 5 [16]. The temperature of the right wall is twice the left wall, while the temperature of the top and bottom walls increase linearly from left to right. Choosing the height as the characteristic flow length, the mean density as the reference density, and the

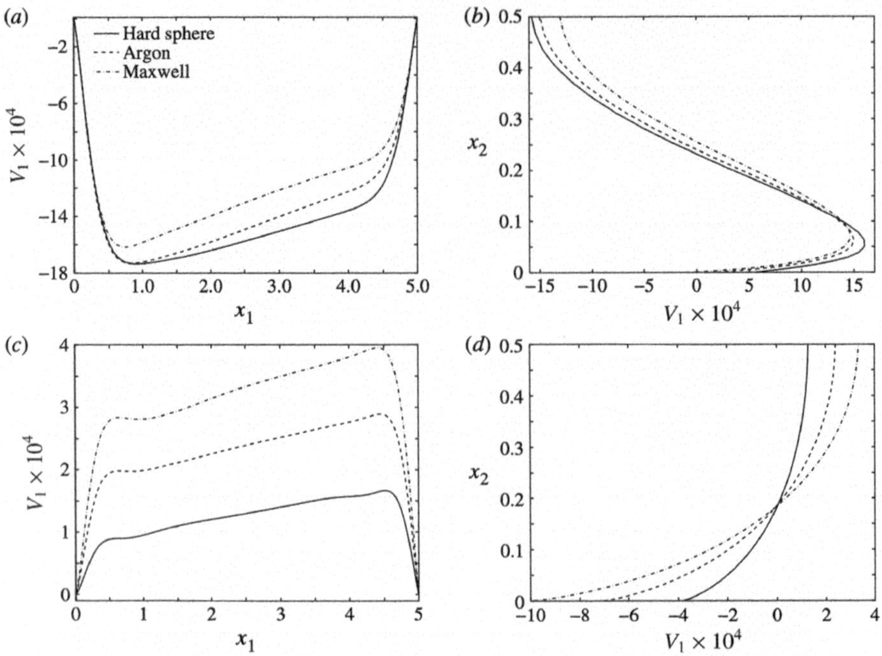

Fig. 4.11 Velocity in the thermal transpiration in a cavity [16]. **a, b** Kn = 0.08. **c, d** Kn = 10. (Left column) $x_2 = 0.5$. (Right column) $x_1 = 2.5$

temperature of the left wall as the reference temperature, rarefied gas flows with the Knudsen number Kn = 0.08, 0.2, 0.25, 0.6, 2, and 10 are simulated.

Figure 4.10 draws the velocity streamlines on top of the temperature contours. When the Knudsen number changes, different flow patterns are seen, as a competition between the thermal transpiration (which drives the gas from left to right) and its induced Poiseuille flow (which drives the gas from right to left). Note that in order to capture the flow velocity accurately, the non-uniform velocity discretization (4.34) is used in the $v_1 v_2$ directions; the reason is that the VDF has discontinuities and rapid variations, and the flow velocity is rather small: if the VDF is not properly resolved, there will be huge errors in the velocity field.

Figure 4.11 compares the flow velocities obtained at the same Knudsen number but different molecular collision kernels controlled by the viscosity index (4.3). It is observed that the viscosity index ω affects the velocity magnitudes significantly.

Many more challenging numerical simulations of the microflows are presented in Ref. [16], where the FSM is demonstrated to be faster than the DSMC [17] and low-noise DSMC [18].

4.6 Concluding Remarks

Collision kernels in the Boltzmann equation are designed to recover the shear viscosity of monatomic gas, and the fast spectral method approximation of the Boltzmann collision operator is elaborated. Many canonical rarefied gas flows are used to validate the efficiency and accuracy of FSM. Specially, we have simulated a number of microflows which is much more efficient than DSMC. We briefly analyze why the FSM is suitable for the simulation of rarefied microflows. The VDFs have large discontinuities at large Kn, hence a significant number of velocity grids are needed. This poses an extremely difficult problem for other deterministic methods that handle binary collisions in velocity space. However, this problem becomes amenable for the FSM because the collisions are treated in frequency space. The FSM approximates the collision operator with spectral accuracy [2]; the number of frequency components does not need to be as large as the velocity grids. One reason for this is that discontinuities in the VDF produce high-frequency components in its spectrum (and this is usually smooth, or at least smoother than the VDF); in calculating the spectrum of the collision operator, the spectrum of VDF is multiplied by a weight function which is very small for high-frequency components, see Fig. 4.3. Therefore, very high frequencies can be safely ignored: in the transition flow regime, we have shown that 32 frequency components in each direction are adequate.

References

1. Bird GA (1994) Molecular gas dynamics and the direct simulation of gas flows. Oxford Science Publications, Oxford University Press Inc, New York
2. Mouhot C, Pareschi L (2006) Fast algorithms for computing the Boltzmann collision operator. Math Comput 75(256):1833–1852
3. Wu L, White C, Scanlon TJ, Reese JM, Zhang YH (2013) Deterministic numerical solutions of the Boltzmann equation using the fast spectral method. J Comput Phys 250:27–52
4. Su W, Wang P, Liu HH, Wu L (2019) Accurate and efficient computation of the Boltzmann equation for Couette flow: influence of intermolecular potentials on Knudsen layer function and viscous slip coefficient. J Comput Phys 378:573–590
5. Wu L, Zhang YH, Li ZH (2017) Computable model on the collision integral of Boltzmann equation and application to rarefied aerodynamics (in Chinese). Sin-Phys Mech Astron 47:070004
6. Gamba IM, Tharkabhushanam SH (2009) Spectral-Lagrangian methods for collisional models of non-equilibrium statistical states. J Comput Phys 228(6):2012–2036
7. Takata S, Funagane H (2011) Poiseuille and thermal transpiration flows of a highly rarefied gas: over-concentration in the velocity distribution function. J Fluid Mech 669:242–259
8. Krook M, Wu TT (1977) Exact solutions of the Boltzmann equation. Phys Fluids 20:1589
9. Filbet F, Mouhot C, Pareschi L (2006) Solving the Boltzmann equation in NlogN. SIAM J Sci Comput 28(3):1029–1053
10. Pareschi L, Russo G (2000) Numerical solution of the Boltzmann equation I: spectrally accurate approximation of the collision operator. SIAM J Numer Anal 37(4):1217–1245
11. Ohwada T (1993) Structure of normal shock-waves: direct numerical analysis of the Boltzmann equation for hard-sphere molecules. Phys Fluids A-Fluid Dyn 5(1):217–234

12. Valentini P, Schwartzentruber TE (2009) Large-scale molecular dynamics simulations of normal shock waves in dilute argon. Phys Fluids 21(6)
13. Meng JP, Wu L, Reese JM, Zhang YH (2013) Assessment of the ellipsoidal-statistical Bhatnagar–Gross–Krook model for force-driven Poiseuille flows. J Comput Phys 251:383–395
14. Mieussens L, Struchtrup H (2004) Numerical comparison of Bhatnagar-Gross-Krook models with proper Prandtl number. Phys Fluids 16(8):2297–2813
15. Zheng Y, Garcia AL, Alder BJ (2002) Comparison of kinetic theory and hydrodynamics for Poiseuille flow. J Stat Phys 109:495–505
16. Wu L, Reese JM, Zhang YH (2014) Solving the Boltzmann equation by the fast spectral method: application to microflows. J Fluid Mech 746:53–84
17. John B, Gu XJ, Emerson DR (2010) Investigation of heat and mass transfer in a lid-driven cavity under nonequilibrium flow conditions. Num Heat Transfer B 52:287–303
18. Radtke GA, Hadjiconstantinou NG, Wagner W (2011) Low-noise Monte Carlo simulation of the variable hard sphere gas. Phys Fluids 23(3):030606

Chapter 5
Fast Spectral Method for Linear Gas Flow

The fast spectral method is applied to solve the linearized Boltzmann equation. First, two canonical problems in rarefied gas dynamics, the Poiseuille flow and thermal transpiration, are simulated and compared to solutions of the numerical kernel method. Then, the linearized Boltzmann equation with collision kernels directly calculated from the Lennard-Jones potential is solved, and the role of intermolecular potential is analyzed. The mass and heat flow rates are presented for different boundary conditions over a wide range of Knudsen numbers, which can be used in the upscaling method (like the Reynolds lubrication theory) to predict the rarefied gas flows in geometries with large aspect ratios.

5.1 Linearization

The linearized BCO (2.41) cannot be solved by FSM, because in general the perturbation function is not periodic in the velocity space, i.e., $\phi(-L_v) \neq \phi(L_v)$. However, the product of the equilibrium distribution function and ϕ vanishes at a large molecular speed. Therefore, the Boltzmann equation is linearized by expressing the VDF as (the normalization in Sect. 4.2 is used)

$$f(t, \boldsymbol{x}, \boldsymbol{v}) = f_{eq}(\boldsymbol{v}) + \beta h(t, \boldsymbol{x}, \boldsymbol{v}) \tag{5.1}$$

with the equilibrium distribution function

$$f_{eq}(\boldsymbol{v}) = \pi^{-3/2} \exp(-v^2), \tag{5.2}$$

L. Wu, *Rarefied Gas Dynamics*,
https://doi.org/10.1007/978-981-19-2872-7_5

where β is a small constant related to the dimensionless strength of perturbation (e.g., in the linear Couette flow β is the wall velocity divided by the most probable speed), and βh is the VDF for small perturbation satisfying $|\beta h/f_{eq}| \ll 1$. The Boltzmann equation (2.19) is linearized to

$$\frac{\partial h}{\partial t} + \boldsymbol{v} \cdot \frac{\partial h}{\partial \boldsymbol{x}} - 2\boldsymbol{a} \cdot \boldsymbol{v} f_{eq} = \mathcal{L}(h) \equiv \mathcal{L}^+ - \nu_{eq}(\boldsymbol{v})h, \tag{5.3}$$

where

$$\mathcal{L}^+(h) = \iint B(\theta, v_r)[f_{eq}(\boldsymbol{v}')h(\boldsymbol{v}'_*) + f_{eq}(\boldsymbol{v}'_*)h(\boldsymbol{v}') - f_{eq}(\boldsymbol{v})h(\boldsymbol{v}_*)]d\Omega d\boldsymbol{v}_* \tag{5.4}$$

can be viewed as a gain term of the linearized BCO, while $\nu_{eq}(\boldsymbol{v})h$ is the loss term, with the equilibrium collision frequency being

$$\nu_{eq}(\boldsymbol{v}) = \iint B(\theta, v_r) f_{eq}(\boldsymbol{v}_*) d\Omega d\boldsymbol{v}_*. \tag{5.5}$$

The macroscopic quantities deviated from their corresponding values in equilibrium state, such as the number density ϱ, bulk velocity $\boldsymbol{u}$, temperature T, stress tensor σ_{ij}, and heat flux $\boldsymbol{q}$ can be calculated as

$$[\varrho, \boldsymbol{u}, T, \sigma_{ij}, \boldsymbol{q}] = \int \left[1, \boldsymbol{v}, \frac{2}{3}\left(v^2 - \frac{3}{2}\right), 2v_{\langle i}v_{j\rangle}, \boldsymbol{v}\left(v^2 - \frac{5}{2}\right)\right] h d\boldsymbol{v}, \tag{5.6}$$

which, on top of the normalization (4.8), are further normalized by the dimensionless constant β.

The equilibrium collision frequency can be calculated analytically or approximated by the algorithm in Appendix D, if $\hat{f}$ is replaced by the spectrum of equilibrium distribution function $\hat{f}_{eq}$. This term only needs to be calculated once since each spatial cell uses the same equilibrium collision frequency. Typical profiles are shown in Fig. 2.2.

The gain term is calculated similarly by the method introduced in the previous chapter. Let

$$\phi_{loss} = \sum_{p,q=1}^{M} \omega_p \omega_q \sin\theta_p \phi_{\alpha+\gamma}(\xi_m, \theta_p, \varphi_q)\psi_\gamma(\xi_m, \theta_p, \varphi_q), \tag{5.7}$$

and $\hat{h}$ the spectrum of h. If Eq. (4.27) is approximated by Gauss-Legendre quadrature, its jth Fourier mode of the gain term $\mathcal{L}^+$ is

$$\begin{aligned}\widehat{\mathcal{L}}_j^+ \approx & \frac{4}{\text{Kn}'} \sum_{p,q=1}^{M} \sum_{l+m=j} \omega_p\omega_q \sin\theta_p [\hat{f}_{eq}(\boldsymbol{l})\phi_{\alpha+\gamma}(\xi_{\boldsymbol{l}}, \theta_p, \varphi_q)] \cdot [\hat{h}_{\boldsymbol{m}}\psi_\gamma(\xi_{\boldsymbol{m}}, \theta_p, \varphi_q)] \\ & + \frac{4}{\text{Kn}'} \sum_{p,q=1}^{M} \sum_{l+m=j} \omega_p\omega_q \sin\theta_p [\hat{h}_{\boldsymbol{l}}\phi_{\alpha+\gamma}(\xi_{\boldsymbol{l}}, \theta_p, \varphi_q)] \cdot [\hat{f}_{eq}(\boldsymbol{m})\psi_\gamma(\xi_{\boldsymbol{m}}, \theta_p, \varphi_q)] \\ & - \frac{4}{\text{Kn}'} \sum_{l+m=j} \hat{f}_{eq}(\boldsymbol{l}) \cdot [\hat{h}_{\boldsymbol{m}}\phi_{loss}]. \end{aligned} \tag{5.8}$$

The gain term $\mathcal{L}^+$ can be calculated by the second algorithm in Appendix D. Since the Fourier transform of $\hat{f}_{eq}(\boldsymbol{l})\phi_{\alpha+\gamma}(\xi_{\boldsymbol{l}}, \theta_p, \varphi_q)$ and $\hat{f}_{eq}(\boldsymbol{m})\psi_\gamma(\xi_{\boldsymbol{m}}, \theta_p, \varphi_q)$ can be pre-computed and stored, the computational time of the linearized BCO is nearly the same as that of the full BCO. Therefore, it seems that there is no need to consider the Boltzmann equation in linearized form. However, the following three factors indicate that the linearization is necessary and beneficial:

- In LBE the small signal is magnified and thus can be better resolved, while in the full Boltzmann equation this signal might be contaminated by numerical error.
- In oscillatory flows the time derivative can be removed if one are interested in the "steady-state" solution, where the flow oscillation has been fully established (see Chap. 9); the use of FSM for LBE facilitates the fast convergence to the final solution, without running time-explicit numerical solvers.
- For inverse power-law potentials, if we choose $\gamma = (1-\alpha)/2$ in the collision kernel (4.6), the linearized gain term $\mathcal{L}^+(h)$ becomes

$$\begin{aligned} & \frac{4}{\text{Kn}'} \iint d\boldsymbol{y}d\boldsymbol{z}\delta(\boldsymbol{y}\cdot\boldsymbol{z})(|\boldsymbol{y}||\boldsymbol{z}|)^{-\gamma} \\ & \times [f_{eq}(\boldsymbol{v}+\boldsymbol{z})h(\boldsymbol{v}+\boldsymbol{y}) + h(\boldsymbol{v}+\boldsymbol{z})f_{eq}(\boldsymbol{v}+\boldsymbol{y}) - h(\boldsymbol{v}+\boldsymbol{y}+\boldsymbol{z})f_{eq}(\boldsymbol{v})] \end{aligned} \tag{5.9}$$

after the Carleman representation. Since the interchange of $\boldsymbol{y}$ and $\boldsymbol{z}$ does not change the linearized gain term, Eq. (5.8) can be simplified to

$$\begin{aligned}\widehat{\mathcal{L}}_j^+ \approx & \frac{8}{\text{Kn}'} \sum_{p,q=1}^{M} \sum_{l+m=j} \omega_p\omega_q [\hat{f}_{eq}(\boldsymbol{l})\phi_{\alpha+\gamma}(\xi_{\boldsymbol{l}}, \theta_p, \varphi_q)][\hat{h}_{\boldsymbol{m}}\psi_\gamma(\xi_{\boldsymbol{m}}, \theta_p, \varphi_q)] \sin\theta_p \\ & - \frac{4}{\text{Kn}'} \sum_{l+m=j} \hat{f}_{eq}(\boldsymbol{l})[\hat{h}_{\boldsymbol{m}}\phi_{loss}]. \end{aligned} \tag{5.10}$$

Therefore, the computational cost will be reduced by half when compared to that of the full BCO.

5.2 Poiseuille Flow

Consider an infinite long channel of a cross-section area S located in the x_1x_2 plane. The gas inside is subject to a uniform pressure gradient in the x_3 direction:

$$p = p_0\left(1 + \beta_p \frac{x_3}{L}\right), \quad \beta_p = \frac{L}{p_0}\frac{dp}{dx_3}, \tag{5.11}$$

and the dimensionless pressure gradient β_p is very small. Both the wall and gas temperature are kept at T_0. The Boltzmann equation can be linearized around the equilibrium state, when h in Eq. (5.1) is replaced by $h + x_3 f_{eq}$, resulting in

$$v_1 \frac{\partial h}{\partial x_1} + v_2 \frac{\partial h}{\partial x_2} = \mathcal{L}^+(h) - \nu_{eq} h - v_3 f_{eq}. \tag{5.12}$$

Note that here the macroscopic variables and VDF have been normalized so that the MFR and heat flow rate (HFR) are respectively expressed as

$$\dot{M} = -\frac{2p_0 S}{v_m}\beta_p G_P, \quad \dot{E} = p_0 v_m S \beta_p Q_P, \tag{5.13}$$

with the normalized MFR and HFR

$$\begin{aligned} G_P &= -\frac{1}{A}\iint u_3 dx_1 dx_2 \equiv -\frac{1}{A}\iiint v_3 h dx_1 dx_2 d\boldsymbol{v}, \\ Q_p &= \frac{1}{A}\iint q_3 dx_1 dx_2 \equiv \frac{1}{A}\iiint \left(v^2 - \frac{5}{2}\right) v_3 h dx_1 dx_2 d\boldsymbol{v}, \end{aligned} \tag{5.14}$$

where $A = S/L^2$.

5.2.1 *Poiseuille Flow Between Parallel Plates*

Consider the Poiseuille flow between two plates located at $x_2 = \pm 1/2$. The spatial region is divided into $N_s = 50$ non-uniform cells, as

$$x = (10 - 15s + 6s^2)s^3, \tag{5.15}$$

with $s = (0, 1, \ldots, N_s)/N_s$. Such a non-uniform discretization has refined grids in the vicinity of solid walls, hence helps to capture the Knudsen layer structure. Because of symmetry, we only consider the half spatial region $-1/2 \le x_2 \le 0$ with the specular-reflection BC being imposed at $x_2 = 0$. The diffuse BC is adopted at the wall:

$$h\,(x_2 = -1/2, v_2 > 0) = 0. \tag{5.16}$$

The maximum molecular velocity is $L_v = 6$. Because of the symmetry and smoothness of the VDF in $v_1(>0)$ and $v_3(>0)$ directions, 12×12 uniform grids are used. In the discretization of v_2, $N_v = 64$ non-uniform velocity grids (4.34) with $\iota = 3$ are used to resolve the over-concentration of VDF at large Knudsen numbers. The number of frequency components in the ξ_1 and ξ_3 directions is 24. Although we use N_v non-uniform velocity grids in v_2 direction, the corresponding frequency domain is divided into 32 equidistant points. In the approximation of the kernel mode (4.27), the Gauss-Legendre quadrature with $M = 10$ is used. The MATLAB code is given in Appendix F, where besides the CIS, the general synthetic iterative scheme [1, 2] detailed in Chap. 8 is implemented to find the steady-state solution efficiently and accurately.

Typical profiles of the velocity and heat flux are shown in Fig. 5.1. When Kn $= 0.1$, the velocity is large. The velocity profile becomes flatter when the Knudsen number increases, due to the increased friction between gas molecules. When the Knudsen number is larger than 1, however, the velocity slip at the solid wall increases, which compensates the friction. As a result, a minimum MFR around Kn ~ 1 appears, which is known as the Knudsen minimum. The HFR always increases with the Knudsen number.

At large Knudsen number, the MFR and HFR are sensitive to the intermolecular potential (reflected through the viscosity index ω), even when the Knudsen number is fixed. That is, the smaller the viscosity index, the larger the mass and heat flow rates.

It should be noted that, Because of the "over-concentration" in VDF [4], numerical simulation of highly rarefied gas flows is a difficult task; for a long time the accurate numerical results have been limited to HS molecules when Kn $\lesssim 20$ [5, 6]. With FSM, the LBE can be solved accurately and efficiently up to Kn $\sim 10^6$. Typical VDF profiles demonstrating the over-concentration phenomena are shown in Fig. 5.2. When Kn $= 1$, the marginal VDF is roughly proportional to $\exp(-v_2^2)$. However, in the free-molecular regime, the marginal VDF has a long tail, and its width shrinks drastically as $1/\text{Kn}$, while its amplitude scales as Kn. If the velocity discretization is not refined around $v_2 \sim 0$, the calculated flow rates will be wrong.

5.2.2 *Poiseuille Flow Through a Long Duct*

Consider the Poiseuille flow in a long duct with the aspect ratio $A = S/L^2$, under the diffuse BC. Unlike the Poiseuille flow between parallel plates where the flow rates increase logarithmically at large Kn, here they saturate at $G_P = \mathcal{M}$ and $Q_P = \mathcal{M}/2$ when Kn $\to \infty$ [7], where

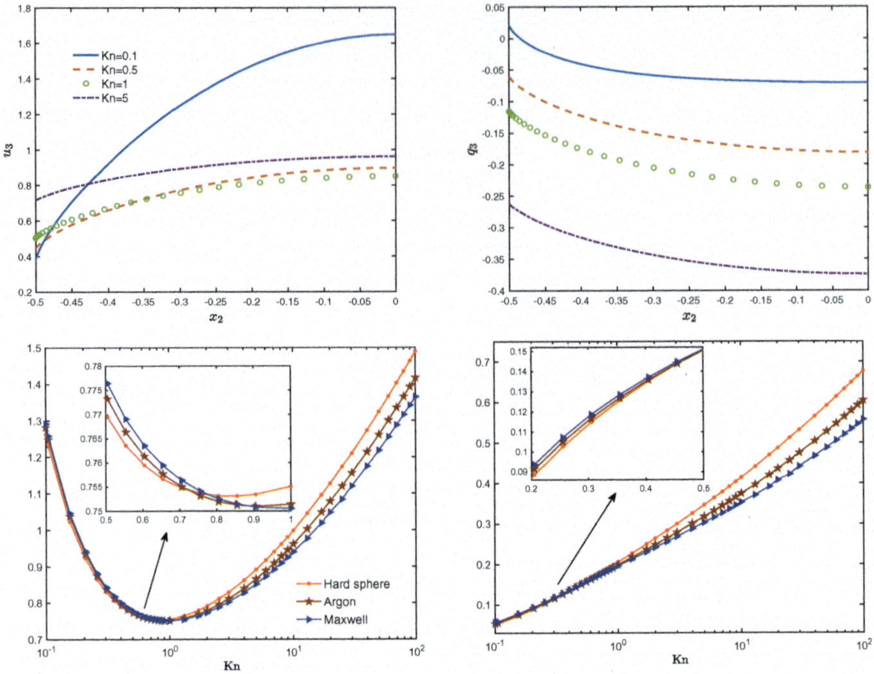

Fig. 5.1 Poiseuille gas flow between parallel plates. (Top row) Velocity and heat flux in the HS gas. (Bottom row) Comparisons in MFR (left) and HFR (right) for different inverse power-law potentials [3]. The viscosity index for argon is $\omega = 0.81$

$$\mathcal{M} = \frac{1}{4\sqrt{\pi}}\left[\frac{2(A^3+1)}{3A} - \frac{2(A^2+1)^{3/2}}{3A} + \ln\frac{(A^2+1)^{1/2}+A}{(A^2+1)^{1/2}-A} + A\ln\frac{(A^2+1)^{1/2}+1}{(A^2+1)^{1/2}-1}\right]. \tag{5.17}$$

Nice agreement in the mass and heat flow rates of HS gas between the FSM and numerical kernel method are observed in Table 5.1. To show the numerical efficiency we consider the HS gas inside a square channels; with a 25×25 spatial cells, $32 \times 32 \times 24$ frequency components, and $M = 6$, the flow rates at $\text{Kn}_{vhs} = 1$ and 10 are obtained in about 1 min, when our Fortran code runs on a single core of an Intel Q9650 (3.0 GHz Core 2 Quad processor), while the low-noise DSMC takes respectively 66 and 12 min [6, 9].

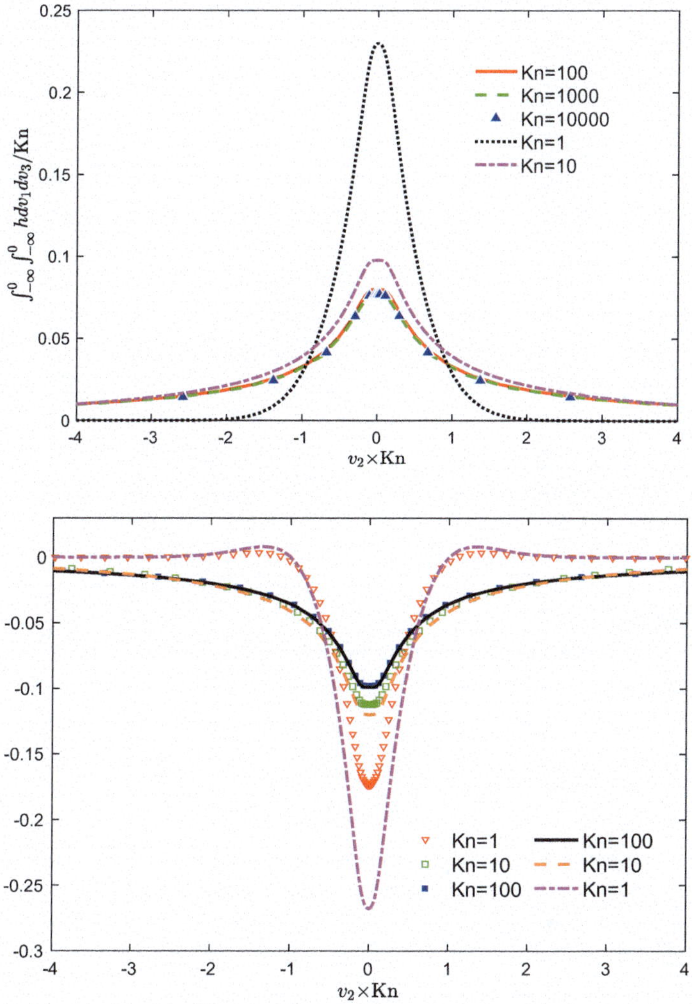

Fig. 5.2 (Top) The marginal VDF in the Poiseuille flow at the channel center. When $\mathrm{Kn} \geq 100$, the velocity distribution perpendicular to the plate shrinks as $1/\mathrm{Kn}$, which is known as the "over-concentration" [4]. (Bottom) The Onsager-Casimir relation at the mesoscopic level. Symbols: the marginal VDFs $\iint_{-\infty}^{0} hdv_1dv_3/\mathrm{Kn}$, where h is obtained in the thermal transpiration. Lines: the marginal VDFs $\iint_{-\infty}^{0} \left(v^2 - \frac{5}{2}\right) hdv_1dv_3/\mathrm{Kn}$, where h is obtained in the Poiseuille flow

Table 5.1 MFR and HFR in the Poiseuille flow of HS gas through a rectangular channel [8]. Note that $\mathrm{Kn} = 5\pi k'/16$, and the numerical kernel method is used in Doi's paper

k'	$A = 1$				$A = 2$			
	FSM		Doi [6]		FSM		Doi [6]	
	G_P	Q_P	G_P	Q_P	G_P	Q_P	G_P	Q_P
0.1	0.6116	0.0445	0.613	0.045	0.9009	0.0476	0.905	0.048
0.3	0.4260	0.0889	0.426	0.089	0.5950	0.1013	0.595	0.101
0.5	0.3960	0.1103	0.396	0.110	0.5433	0.1301	0.544	0.130
0.8	0.3833	0.1280	0.383	0.128	0.5203	0.1554	0.520	0.156
1	0.3811	0.1356	0.381	0.136	0.5141	0.1664	0.515	0.167
3	0.3835	0.1656	0.383	0.166	0.5148	0.2131	0.516	0.214
5	0.3886	0.1760	0.388	0.176	0.5228	0.2300	0.523	0.230
8	0.3938	0.1837	0.394	0.184	0.5314	0.2428	0.532	0.243
10	0.3963	0.1869	0.396	0.187	0.5354	0.2481	0.536	0.248
20	0.4033	0.1948	–	–	0.5479	0.2613	–	–
50	0.4102	0.2016	–	–	0.5596	0.2734	–	–
10^2	0.4136	0.2048	–	–	0.5657	0.2791	–	–
10^3	0.4183	0.2089	–	–	0.5742	0.2865	–	–
10^4	0.4191	0.2095	–	–	0.5758	0.2878	–	–
∞[a]	0.4194	0.2097	0.4194	0.2097	0.5762	0.2881	0.5762	0.2881

[a] Data in columns 4, 5, 8, and 9 are obtained from Eq. (5.17)

5.3 Thermal Transpiration

Thermal transpiration, where the gas moves toward a hotter region even in the absence of a pressure gradient, is one of the fundamental problems in RGD [10, 11]. The gas pressure is maintained constant, while the wall temperature varies linearly in the x_3 direction as

$$T = T_0\left(1 + \beta_T \frac{x_3}{L}\right), \quad \beta_T = \frac{L}{T_0}\frac{dT}{dx_3}, \tag{5.18}$$

where the dimensionless temperature gradient β_T is very small. The Boltzmann equation can be linearized around the equilibrium state, when h in Eq. (5.1) is replaced by $h + x_3 f_{eq}(v^2 - \frac{5}{2})$, resulting in

$$v_1\frac{\partial h}{\partial x_1} + v_2\frac{\partial h}{\partial x_2} = \mathcal{L}^+(h) - \nu_{eq}h - v_3\left(v^2 - \frac{5}{2}\right) f_{eq}. \tag{5.19}$$

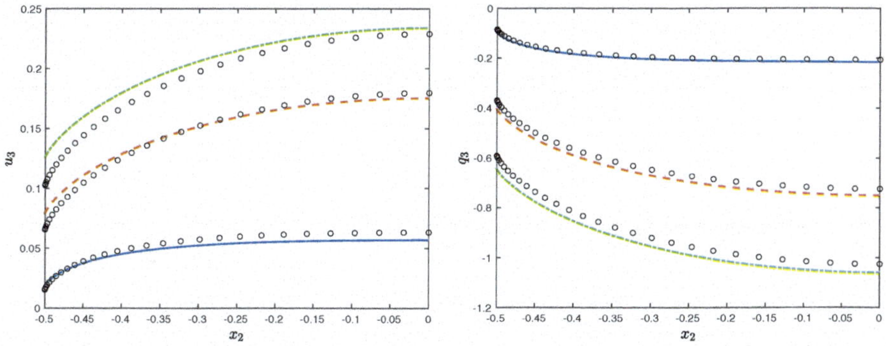

Fig. 5.3 Profiles of velocity and heat flux in the thermal transpiration between two parallel plates. Solid, dashed, dash-dotted lines are the results of HS gas, when Kn = 0.1, 0.5, and 1, respectively, while circles are the corresponding results of Maxwell gas

The MFR and HFR are expressed as

$$\dot{M} = \frac{2p_0 S}{v_m}\beta_T G_T, \quad G_T = \frac{1}{A}\iiint v_3 h dx_1 dx_2 d\boldsymbol{v},$$
$$\dot{E} = -p_0 v_m S \beta_T Q_T, \quad Q_T = -\frac{1}{A}\iiint \left(v^2 - \frac{5}{2}\right) v_3 h dx_1 dx_2 d\boldsymbol{v}. \tag{5.20}$$

Typical profiles of the velocity and heat flux are shown in Fig. 5.3, whose magnitudes increase with the Knudsen number. Again, it is observed that although the Knudsen number is the same, different intermolecular potential (viscosity index) has different macroscopic profiles.

For the rectangular cross-section, the MFR is $G_T = \mathcal{M}/2$ and the HFR in the free-molecular flow regime [7] is $G_T = 9\mathcal{M}/4$, where $\mathcal{M}$ is given by Eq. (5.17).

5.4 Onsager-Casimir Relation

The Onsager-Casimir relation states that the MFR in the thermal transpiration is equal to the HFR in the Poiseuille flow [12]. In the numerical simulation we find that this relation is held with the absolute error smaller than 10^{-4}.

Takata and Funagane [4] made the important observation that at large Kn, $u_3[h_T]$ and $q_3[h_P]$ are even identical at the level of spatial profile:

$$u_3[h_T] = q_3[h_P] + O\left(\frac{(\ln \text{Kn})^2}{\text{Kn}}\right). \tag{5.21}$$

Our numerical results in Fig. 5.2 further show that the agreement is even at the mesoscopic level:

$$h_T \approx \left(v^2 - \frac{5}{2}\right) h_P. \tag{5.22}$$

5.5 Influence of Intermolecular Potential

Note that except for the HS gas, the collision kernels in the previous chapter are modeled to recover the shear viscosity, rather than calculated directly from the intermolecular potential. Here we extend the FSM to solve the LBE with the collision kernel directly calculated from the Lennard-Jones potential (2.10), rather than using the modeled kernel (4.7).

5.5.1 *Lennard-Jones Potential*

When the Lennard-Jones potential is considered, after the normalization in Sect. 4.2, the collision kernel $B(\theta, v_r)$ in Eqs. (5.4) and (5.5) becomes [13]:

$$B(\theta, v_r) = (n_0 d_{LJ}^2 L)\sigma_D v_r, \tag{5.23}$$

where the differential cross-section, after the Carleman representation, is

$$\sigma_D = \sigma_D\left(2\arctan\frac{|y|}{|z|}, v_m\sqrt{y^2+z^2}\right) \equiv \sigma_D'(|y|, |z|). \tag{5.24}$$

In general, y and z in $\sigma_D'(|y|, |z|)$ cannot be separated as $\sigma_1(|y|)\sigma_2(|z|)$, and the kernel mode $\beta(\boldsymbol{l}, \boldsymbol{m})$ can only be simplified to

$$\begin{aligned}\beta(\boldsymbol{l}, \boldsymbol{m}) =& (n_0 d_{LJ}^2 L)\iint d\boldsymbol{e}' d\boldsymbol{e}\delta(\boldsymbol{e}\cdot\boldsymbol{e}') \\ &\times \int_{-R}^{R}\int_{-R}^{R}|\rho\rho'|\sigma_D'(|\rho|, |\rho'|)\exp(i\rho\xi_l\cdot\boldsymbol{e} + i\rho'\xi_m\cdot\boldsymbol{e}')d\rho d\rho'.\end{aligned} \tag{5.25}$$

To enable the FFT-based convolution, the integration with respect to ρ is approximated by a numerical quadrature [13, 14], and the integration with respect to ρ' is calculated accurately, or vice versa. Suppose ρ_r and ω_r with $r = 1, 2, \ldots, M_r$ are the abscissas and weights of a quadrature in the region $[0, R]$, Eq. (5.25) becomes

$$\beta(\boldsymbol{l}, \boldsymbol{m}) = (n_0 d_{LJ}^2 L)\sum_r \omega_r \iint \delta(\boldsymbol{e}\cdot\boldsymbol{e}')\phi(\rho_r, \xi_l\cdot\boldsymbol{e})\psi(\rho_r, \xi_m\cdot\boldsymbol{e}')d\boldsymbol{e}'d\boldsymbol{e}, \tag{5.26}$$

where $\psi(\rho_r, s) = 2\int_0^R \rho' \sigma'_D(\rho_r, \rho') \cos(\rho' s) d\rho'$ and $\phi(\rho_r, s) = 2\rho_r \cos(\rho_r s)$. Then, following the straightforward algebraic calculation in Sect. 4.3.2, we have

$$\beta(\boldsymbol{l}, \boldsymbol{m}) = 4(n_0 d_{LJ}^2 L) \sum_{p,q,r=1}^{M,M,M_r} \omega_p \omega_q \omega_r \phi(\rho_r, \xi_l \cdot \boldsymbol{e}_{\theta_p, \varphi_q}) \psi' \left(\rho_r, \xi_l^{\perp}\right) \sin\theta_p, \quad (5.27)$$

where

$$\psi'(\rho_r, s) = 2\pi \int_0^R \rho' \sigma'_D(\rho_r, \rho') J_0(\rho' s) d\rho'. \quad (5.28)$$

Thus, the BCO for the Lennard-Jones potential can be calculated through the FFT-based convolution, with a computational cost of $O(M^2 M_r N_v^3 \log N_v)$. Normally $M_r \sim N_v$, so the computational cost for the Lennard-Jones potential is higher than the modeled collision kernel (4.6) by about one order of magnitude. A detailed numerical method to calculate the differential cross-section and the corresponding kernel mode is given in Ref. [13].

5.5.2 Accurate Transport Coefficients

Note that the shear viscosity (3.28) is derived based on the eigenvalues and eigenfunctions of the LBE for Maxwellian molecules. For general intermolecular potentials, however, Eq. (3.28) is not accurate, and its high-order correction is cumbersome. Here we show how to obtain accurate transport coefficients of the Boltzmann equation via FSM.

According to the Chapman-Enskog expansion, the exact shear viscosity μ and thermal conductivity κ of the Boltzmann equation are calculated by

$$\begin{aligned} \mu &= \frac{m v_m}{d_{LJ}^2} \int h_\mu(v) v_1 v_2 dv \equiv \frac{m v_m}{d_{LJ}^2} \mu', \\ \kappa &= \frac{k_B v_m}{d_{LJ}^2} \int h_\kappa(v) v_1 \left(v^2 - \frac{5}{2}\right) dv \equiv \frac{k_B v_m}{d_{LJ}^2} \kappa', \end{aligned} \quad (5.29)$$

where μ' and κ' are the reduced viscosity and thermal conductivity, respectively. The two functions $h_\mu(v)$ and $h_\kappa(v)$ satisfy the following integral equations if we choose $n_0 d_{LJ}^2 L = 1$ in the collision kernel:

$$\begin{aligned} \mathcal{L}(h_\mu) &= -2 f_{eq} v_1 v_2, \\ \mathcal{L}(h_\kappa) &= -f_{eq} v_1 \left(v^2 - \frac{5}{2}\right), \quad \text{and} \int h_\kappa v_1 dv = 0. \end{aligned} \quad (5.30)$$

Table 5.2 Comparisons of reduced transport coefficients obtained from the FSM at temperature 300 K [13] with those from the variational method with first- and third-order Chapman-Cowling approximation [15]

Gas	$\mu'^{[1]}$	$\mu'^{[3]}$	FSM μ'	$\kappa'^{[1]}$	$\kappa'^{[3]}$	FSM κ'
He	0.1773	0.1787	0.1789	0.6650	0.6732	0.6742
Ne	0.1477	0.1488	0.1486	0.5539	0.5602	0.5596
Ar	0.1129	0.1131	0.1132	0.4234	0.4248	0.4251
Kr	0.0969	0.0969	0.0967	0.3632	0.3635	0.3629
Xe	0.0892	0.0893	0.0894	0.3348	0.3349	0.3354

To find h_μ, we use the following iterative scheme:

$$h_\mu^{(k+1)} = \frac{\mathcal{L}^+(h_\mu^{(k)}) + 2 f_{eq} v_1 v_2}{\nu_{eq}}, \tag{5.31}$$

while to find h_κ, we use

$$\begin{aligned} \widetilde{h}_\kappa^{(k+1)} &= \frac{\mathcal{L}^+(h_\kappa^{(k)}) + f_{eq} v_1 \left(v^2 - \frac{5}{2}\right)}{\nu_{eq}}, \\ h_\kappa^{(k+1)} &= \widetilde{h}_\kappa^{(k+1)} - 2 f_{eq} v_1 \int \widetilde{h}_\kappa^{(k+1)} v_1 d\boldsymbol{v}. \end{aligned} \tag{5.32}$$

The Lennard-Jones potential for five noble gases, He, Ne, Ar, Kr, and Xe, are considered at $T_0 = 300$ K, with the potential depths $k_B T_0/\epsilon = 29.35$, 8.403, 2.419, 1.579, and 1.310, respectively. The numerical results in Table 5.2 confirm the accuracy of FSM in the approximation of BCO when the collision kernel is directly calculated from the Lennard-Jones potential.

5.5.3 *Poiseuille Flow*

The LBE is solved with the Lennard-Jones potential. It is noted that the spatial and temporal Knudsen numbers, as well as the rarefaction parameter, are defined in terms of the shear viscosity. To obtain accurate numerical results from the Boltzmann equation, accurate transport coefficients should be used. In this case, in the simulation of Lennard-Jones potential, the term $n_0 d_{LJ}^2 L$ in the collision kernel (5.23) should be modified as

$$n_0 d_{LJ}^2 L \to 2\mu' \delta_{rp}, \tag{5.33}$$

Table 5.3 Mass flow rate G_P and heat flow rate Q_P in the Poiseuille flow between parallel infinite plates when the Lennard-Jones potential is considered [13]

δ_{rp}	He		Ne		Ar		Kr		Xe	
	G_P	Q_P	G_P	Q_P	G_P	Q_P	G_P	Q_P	G_P	Q_P
0.01	2.713	1.152	2.607	1.070	2.535	1.057	2.538	1.065	2.547	1.072
0.02	2.432	1.021	2.362	0.961	2.292	0.930	2.286	0.933	2.290	0.938
0.025	2.348	0.981	2.289	0.929	2.219	0.893	2.211	0.894	2.214	0.898
0.04	2.180	0.900	2.141	0.862	2.076	0.819	2.063	0.815	2.064	0.818
0.05	2.105	0.863	2.074	0.831	2.011	0.785	1.998	0.780	1.998	0.782
0.10	1.892	0.749	1.879	0.732	1.829	0.688	1.815	0.679	1.813	0.679
0.20	1.713	0.637	1.708	0.629	1.677	0.596	1.666	0.586	1.663	0.584
0.25	1.665	0.601	1.661	0.595	1.636	0.566	1.626	0.557	1.624	0.555
0.40	1.581	0.526	1.579	0.523	1.564	0.504	1.558	0.497	1.556	0.495
0.50	1.550	0.491	1.549	0.489	1.539	0.474	1.534	0.468	1.532	0.466
1.00	1.505	0.385	1.505	0.384	1.505	0.380	1.505	0.378	1.505	0.377
1.60	1.532	0.315	1.533	0.315	1.537	0.315	1.540	0.316	1.541	0.315
2.00	1.568	0.282	1.568	0.283	1.575	0.285	1.579	0.286	1.580	0.286
2.50	1.622	0.251	1.623	0.251	1.630	0.254	1.636	0.256	1.637	0.257
4.00	1.817	0.188	1.818	0.189	1.828	0.193	1.835	0.196	1.838	0.197
5.00	1.960	0.161	1.961	0.162	1.972	0.166	1.980	0.169	1.983	0.170
10.0	2.732	0.093	2.732	0.093	2.743	0.097	2.752	0.099	2.756	0.100

while in the power-law potential with the modeled collision kernel (4.6), after obtaining the reduced shear viscosity μ', the term $1/\text{Kn}'$ in Eqs. (4.11) and (4.12) should be modified as

$$\frac{1}{\text{Kn}'} \rightarrow \frac{2\mu'\delta_{rp}}{\text{Kn}'}. \tag{5.34}$$

Such a modification, although only introduces a relative difference of less than 2.5%, becomes important when calculating the Knudsen layer function.

The numerical results in Table 5.3 show that different parameters in the Lennard-Jones potential (2.10) lead to different flow rates, especially when the rarefaction parameter is small.

5.5.4 *Planar Fourier Flow*

Consider the Fourier heat conduction between two plates located at $x_2 = -0.5$ and $x_2 = 0.5$, where the temperature is $T_0 - \Delta T/2$ and $T_0 + \Delta T/2$, respectively. The temperature difference ΔT is negligible compared to T_0 so that the Boltzmann

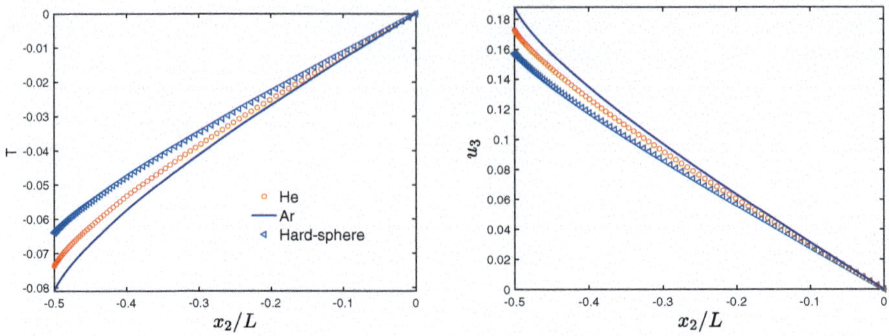

Fig. 5.4 Normalized temperature (velocity) in the planar Fourier (Couette) flow [13] obtained from the LBE with the collision kernel directly calculated from the Lennard-Jones potential. Results from the HS gas are shown for comparison. The rarefaction parameter is $\delta_{rp} = 0.1$

equation (4.10) is linearized to Eq. (5.3) by choosing $\beta = \Delta T/T_0$. Assuming diffuse BC, the VDF reflected from the wall reads

$$h = \left[1 - \frac{v^2}{2} - 2\sqrt{\pi}\int_{v_2<0} v_2 h\left(v, x_2 = -0.5\right) dv\right] f_{eq}, \tag{5.35}$$

while at $x_2 = 0$, the symmetry leads to $h(v_1, v_2, v_3) = -h(v_1, -v_2, v_3)$.

Figure 5.4 shows the temperature when $\delta_{rp} = 0.1$ and $T_0 = 300$ K where significant temperature jump at the wall is seem. The influence of intermolecular potential is clearly seen, e.g., the temperature of Ar at the wall is about 40% larger than that of the HS gas. As δ_{rp} increases, the difference decreases, which eventually disappears in near-continuum flow. Interestingly, the difference in heat flux between various gases is small over the whole range of rarefaction parameters.

5.5.5 Planar Couette Flow

The geometry of planar Couette flow is the same as that of the Fourier flow, except that the plate at $x_2 = -0.5$ moves with a speed u_w in the x_3 direction, while the plate at $x_2 = 0.5$ moves in the opposite direction with the same speed. When the wall speed is far less than the most probable speed v_m, the Boltzmann equation is linearized to Eq. (5.3) by choosing $\beta = u_w/v_m$. The diffuse BC for the perturbation VDF reflected from the wall is

$$h(\boldsymbol{v}, x_2 = -0.5) = 2v_3 f_{eq}, \tag{5.36}$$

while at the channel center $h(v_1, v_2, v_3, x_2 = 0) = h(v_1, -v_2, -v_3, x_2 = 0)$.

Figure 5.4 shows the velocity when $\delta_{rp} = 0.1$. The intermolecular potential has a strong influence on the velocity profile, but little influence on the wall shear stress.

5.6 Cercignani-Lampis Boundary Condition

We present the LBE solutions for the Poiseuille flow between two plates and through a circular cross section [16]. Although this classical problem has been investigated extensively, accurate numerical results based on the LBE and the Cercignani-Lampis BC are scarce.

5.6.1 *Poiseuille Flow Through Parallel Plates*

Table 5.4 shows the flow rates for HS molecules. When the rarefaction parameter δ_{rp} and the EAC α_n are fixed, the MFR increases rapidly when the TMAC α_t decreases. However, the influence of α_n on MFR is limited.

The variation of HFR with respect to α_n and α_t is complicated. First, when α_n and δ_{rp} are fixed, the HFR increases slightly with α_t when δ_{rp} is large, while it increases with decreasing α_t when δ_{rp} is small. Second, when $\alpha_t = 1$ and δ_{rp} is fixed, the HFR does not change with α_n: in fact, in this case the Cercignani-Lampis BC is reduced to the diffuse BC [17]. Third, when $\alpha_t (\neq 1)$ and δ_{rp} are fixed, the HFR increases slightly with α_n when δ_{rp} is large, but it increases with decreasing α_n when δ_{rp} is small.

The influence of intermolecular potential is visualized in Fig. 5.5, where we choose xenon and helium, as the results of other noble gases are similar. The large difference is found at small rarefaction parameter. For example, when $\delta_{rp} = 0.01$, the relative differences in the MFR and HFR between HS gas and xenon are about 12% and 23%, respectively, when the Cercignani-Lampis BC with $\alpha_n = 1$ and $\alpha_t = 0.75$ is used.

Figure 5.5c shows the exponent of thermalmolecular pressure difference (TPD), which is an important parameter determining the performance of a Knudsen pump. In the range of δ_{rp} considered, the HS gas has the largest TPD, while xenon has the smallest. This difference increases when δ_{rp} decreases. In the Cercignani-Lampis BC, when δ_{rp} is fixed, the TPD exponent decreases with α_t. In the diffuse BC, when the TMAC α_M decreases, the TPD exponent decreases at large values of δ_{rp}, but at free-molecular flow regime, α_M does not have any influence on the TPD exponent.

5.6.2 *Poiseuille Flow Through Long Tube*

For flows through circular cross sections, the polar coordinates can be applied to reduce the computational cost. Introducing the polar coordinates in the spatial space

Table 5.4 Dimensionless flow rates in the Poiseuille flow of HS molecules between two parallel plates [16]. The Cercignani-Lampis BC is used, but the data in the last two columns are from the Maxwell BC

δ_{rp}	α_t	G_P	Q_P	G_P	Q_P	G_P	Q_P	G_P	Q_P	G_P	Q_P
		$\alpha_n = 0.25$		$\alpha_n = 0.5$		$\alpha_n = 0.75$		$\alpha_n = 1$		$\alpha_M = \alpha_t$	
0.01	0.5	5.115	1.699	4.871	1.499	4.752	1.390	4.684	1.322	6.844	2.991
	1	2.911	1.320	2.911	1.320	2.911	1.320	2.911	1.320	2.911	1.320
	1.5	2.084	1.122	2.196	1.205	2.265	1.265	2.320	1.319		
0.1	0.5	3.996	1.098	3.854	0.953	3.774	0.866	3.724	0.807	4.389	1.570
	1	1.951	0.801	1.951	0.801	1.951	0.801	1.951	0.801	1.951	0.801
	1.5	1.200	0.634	1.270	0.699	1.319	0.749	1.360	0.796		
0.2	0.5	3.710	0.906	3.616	0.798	3.558	0.726	3.519	0.675	3.907	1.211
	1	1.747	0.667	1.747	0.667	1.747	0.667	1.747	0.667	1.747	0.667
	1.5	1.037	0.525	1.086	0.577	1.123	0.620	1.156	0.661		
1	0.5	3.308	0.458	3.297	0.435	3.288	0.415	3.280	0.398	3.327	0.529
	1	1.507	0.389	1.507	0.389	1.507	0.389	1.507	0.389	1.507	0.389
	1.5	0.894	0.336	0.901	0.351	0.908	0.366	0.915	0.381		
2	0.5	3.340	0.300	3.339	0.296	3.338	0.292	3.337	0.288	3.347	0.328
	1	1.564	0.281	1.564	0.281	1.564	0.281	1.564	0.281	1.564	0.281
	1.5	0.970	0.265	0.971	0.268	0.971	0.272	0.972	0.275		
10	0.5	4.522	0.0834	4.514	0.0861	4.508	0.0887	4.503	0.0912	4.535	0.0843
	1	2.729	0.0900	2.729	0.0900	2.729	0.0900	2.729	0.0900	2.729	0.0900
	1.5	2.120	0.0962	2.127	0.0938	2.133	0.0913	2.138	0.0889		
20	0.5	6.162	0.0437	6.151	0.0454	6.142	0.0470	6.133	0.0485	6.177	0.0437
	1	4.360	0.0480	4.360	0.0480	4.360	0.0480	4.360	0.0480	4.360	0.0480
	1.5	3.743	0.0519	3.752	0.0505	3.761	0.0490	3.769	0.0474		
100	0.5	19.47	0.0091	19.45	0.0094	19.44	0.0098	19.43	0.0102	19.49	0.0090
	1	17.66	0.0101	17.66	0.0101	17.66	0.0101	17.66	0.0101	17.66	0.0101
	1.5	17.03	0.0110	17.05	0.0107	17.06	0.0103	17.07	0.0100		

$$x_1 = r\cos\theta, \quad x_2 = r\sin\theta, \tag{5.37}$$

the cylindrical coordinates in the molecular velocity space

$$v_1 = v_r\cos\theta, \quad v_2 = v_r\sin\theta, \tag{5.38}$$

and defining the VDF as $h = h(r, \theta, v_r, v_3)$, the LBE can be written as

$$v_1\frac{\partial h}{\partial r} - \frac{v_2}{r}\frac{\partial h}{\partial \theta} = \mathcal{L}^+(h) - \nu_{eq}h - v_3 f_{eq}. \tag{5.39}$$

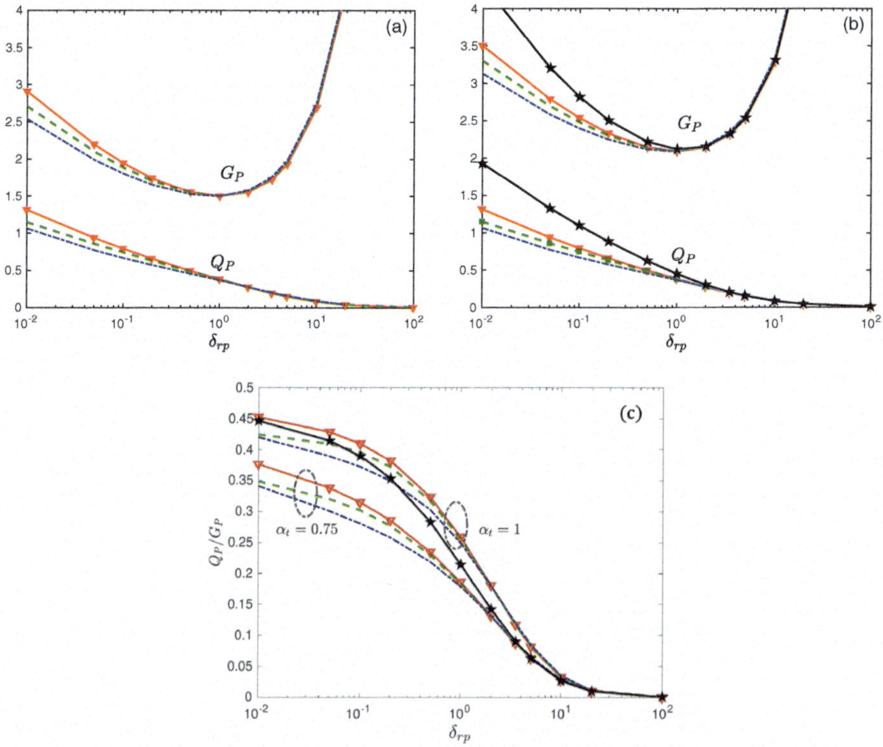

Fig. 5.5 Flow rates in the Poiseuille flow between two parallel plates [16]. Triangles: HS gas. Dashed lines: He. Dash-dotted lines: Xe. The Lennard-Jones potential is used, the wall temperature is $T_w = 300$ K. In the Cercignani-Lampis BC, $\alpha_n = 1$, **a** $\alpha_t = 1$ and **b** $\alpha_t = 0.75$. In the Maxwell BC, $\alpha_M = 0.75$ is used (Pentagrams). **c** The exponent of thermal pressure difference, i.e., the ratio of MFR G_T/G_P, or Q_P/G_P as per the Onsager-Casimir relation

In the numerical simulation, r is discretized by 150 non-uniform points, with most points located near the pipe surface $r = 1$. Due to symmetry, the truncated velocity $v_r \in (0, 4)$ is discretized by 64 non-uniform points, with most points located near $v_r = 0$ to capture the discontinuities in the VDF, while $\theta \in [0, \pi]$ and $v_3 \in (0, 6)$ are discretized by 40 and 12 uniform points, respectively.

The FSM is implemented in the following way [1]: first, the spectrum of the VDF is calculated by Fourier transform from the cylindrical molecular velocity space to the Cartesian frequency space, where the frequency is uniformly discretized. This is more expansive than the use of non-uniform velocity grids in Cartesian coordinates (see Sect. 4.3.4). Second, the FSM is applied to find the spectrum of the linearized BCO in the Cartesian coordinate. Finally, the inverse Fourier transform is used to find the collision operator in the cylindrical space.

The MFR and HFR of the HS molecules through a long tube are shown in Table 5.5. The influence of BC exists, but is not as large as that between parallel plates. This is

Table 5.5 Dimensionless flow rates in the Poiseuille flow of HS molecules through a circular tube, using the Cercignani-Lampis BC. The characteristic length L is the radius of circular cross section [16]

δ_{rp}	α_t	G_P	Q_P	G_P	Q_P	G_P	Q_P	G_P	Q_P
		$\alpha_n = 0.25$		$\alpha_n = 0.5$		$\alpha_n = 0.75$		$\alpha_n = 1$	
0.01	0.5	3.363	0.989	3.321	0.881	3.295	0.809	3.277	0.759
	1	1.472	0.725	1.472	0.725	1.472	0.725	1.472	0.725
	1.5	0.809	0.584	0.825	0.620	0.840	0.657	0.855	0.697
0.1	0.5	3.251	0.837	3.227	0.763	3.210	0.709	3.198	0.669
	1	1.397	0.634	1.397	0.634	1.397	0.634	1.397	0.634
	1.5	0.753	0.516	0.763	0.545	0.773	0.574	0.784	0.607
0.5	0.5	3.181	0.574	3.176	0.553	3.172	0.536	3.169	0.520
	1	1.381	0.492	1.381	0.492	1.381	0.492	1.381	0.492
	1.5	0.767	0.432	0.770	0.443	0.772	0.455	0.775	0.468
1	0.5	3.233	0.432	3.232	0.429	3.231	0.426	3.230	0.423
	1	1.448	0.403	1.448	0.403	1.448	0.403	1.448	0.403
	1.5	0.845	0.379	0.847	0.380	0.847	0.383	0.848	0.385
2	0.5	3.423	0.295	3.420	0.300	3.418	0.305	3.417	0.309
	1	1.639	0.298	1.639	0.298	1.639	0.298	1.639	0.298
	1.5	1.038	0.301	1.040	0.297	1.042	0.293	1.043	0.288
5	0.5	4.113	0.152	4.105	0.157	4.098	0.162	4.093	0.167
	1	2.319	0.164	2.319	0.164	2.319	0.164	2.319	0.164
	1.5	1.708	0.175	1.715	0.170	1.721	0.165	1.726	0.161
10	0.5	5.333	0.0835	5.322	0.0868	5.313	0.0899	5.305	0.0929
	1	3.531	0.0917	3.531	0.0917	3.531	0.0917	3.531	0.0917
	1.5	2.913	0.0992	2.923	0.0963	2.932	0.0934	2.939	0.0904
20	0.5	7.815	0.0437	7.802	0.0456	7.791	0.0472	7.782	0.0489
	1	6.007	0.0484	6.007	0.0484	6.007	0.0484	6.007	0.0484
	1.5	5.385	0.0527	5.396	0.0511	5.406	0.0495	5.416	0.0479
50	0.5	15.30	0.0179	15.28	0.0187	15.27	0.0194	15.26	0.0201
	1	13.49	0.0200	13.49	0.0200	13.49	0.0200	13.49	0.0200
	1.5	12.86	0.0218	12.87	0.0212	12.89	0.0205	12.89	0.0198
100	0.5	27.78	0.0090	27.77	0.0094	27.75	0.0098	27.74	0.0101
	1	25.97	0.0101	25.97	0.0101	25.97	0.0101	25.97	0.0101
	1.5	25.34	0.0110	25.35	0.0107	25.37	0.0103	25.38	0.0100

because, in the free-molecular flow regime, the flow rates in tube flow are constant, while that between parallel plates increase logarithmically with the Knudsen number.

References

1. Wu L, Zhang J, Liu HH, Zhang YH, Reese JM (2017) A fast iterative scheme for the linearized Boltzmann equation. J Comput Phys 338:431–451
2. Su W, Zhu LH, Wang P, Zhang YH, Wu L (2020) Can we find steady-state solutions to multiscale rarefied gas flows within dozens of iterations? J Comput Phys 407:109245
3. Wu L, Reese JM, Zhang YH (2014) Solving the Boltzmann equation by the fast spectral method: application to microflows. J Fluid Mech 746:53–84
4. Takata S, Funagane H (2011) Poiseuille and thermal transpiration flows of a highly rarefied gas: over-concentration in the velocity distribution function. J Fluid Mech 669:242–259
5. Ohwada T, Sone Y, Aoki K (1989) Numerical analysis of the Poiseuille and thermal transpiration flows between two parallel plates on the basis of the Boltzmann equation for hard sphere molecules. Phys Fluids 1:2042
6. Doi T (2010) Numerical analysis of the Poiseuille flow and thermal transpiration of a rarefied gas through a pipe with a rectangular cross section based on the linearized Boltzmann equation for a hard sphere molecular gas. J Vac Sci Technol, A 28:603–612
7. Loyalka SK, Storvick TS, Park HS (1976) Poiseuille flow and thermal creep flow in long, rectangular channels in the molecular and transition flow regimes. J Vac Sci Technol 13:1188
8. Wu L (2013) Deterministic numerical simulation of the Boltzmann and kinetic model equations for classical and quantum dilute gases. PhD thesis, University of Strathclyde, Glasgow
9. Radtke GA, Hadjiconstantinou NG, Wagner W (2011) Low-noise Monte Carlo simulation of the variable hard sphere gas. Phys Fluids 23(3):030606
10. Reynolds O (1879) On certain dimensional properties of matter in the gaseous state. Philos Trans R Soc Part 1(170):727–845
11. Maxwell JC (1879) VII. On stresses in rarified gases arising from inequalities of temperature. Proc R Soc Lond 170:231–256
12. Sharipov F (1994) Onsager-Casimir reciprocity relations for open gaseous systems at arbitrary rarefaction. I. General theory for single gas. Physica A 203:437–456
13. Wu L, Liu HH, Zhang YH, Reese JM (2015) Influence of intermolecular potentials on rarefied gas flows: fast spectral solutions of the Boltzmann equation. Phys Fluids 27:082002
14. Hu JW, Ying LX (2012) A fast spectral algorithm for the quantum Boltzmann collision operator. Commun Math Sci 10(3):989–999
15. Ferziger JH, Kaper HG (1972) Mathematical theory of transport processes in gases. North-Holland Publishing Company, Amsterdam
16. Wu L, Struchtrup H (2017) Assessment and development of the gas kinetic boundary condition for the Boltzmann equation. J Fluid Mech 823:511–537
17. Sharipov F (2002) Application of the Cercignani–Lampis scattering kernel to calculations of rarefied gas flows. I. Plane flow between two parallel plates. Eur J Mech B/Fluids 21:113–123

Chapter 6
Kinetic Modeling of Monatomic Gas Flow

The complicated nature of the Boltzmann collision operator has stimulated the development of simpler kinetic models that try to imitate its behavior as closely as possible. In this chapter we introduce and assess the accuracy of typical gas kinetic models for single-species monatomic gas.

6.1 Basic Rules

Several rules are considered when simplifying the BCO [1]. First of all, the mass, momentum, and energy conservation must be guaranteed. Second, in the equilibrium state, the VDF must be reduced to the Maxwellian distribution. Third, the transport coefficients of the kinetic model equation should be equal to those of the Boltzmann equation. Fourth, the H-theorem, which states that the entropy production is always positive and vanishes only if the system is in equilibrium, should be satisfied.

While the first and second requirements are fundamental, and the third one yields consistent solutions with the NSF equations in the continuum flow regime, the fourth one is just a physical requirement which does not guarantee accurate numerical results. If a kinetic model is exactly the same as the Boltzmann equation, the entropy production rate should be the same as well; however, this is in general impossible. In fact, as we shall see later, in most cases, the Shakhov kinetic model, where the H-theorem has been proven only in linearized cases, performs better than the ellipsoidal-statistical (ES) BGK model that fulfills the H-theorem.

Since the BCO can be written as $Q = Q^+ - \nu f$, the general and popular form of the modeled collision operator is formally written as a relaxation approximation:

$$Q(f, f_*) = \nu(f_r - f), \tag{6.1}$$

L. Wu, *Rarefied Gas Dynamics*,
https://doi.org/10.1007/978-981-19-2872-7_6

where ν is the collision frequency and f_r is the reference VDF. Note that here ν differs from that of the Boltzmann equation. On the other hand, since in the grazing collision limit the BCO can be approximated by the Fokker-Planck operator, the Fokker-Planck equation can also be used to model the rarefied gas flow.

For the relaxation-type kinetic model (6.1), there are two categories of gas kinetic models, classified by whether the collision frequency is molecular velocity-independent [2–7] or dependent [8–13]. In this book, if without specification, we focus on the kinetic models with velocity-independent collision frequencies, since they are frequently used due to relative simplicity, although except for the Maxwellian molecules the collision frequency in the Boltzmann equation is velocity-dependent, see the equilibrium collision frequency in Sect. 2.3.2 for some examples.

6.2 Velocity-Independent Collision Frequency

6.2.1 *BGK Model*

The model proposed by Bhatnagar, Gross, and Krook [2] and independently by Welander [14] is the simplest kinetic model ever known, since the reference VDF is the local Maxwellian distribution (2.31). One can verify that the BGK model satisfies the conservation laws. In the equilibrium where the collision operator vanishes, $f = F_{eq}$, which fulfills the second requirement of kinetic modeling. The H-theorem is easy to prove. For simplicity, we consider the spatial homogeneous case. Since $\ln F_{eq}$ is the linear combination of the collision invariants, we have $\int (F_{eq} - f) \ln F_{eq} d\boldsymbol{v} = 0$, and

$$\begin{aligned} \frac{dH}{dt} &= -\nu \int (F_{eq} - f) \ln f d\boldsymbol{v} \\ &= -\nu \int (F_{eq} - f)(\ln F_{eq} - \ln f) d\boldsymbol{v} \geq 0. \end{aligned} \tag{6.2}$$

Now let us have a look at the shear viscosity and thermal conductivity of the BGK model. Although rigorous steps to derive the two transport coefficients from the Chapman-Enskog expansion are detailed in Sect. 3.2, here we introduce a much simpler derivation.

6.2.1.1 Shear Viscosity

Consider a simple Couette flow between two parallel plates perpendicular to the y-axis. The flow velocity $u_x(y)$ is so small that the viscous heating is negligible, hence the gas temperature and pressure are uniform in the whole domain. The local equilibrium VDF is

$$F_{eq} = n \left(\frac{m}{2\pi k_B T} \right)^{3/2} \exp \left(-\frac{(v_x - u_x(y))^2 + v_y^2 + v_z^2)}{2k_B T/m} \right). \tag{6.3}$$

Expand the VDF as $f = F_{eq} + f^{(1)} + f^{(2)} + \cdots$, where $f^{(1)}$ is of order ν^{-1}. We found that, in the steady-state, the first-order approximation of the VDF is

$$f^{(1)} = -\frac{v_y}{\nu} \frac{\partial F_{eq}}{\partial y} = -\frac{v_y}{\nu} \frac{v_x - u_x(y)}{k_B T/m} \frac{\partial u_x}{\partial y} F_{eq}. \tag{6.4}$$

Therefore, to the first-order approximation, the shear stress is

$$p_{xy} = m \int (v_x - u_x) v_y \left[F_{eq} + f^{(1)} \right] d\boldsymbol{v} = -\frac{p}{\nu} \frac{\partial u_x}{\partial y}. \tag{6.5}$$

Obviously, if one wants to recover the shear viscosity in the BGK model, one should choose the velocity-independent collision frequency as

$$\nu = \frac{p}{\mu}. \tag{6.6}$$

6.2.1.2 Thermal Conductivity

To find the thermal conductivity we consider the simple Fourier flow, where the flow velocity is zero but a temperature gradient is imposed in the y-direction. The local equilibrium VDF is given by

$$F_{eq} = \frac{p}{k_B T(y)} \left(\frac{m}{2\pi k_B T(y)} \right)^{3/2} \exp \left(-\frac{m v^2}{2 k_B T(y)} \right), \tag{6.7}$$

where the pressure p is a constant in the steady state.

Applying the Chapman-Enskog expansion we find

$$f^{(1)} = -\frac{v_y}{\nu} \frac{\partial F_{eq}}{\partial y} = -\frac{v_y}{\nu} \left(\frac{m v^2}{2 k_B T} - \frac{5}{2} \right) \frac{1}{T} \frac{\partial T}{\partial y} F_{eq}, \tag{6.8}$$

so the heat flux is

$$q_y = \frac{m}{2} \int v_y v^2 f^{(1)} d\boldsymbol{v} = -\frac{p}{\nu} \frac{5 k_B}{2m} \frac{\partial T}{\partial y}, \tag{6.9}$$

which produces a thermal conductivity of

$$\kappa = \frac{p}{\nu} \frac{5 k_B}{2m}. \tag{6.10}$$

Thus, the Prandtl number (3.30) in the BGK model is always unity, while that from the Boltzmann equation is about 2/3 for monatomic gases. This means that the BGK model cannot recover the shear viscosity and thermal conductivity simultaneously. For problems dominated by the heat conduction, one should choose

$$\nu = \frac{2p}{3\mu} \tag{6.11}$$

to recover the correct thermal conductivity in the BGK model.

6.2.1.3 Linearized BGK Model

If the same linearization as in Sect. 5.1 is adopted, and the collision frequency is chosen as per Eq. (6.6), the linearized BGK collision operator becomes

$$\begin{aligned} \mathcal{L}_{BGK}(h) &= \frac{p}{\mu}\left[\varrho + 2\boldsymbol{u}\cdot\boldsymbol{v} + T\left(v^2 - \frac{3}{2}\right)\right] f_{eq} - \frac{p}{\mu}h \\ &\equiv \mathcal{L}^{+}_{BGK} - \frac{p}{\mu}h, \end{aligned} \tag{6.12}$$

where ϱ, $\boldsymbol{u}$, and T are the perturbed number density, flow velocity, and temperature that deviate from their equilibrium values, see definition in Eq. (5.6); $\mathcal{L}^{+}_{BGK}$ is regarded as the gain part of the linearized collision operator.

6.2.2 *Ellipsoidal-Statistical BGK Model*

To correct the Prandtl number in the BGK model, Holway proposed the ESBGK model [4], where the reference VDF is obtained by maximizing the entropy (2.25) under the given information of mass, momentum, energy, and the stress tensor (or precisely, the stress tensor times a constant b):

$$f_r^{ES} = \frac{n}{\sqrt{\det[2\pi\lambda_{ij}]}} \exp\left(-\frac{1}{2}\lambda_{ij}^{-1}c_i c_j\right), \tag{6.13}$$

where

$$\lambda_{ij} = \frac{k_B T(1-b)}{m}\delta_{ij} + \frac{bp_{ij}}{nm} = \frac{p\delta_{ij} + b\sigma_{ij}}{nm}. \tag{6.14}$$

If $b = 0$, λ_{ij} becomes diagonal, and the ESBGK model reduces to the BGK model. The presence of pressure tensor does not affect the thermal conduction, therefore, the thermal conductivity is still given by Eq. (6.10). However, since when σ_{ij} is small, f_r^{ES} can be expanded as

$$f_r^{ES} = F_{eq} + b\frac{\sigma_{ij}}{p}\frac{c_{\langle i}c_{j\rangle}}{2k_BT/m}F_{eq}, \tag{6.15}$$

the first-order approximation of the VDF in the simple Couette flow becomes

$$f^{(1)} = -\frac{v_y}{\nu}\frac{v_x - u_x(y)}{k_BT/m}\frac{\partial u_x}{\partial y}F_{eq} + b\frac{\sigma_{ij}}{p}\frac{c_{\langle i}c_{j\rangle}}{2k_BT/m}F_{eq}, \tag{6.16}$$

which results in the shear viscosity

$$\mu = \frac{1}{1-b}\frac{p}{\nu}, \tag{6.17}$$

and the Prandtl number

$$\mathrm{Pr} = \frac{1}{1-b}. \tag{6.18}$$

Clearly, b should be smaller than unity to keep the collision frequency positive. There are, however, low bound for b: for f_r^{ES} to have finite norm, one must choose $b \geq -1/2$. Therefore, the Prandtl number of the ESBGK model varies from $2/3$ to positive infinity as b increases from $-1/2$ to 1. Fortunately, to get the correct Prandtl number for monatomic gas, one chooses the minimum available value

$$b = -\frac{1}{2}. \tag{6.19}$$

When the collision operator vanishes, we have $f = f_r^{ES}$, which is not Maxwellian if $\sigma_{ij} \neq 0$. This, however, does not violate the second requirement of kinetic modeling. This is because even when $f = f_r^{ES}$, the evolution of the deviatoric stress follows

$$\frac{\partial \sigma_{ij}}{\partial t} = -\frac{p}{\mu}\sigma_{ij}, \tag{6.20}$$

which gradually decays to zero. Therefore, the route to equilibrium in the ESBGK model can be understood as follows: while the VDF relaxes toward the ellipsoidal distribution, the ellipsoidal distribution itself evolves toward the Maxwellian distribution; and eventually we have $f = F_{eq}$ in equilibrium.

The H-theorem has been proved by Andries et al., making this model the only kinetic model satisfying the H-theorem and having the correct transport coefficients [15]. Since then, the ESBGK model has attracted great attention.

The linearized ESBGK collision operator takes the following form:

$$\mathcal{L}_{ES}(h) = \frac{2p}{3\mu}\left[\varrho + 2\boldsymbol{u}\cdot\boldsymbol{v} + T\left(v^2 - \frac{3}{2}\right) - \sigma_{ij}\frac{v_{\langle i}v_{j\rangle}}{2}\right]f_{eq} - \frac{2p}{3\mu}h. \tag{6.21}$$

6.2.3 Shakhov Model

In contrast to the ESBGK model where the stress tensor enters the reference VDF, in the Shakhov model the heat flux is introduced on top of the BGK model [5, 6]:

$$f_r^S = F_{eq}\left[1 + (1-\mathrm{Pr})\frac{2m\boldsymbol{q}\cdot\boldsymbol{c}}{5n(k_BT)^2}\left(\frac{mc^2}{2k_BT}-\frac{5}{2}\right)\right]. \tag{6.22}$$

The conservation laws are fulfilled. Also, the additional term related to the heat flux does not affect the shear viscosity, so the collision frequency is given by Eq. (6.6). For the calculation of thermal conductivity, the first-order approximation of the VDF in the simple Fourier flow is changed to

$$f^{(1)} = -\frac{v_y}{\nu}\left(\frac{mv^2}{2k_BT}-\frac{5}{2}\right)\frac{1}{T}\frac{\partial T}{\partial y}F_{eq} + (1-\mathrm{Pr})\frac{2m\boldsymbol{q}\cdot\boldsymbol{c}}{5n(k_BT)^2}\left(\frac{mc^2}{2k_BT}-\frac{5}{2}\right)F_{eq}, \tag{6.23}$$

so that the thermal conductivity becomes

$$\kappa = \frac{1}{\mathrm{Pr}}\frac{p}{\nu}\frac{5k_B}{2m}. \tag{6.24}$$

Thus, the Prandtl number is recovered. Also, as the VDF is relaxed toward f_r^S, the reference VDF itself evolves toward the Maxwellian distribution, due to the fact that the heat flux gradually decays to zero as per

$$\frac{\partial \boldsymbol{q}}{\partial t} = -\frac{2p}{3\mu}\boldsymbol{q}. \tag{6.25}$$

Therefore, in the equilibrium one has $f = F_{eq}$.

Two deficiencies exist in the Shakhov model. First, the H-theorem is only proved in linearized flows,[1] while one can neither prove nor disprove the H-theorem in nonlinear flow. Second, the VDF might become negative in strong non-equilibrium flow, which is not physical. Despite these, the Shakhov model has been widely used, as in most cases it gives satisfactory results when compared to those of the Boltzmann equation.

---

[1] For simplicity we consider the spatial homogeneous case, thus

$$\frac{dH}{dt} = -\nu\int(f_r^S - f)[\ln f - \ln f_r^S]d\boldsymbol{v} - \nu\int(f_r^S - f)\ln f_r^S d\boldsymbol{v}, \tag{6.26}$$

where the first term on the right-hand side is non-negative, while in the linearized case the second term can be simplified to $8\nu q^2/45p^2v_m^2$, which is also non-negative. Therefore, the H-theorem is proved in the linearized case.

The Shakhov linearized collision operator is

$$\mathcal{L}_S(h) = \mathcal{L}_{BGK}(h) + \frac{4p(1-\mathrm{Pr})}{5\mu}\boldsymbol{q}\cdot\boldsymbol{v}\left(v^2-\frac{5}{2}\right)f_{eq}. \tag{6.27}$$

6.2.4 Gross-Jackson Model

Although various kinetic models with the velocity-independent collision frequency are proposed, they can be generalized by the Gross-Jackson model [3] in the linearized case. In fact, the omnipotent thing in kinetic modeling is to get the relaxation rates of the corresponding macroscopic quantities right in spatial homogeneous problem. For shear viscosity and thermal conductivity the corresponding quantities are stress and heat flux. And one can verify that the Chapman-Enskog expansion, when applied to a kinetic model with velocity-independent collision frequency, yields the following relation:

$$\begin{aligned} \frac{\partial \sigma_{ij}}{\partial t} &= -\frac{\sigma_{ij}}{\tau_s}, \quad &\rightarrow \mu = p\tau_s, \\ \frac{\partial \boldsymbol{q}}{\partial t} &= -\frac{\boldsymbol{q}}{\tau_h}, \quad &\rightarrow \kappa = \frac{5k_B}{2m}p\tau_h. \end{aligned} \tag{6.28}$$

To recover the shear viscosity and thermal conductivity, the relaxation rates of stress and heat flux must be exactly the same as that of the Boltzmann equation, i.e., Eq. (2.49).

It should be emphasized that, while the transport coefficients are only defined in the continuum flow regime, the relaxation processes, which are intrinsic to the kinetic equations, are valid in the whole flow regime. In other words, the transport coefficients and the corresponding relaxation processes are the appearance and essence of RGD.

Based on the eigenvalues and eigenfunctions of the linearized BCO for Maxwellian molecule (see Sect. 2.3.3), Gross and Jackson proposed the following kinetic model:

$$\begin{aligned} \mathcal{L}(h) = \lambda_{st}h + f_{eq}\sum_{nl}(\lambda_{nl}-\lambda_{st}) \\ \times \frac{2l+1}{4\pi}\int P_l(\cos\phi')g_{nl}(v)g_{nl}(v_1)h(\boldsymbol{v}_1)d\boldsymbol{v}_1, \end{aligned} \tag{6.29}$$

where ϕ' is the angle between $\boldsymbol{v}$ and $\boldsymbol{v}'$, λ_{st} is an arbitrary negative number, λ_{nl} is the eigenvalue of the linearized Boltzmann equation for Maxwellian molecules given in Eq. (2.43), and the molecular velocity has been normalized by the most probable speed. The Nth approximation of $\mathcal{L}(h)$ is

$$\mathcal{L}^{(N)}(h) = \lambda_{0N}h - \lambda_{0N} f_{eq} \sum_{2n+l\le N} \frac{2l+1}{4\pi}\left(1-\frac{\lambda_{nl}}{\lambda_{0N}}\right) g_{nl}(v) \times \int P_l(\cos\phi') g_{nl}(v_1) h(\boldsymbol{v}_1) d\boldsymbol{v}_1, \tag{6.30}$$

where all polynomial $g_{nl}(v)$ of order N or less have been retained, and λ_{st} has been set equal to λ_{0N}. Thus, the highest eigenvalues of $\mathcal{L}(h)$ are all approximated by one eigenvalue λ_{0N}.

As a matter of fact, the BGK, ESBGK, and Shakhov models, when linearized, are special cases of the Gross-Jackson model. For instance, if $N=2$, from Eq. (6.30) we have $\mathcal{L}^{(2)} = \mathcal{L}_{BGK}$. If $N=3$, we have

$$\mathcal{L}^{(3)} = \frac{3}{2}L^{(2)}(h) + \frac{p}{\mu}\left[\frac{\sigma_{ij}}{2}v_{\langle i}v_{j\rangle} + \frac{2}{3}\boldsymbol{q}\cdot\boldsymbol{v}\left(v^2-\frac{5}{2}\right)\right] f_{eq}, \tag{6.31}$$

which does not coincide with either the linearized ESBGK model (6.21) nor the Shakhov model (6.27). However, it can be viewed as a linear combination of the ESBGK and Shakhov models:

$$\mathcal{L}^{(3)} = -\frac{3}{2}\mathcal{L}_{ES}(h) + \frac{5}{2}\mathcal{L}_S(h). \tag{6.32}$$

On the other hand, if we choose $\lambda_{st} = -2p/3\mu$ and $\lambda_{st} = -p/\mu$ in Eq. (6.29), and retain terms with $2n+l \le 3$, we have $\mathcal{L}^{(3)} = \mathcal{L}_{ES}(h)$ and $\mathcal{L}^{(3)} = \mathcal{L}_S(h)$, which are respectively the linearized ESBGK and Shakhov models.

6.2.5 Nonlinearization

It can be deduced from the relation between the BGK, ESBGK, and Shakhov models and their linearized versions that the linearized collision operators can be "non-linearized" back to nonlinear counterparts in the following manner:

$$\begin{aligned}
\left[n + 2\boldsymbol{u}\cdot\boldsymbol{v} + T\left(v^2-\frac{3}{2}\right)\right] f_{eq} - h &\to F_{eq} - f, \\
\sigma_{ij}\frac{v_{\langle i}v_{j\rangle}}{2} f_{eq} &\to \frac{\sigma_{ij}}{p}\frac{c_{\langle i}c_{j\rangle}}{2k_BT} F_{eq}, \\
\boldsymbol{q}\cdot\boldsymbol{v}\left(v^2-\frac{5}{2}\right) f_{eq} &\to \frac{m\boldsymbol{q}\cdot\boldsymbol{c}}{2n(k_BT)^2}\left(\frac{mc^2}{2k_BT}-\frac{5}{2}\right) F_{eq}.
\end{aligned} \tag{6.33}$$

Then, inspired by Eq. (6.32), one may construct nonlinear kinetic models from the Gross-Jackson model. For example, the reference VDF in nonlinear kinetic model may take the following form (similar kinetic models are given in Refs. [16, 17]):

$$f_r^N = F_{eq}\left[1 + r_1\frac{\sigma_{ij}}{p}\frac{c_{\langle i}c_{j\rangle}}{2k_BT/m} + r\frac{2m\boldsymbol{q}\cdot\boldsymbol{c}}{5n(k_BT)^2}\left(\frac{mc^2}{2k_BT} - \frac{5}{2}\right)\right], \tag{6.34}$$

and in order to recover the shear viscosity and thermal conductivity the collision frequency in Eq. (6.1) is chosen as $\nu = p/\mu(1 - r_1)$, while the Prandtl number is controlled through $\Pr = (1 - r)/(1 - r_1)$.

This model has only one free parameter r_1 as $\Pr = 2/3$. When $r_1 = -1/2$, we have $r = 0$, hence the model mimics the behavior of ESBGK model: in fact, r_1 is equivalent to b in the ESBGK model, but now there is no lower bound for r_1; if the r_1 term is absorbed back into the exponential function, then the ESBGK model will be exactly recovered. When $r_1 = 0$, we have $r = 1 - \Pr$, thus the model is reduced to the Shakhov model.

By considering more terms in the Gross-Jackson model (6.29) and "non-linearizing" them, one may construct nonlinear kinetic models with high accuracy. In this case, the reference VDF takes a similar form of the VDF used in Grad's moment method.

6.3 Velocity-Dependent Collision Frequency

In order to recover the correct Prandtl number while using the Gaussian reference VDF, Struchtrup and Mieussens developed a kinetic model, where the collision frequency ν is a function of the molecular velocity, and the reference VDF in Eq. (6.1) takes the following form [11, 12]:

$$f_r = a\exp(-\Gamma c^2 + \gamma_i c_i), \tag{6.35}$$

where a, Γ, and γ_i are chosen so that the conservation laws are fulfilled:

$$\int \nu(v)(f_r - f)[1, \boldsymbol{v}, c^2]d\boldsymbol{v} = [0, 0, 0]. \tag{6.36}$$

In general, f_r is not Maxwellian, but since $\ln f_r$ is a linear combination of the collisional invariants, the H-theorem can be proved in a similar way to Eq. (6.2). In equilibrium, the collision operator vanishes, so $f = f_r$; and since both VDFs share the same moments it can be concluded that f is the Maxwellian distribution: Eq. (6.35) can be written as

$$f_r = a\exp\left(\frac{\gamma_i\gamma_i}{4\Gamma}\right)\exp\left[-\Gamma\left(v_i - u_i - \frac{\gamma_i}{2\Gamma}\right)\left(v_i - u_i - \frac{\gamma_i}{2\Gamma}\right)\right], \tag{6.37}$$

which should produce the density n, velocity $\boldsymbol{u}$, and temperature T; it can be easily found that $\gamma_i = 0$, $a = n/(2\pi k_BT/m)^{3/2}$, and $\Gamma = m/2k_BT$. That is, $f_r = F_{eq}$.

For a system that deviates slightly from the equilibrium, the reference VDF can be linearized as $f_r = F_{eq}(1 - \hat{a} + \hat{\Gamma}c^2 + \gamma_i c_i)$, where the coefficients $\hat{a}$, γ_i, and Γ measure the deviation from the Maxwellian distribution. After performing the Chapman-Enskog expansion, it is found that the VDF is

$$\begin{aligned} f =&F_{eq}(1-\hat{a}+\hat{\Gamma}c^2+\gamma_i c_i) \\ &-\frac{F_{eq}}{\nu(v)}\left\{\frac{m}{k_BT}\frac{\partial u_{\langle i}}{\partial x_{j\rangle}}c_{\langle i}c_{j\rangle}+\frac{1}{T}\frac{\partial T}{\partial x_i}c_i\left(\frac{mc^2}{2k_BT}-\frac{5}{2}\right)\right\}, \end{aligned} \tag{6.38}$$

where, according to the conservation laws, $\hat{a}=\hat{\Gamma}=0$, and

$$\gamma_i=\frac{8}{3\sqrt{\pi}}\frac{1}{T}\frac{\partial T}{\partial x_i}\int_0^\infty\frac{\xi^4}{\nu(\xi)}\left(\xi^2-\frac{5}{2}\right)\exp(-\xi^2)d\xi, \tag{6.39}$$

with $\xi=c/v_m$ being the normalized peculiar velocity. Hence, the shear viscosity and thermal conductivity are

$$\begin{aligned} \mu&=\frac{16p}{15\sqrt{\pi}}\int_0^\infty\frac{\xi^6}{\nu(\xi)}\exp(-\xi^2)d\xi, \\ \kappa&=\frac{16p}{15\sqrt{\pi}}\frac{5k_B}{2m}\int_0^\infty\frac{\xi^4\left(\xi^2-\frac{5}{2}\right)^2}{\nu(\xi)}\exp(-\xi^2)d\xi. \end{aligned} \tag{6.40}$$

Appropriate forms of the collision frequency can be chosen to recover the shear viscosity and thermal conductivity simultaneously. Introducing the dimensionless collision frequency:

$$\nu(\xi)=\frac{p}{\mu}\hat{\nu}(\xi)\ \text{with}\ \frac{16}{15\sqrt{\pi}}\int_0^\infty\frac{\xi^6}{\hat{\nu}(\xi)}\exp(-\xi^2)d\xi=1, \tag{6.41}$$

Mieussens and Struchtrup constructed five forms for $\hat{\nu}(\xi)$:

$$\begin{aligned} \hat{\nu}_1(\xi)=&0.431587\xi^{1.791288}, \\ \hat{\nu}_2(\xi)=&0.0268351(1+14.2724\xi^2), \\ \hat{\nu}_3(\xi)=&0.0365643(1+10\xi^{2.081754}), \\ \hat{\nu}_4(\xi)=&0.1503991(1+0.92897\xi^4), \end{aligned} \tag{6.42}$$

and

$$\hat{\nu}_5(\xi)=\begin{cases}0.2590894, & \xi\le 1.2,\\ 0.8288236, & \xi>1.2.\end{cases} \tag{6.43}$$

It is noted that although these collision frequencies mimic the velocity-dependent behaviors, they do not agree with the collision frequency (2.35) in the Boltzmann equation: in fact, when the $\hat{\nu}(\xi)$ takes the form of the equilibrium collision frequency of the Boltzmann equation, the Prandtl number of this kinetic model is very close to unity. Therefore, in the simulation of normal shock waves, no significant improvement is seen [12].

If one is not obsessed with the H-theorem in the kinetic modeling, the velocity-dependent collision frequency can be chosen more flexibly. Following the nonlinearization procedure in the previous section, one may introduce the heat flux to the reference VDF:

$$f_r = a\exp(-\Gamma c^2 + \gamma_i c_i) + r\frac{2m\boldsymbol{q}\cdot\boldsymbol{c}}{5n(k_BT)^2}\left(\frac{mc^2}{2k_BT} - \frac{5}{2}\right)F_{eq}, \tag{6.44}$$

where r is independent of the molecular velocity. Applying the Chapman-Enskog expansion, the VDF to the first-order approximation reads

$$\begin{aligned} f =& F_{eq}(1 - \hat{a} + \hat{\Gamma}c^2 + \gamma_i c_i) + r\frac{2m\boldsymbol{q}\cdot\boldsymbol{c}}{5n(k_BT)^2}\left(\frac{mc^2}{2k_BT} - \frac{5}{2}\right)F_{eq} \\ & - \frac{F_{eq}}{\nu(v)}\left\{\frac{m}{k_BT}\frac{\partial u_{\langle i}}{\partial x_{j\rangle}}c_{\langle i}c_{j\rangle} + \frac{1}{T}\frac{\partial T}{\partial x_i}c_i\left(\frac{mc^2}{2k_BT} - \frac{5}{2}\right)\right\}, \end{aligned} \tag{6.45}$$

where $\hat{a} = \hat{\Gamma} = 0$, and γ_i is still given by Eq. (6.39). Therefore, the shear viscosity is

$$\mu = \frac{16p}{15\sqrt{\pi}}\int_0^\infty \frac{\xi^6}{\nu(\xi)}\exp(-\xi^2)d\xi, \tag{6.46}$$

and thermal conductivity is

$$\kappa = \frac{1}{1-r}\frac{16p}{15\sqrt{\pi}}\frac{5k_B}{2m}\int_0^\infty \frac{\xi^4\left(\xi^2 - \frac{5}{2}\right)^2}{\nu(\xi)}\exp(-\xi^2)d\xi. \tag{6.47}$$

Now the positive collision frequency can be chosen arbitrarily while the Prandtl number is controlled by r.

Finally, it is pointed that the parameters a, Γ, and γ_i should be determined by the conversation laws; this usually requires the Newton algorithm for nonlinear equations, say, Eqs. (6.35) and (6.36). To circumvent this problem, we propose to use the following simplified reference VDF:

$$f_r = \left[\hat{\varrho} + \hat{\Gamma}c^2 + \gamma_i c_i + r\frac{2m\boldsymbol{q}\cdot\boldsymbol{c}}{5n(k_BT)^2}\left(\frac{mc^2}{2k_BT} - \frac{5}{2}\right)\right]F_{eq}, \tag{6.48}$$

which does not affect the transport coefficients, but the parameters $\hat{\varrho}$, $\hat{\Gamma}$, and γ_i can be solved easily by the following linear equations:

$$\begin{bmatrix} \int \nu(v)F_{eq}d\boldsymbol{v} & \int \nu(v)c^2F_{eq}d\boldsymbol{v} \\ \int \nu(v)c^2F_{eq}d\boldsymbol{v} & \int \nu(v)c^4F_{eq}d\boldsymbol{v} \end{bmatrix}\begin{bmatrix} \hat{\varrho} \\ \hat{\Gamma} \end{bmatrix} = \begin{bmatrix} \int \nu(v)fd\boldsymbol{v} \\ \int \nu(v)c^2fd\boldsymbol{v} \end{bmatrix}, \tag{6.49}$$

and

$$\gamma_i = \frac{\int \nu(v)c_i\left[f - r\frac{2mq_ic_i}{5n(k_BT)^2}\left(\frac{mc^2}{2k_BT} - \frac{5}{2}\right)F_{eq}\right]d\boldsymbol{v}}{\int \nu(v)c_i^2 F_{eq}d\boldsymbol{v}}. \tag{6.50}$$

For Maxwellian molecules, the collision frequency ν is velocity-independent, so we have $\hat{\varrho} = 1$, and $\hat{\Gamma} = \gamma_i = 0$; and this model is reduced to the Shakhov model. We call this model the ν model [25].

6.4 Fokker-Planck Model

If the total cross-section is infinite, the collision frequency is also infinite. However, it should be noted that such an infinite cross-section arises from the grazing collision, see Eq. (2.17), where the post-collision velocities deviate slightly from the pre-collision ones. Therefore, their contribution to the collision operator is small. In this case, the collision effect can be described by the Fokker-Planck collision operator.

In the grazing collision limit, we have the following approximation:

$$\Psi(\boldsymbol{v}') - \Psi(\boldsymbol{v}) \approx (v_i' - v_i)\frac{\partial\Psi}{\partial v_i} + \frac{(v_i' - v_i)(v_j' - v_j)}{2}\frac{\partial^2\Psi}{\partial v_i \partial v_j}. \tag{6.51}$$

Applying this approximation to the BCO leads to

$$\begin{aligned}\int \Psi(\boldsymbol{v})Qd\boldsymbol{v} &= \iiint \left[\Psi(\boldsymbol{v}') - \Psi(\boldsymbol{v})\right] B(\cos\theta, v_r) f(\boldsymbol{v}_*) f(\boldsymbol{v}) d\Omega d\boldsymbol{v}_* d\boldsymbol{v} \\ &= \int \Psi(\boldsymbol{v})\left[-\frac{\partial}{\partial v_i}(A_i f) + \frac{1}{2}\frac{\partial^2}{\partial v_i \partial v_j}(D_{ij} f)\right] d\boldsymbol{v},\end{aligned} \tag{6.52}$$

where

$$\begin{aligned}A_i &= \iint (v_i' - v_i) B(\cos\theta, v_r) f_* d\boldsymbol{v}_* d\Omega, \\ D_{ij} &= \iint (v_i' - v_i)(v_j' - v_j) B(\cos\theta, v_r) f_* d\boldsymbol{v}_* d\Omega\end{aligned} \tag{6.53}$$

describes the drift and diffusion effects, respectively. Note that the integration with respect to the deflection angle occurs around $\theta = 0$. For detailed calculation of A_i and D_{ij}, we refer to the book of Cercignani [18].

Due to the arbitrariness of $\Psi(\boldsymbol{v})$, the BCO in the grazing limit can be approximated by

$$Q = -\frac{\partial}{\partial v_i}(A_i f) + \frac{1}{2}\frac{\partial^2}{\partial v_i \partial v_j}(D_{ij} f). \tag{6.54}$$

In the original Fokker-Planck model, the drift and diffusion effects are simply modeled as

$$A_i = -\frac{c_i}{\tau_{FP}}, \quad D_{ij} = \frac{2k_B T}{m}\frac{1}{\tau_{FP}}\delta_{ij}. \tag{6.55}$$

The shear viscosity and thermal conductivity can be easily obtained if we consider the relaxation of stress and heat flux in spatially homogeneous system (6.28):

$$\begin{aligned} \frac{\partial \sigma_{ij}}{\partial t} &= -\frac{2}{\tau_{FP}}\sigma_{ij}, \quad \rightarrow \mu = p\frac{\tau_{FP}}{2}, \\ \frac{\partial \boldsymbol{q}}{\partial t} &= -\frac{3}{\tau_{FP}}\boldsymbol{q}, \quad \rightarrow \kappa = \frac{5k_B}{2m}p\frac{\tau_{FP}}{3}. \end{aligned} \tag{6.56}$$

Therefore, the Prandtl number of the Fokker-Planck model is 3/2, rather than 2/3 for monatomic gas.

There are several ways to fix the Prandtl number. In the cubic Fokker-Planck model [19], the drift coefficient is changed to

$$A_i = -\frac{c_i}{\tau_{FP}} + \tilde{c}_{ij}c_j + \gamma_i\left(c^2 - \frac{3k_B T}{m}\right) + \Lambda\left(c^2 c_i - \frac{2q_i}{nm}\right), \tag{6.57}$$

where $\tau_{FP} = 2\mu/p$, and the expressions for $\tilde{c}_{ij}$, γ_i, and Λ are rather complicated, see the appendix of Ref. [19], while in the ES Fokker-Planck model, an anisotropic diffusion coefficient tensor is assumed [20]:

$$D_{ij} = \frac{2}{\tau_{FP}}\left[(1-\beta)\frac{k_B T}{m}\delta_{ij} + \beta\frac{p_{ij}}{nm}\right], \tag{6.58}$$

where $\beta = -5/4$ and the relaxation time is $\tau_{FP} = 2(1-\beta)\mu/p$. However, in the ES model, in order to ensure the positivity of the diffusion coefficient, β is usually defined as $\beta = \max\left(-\frac{5}{4}, -\frac{1}{\lambda-1}\right)$, where λ is the largest eigenvalue of the normalized pressure tensor p_{ij}/p.

The collision operator of the Fokker-Planck model contains the derivatives rather than the integration with respect to the molecular velocity. This causes problems if the deterministic numerical method is applied, as the VDF is usually discontinuous. Fortunately, since the Fokker-Planck equation can also be derived from the Brownian motion of particles, it can be solved by the stochastic method.[2] The method strongly resembles the DSMC method, but in the continuum flow regime it is more efficient, since no collision pair is selected and the computational cost is independent of the Knudsen number [19, 21, 22]. Benchmark studies

---

[2] Suppose the i-th simulation particle has a position $\boldsymbol{X}_i$ and a molecular velocity $\boldsymbol{M}_i$, then its dynamics is governed by the stochastic Langevin process [21]:

$$\frac{d\boldsymbol{X}_i}{dt} = \boldsymbol{M}_i, \quad \frac{d\boldsymbol{M}_i}{dt} = -\frac{\boldsymbol{M}_i'}{\tau_{FP}} + \sqrt{\frac{2k_B T}{m\tau_{FP}}}\frac{d\boldsymbol{W}_i(t)}{dt}, \tag{6.59}$$

where $\boldsymbol{M}_i'$ is the peculiar velocity, and $\boldsymbol{W}(t)$ follows the Wiener process. A Wiener process is a Gaussian process with $d\boldsymbol{W}_i(t) = \boldsymbol{W}_i(t+dt) - \boldsymbol{W}_i(t)$, $\langle d\boldsymbol{W}_i(t)\rangle = 0$ and $\langle d\boldsymbol{W}_i(t)d\boldsymbol{W}_i(t)\rangle = dt\delta_{ij}$.

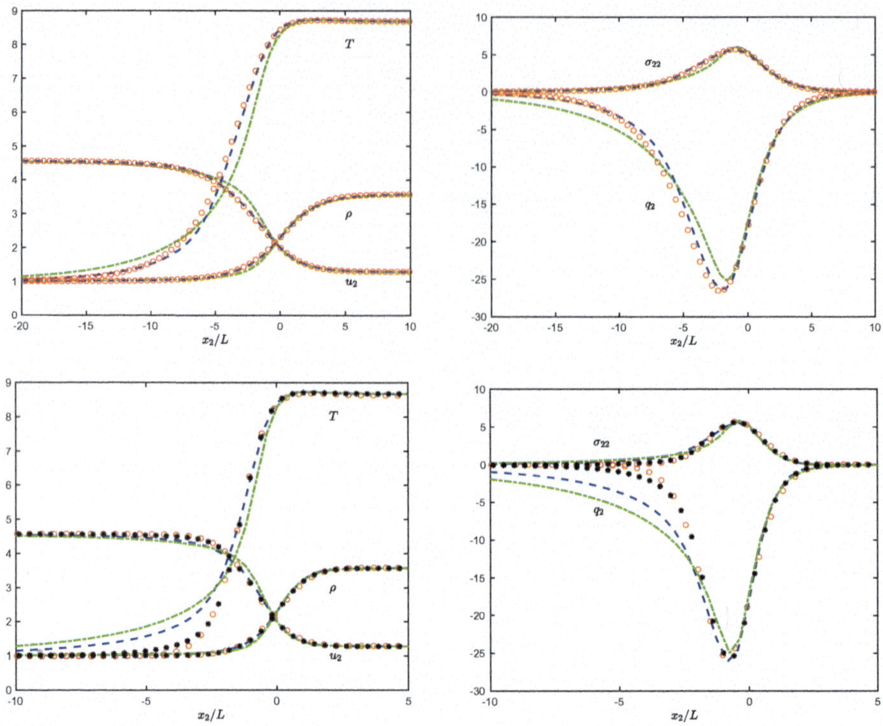

Fig. 6.1 The Ma = 5 shock wave of (First row) Maxwell and (Second row) HS gases [25]. The characteristic length L is chosen to be the MFP in the shock upstream. Shakhov model: blue dashed line; ESBGK model: green dash-dotted lines; Boltzmann equation: red circles. ν-model (6.48) with the collision frequency (6.60): Asterisks

of kinetic models for shock waves are presented in Refs. [23, 24]. It seems that the Fokker-Planck model works for certain type of intermolecular potentials, e.g., the argon gas where the viscosity index is $\omega = 0.81$ [23], but its accuracy deteriorates for HS and Maxwellian molecules [23, 24], where $\omega = 0.5$ and 1, respectively.

6.5 Accuracy of Kinetic Models

In this section we assess the accuracy of the ESBGK model, the Shakhov model, and the ν-model (6.48), by comparing their solutions to those from the Boltzmann equation.

Table 6.1 Numerical fitting of the equilibrium collision frequency by the quartic function $\nu_{eq}(\xi) = \sum_{j=0}^{4} c_j \xi^j$, when $\xi \le 5$, as well as the constants A in Eq. (6.61) and r. When $\xi > 5$, the collision frequency ν_{eq} can be approximated when the first-order Taylor expansion is applied to Eq. (2.35), resulting in $\nu_{eq} = (2-\omega)\sqrt{\pi}\xi^{2(1-\omega)}$. In the calculation of r we take the Prandtl number to be Pr = 2/3, while the collision frequency is given by Eq. (6.60) [25]

ω	c_4	c_3	c_2	c_1	c_0	A	r
0.50	0.0145	−0.2019	1.0561	0.0753	2.9774	0.0871	0.3486
0.55	0.0132	−0.1793	0.8826	0.0863	2.8180	0.0944	0.3470
0.60	0.0117	−0.1557	0.7261	0.0929	2.6691	0.1022	0.3453
0.65	0.0101	−0.1320	0.5859	0.0950	2.5300	0.1107	0.3436
0.70	0.0085	−0.1089	0.4615	0.0926	2.3999	0.1197	0.3419
0.75	0.0069	−0.0867	0.3522	0.0859	2.2782	0.1293	0.3403
0.80	0.0054	−0.0659	0.2571	0.0752	2.1642	0.1395	0.3387
0.85	0.0039	−0.0467	0.1753	0.0608	2.0572	0.1505	0.3372
0.90	0.0025	−0.0292	0.1059	0.0432	1.9566	0.1622	0.3358
0.95	0.0012	−0.0137	0.0477	0.0227	1.8619	0.1747	0.3345

6.5.1 Normal Shock Wave

Structures of the Ma = 5 normal shock wave are compared in Fig. 6.1. For the Maxwellian gas with $\omega = 1$, the Shakhov model predicts the shock structure well, while the ESBGK model overpredicts (underpredicts) the temperature and heat flux in the upstream (downstream) part. For the HS gas with $\omega = 0.5$, the Shakhov model still predicts the density and velocity profiles well but overpredicts the temperature and heat flux in the upstream part.

The ESBGK model is even worse. This numerical example suggests the importance of including the heat flux in the modeled collision operator (6.48). To determine the velocity-dependent collision frequency in the ν-model, we first choose $\nu(\xi) = \nu_{eq}(\xi)$, and find that the upstream temperature is underestimated. Therefore, a flatter collision frequency curve is required; after trial-and-errors we arrive at the following semi-empirical formula [25]:

$$\nu_\omega(\xi) = A\frac{p}{\mu}[\nu_{eq}(\xi) + 2\nu_{eq}(0)], \tag{6.60}$$

where the constant A for inverse power-law potential is

$$A = \frac{16}{15\sqrt{\pi}}\int_0^\infty \frac{\xi^6}{\nu_{eq}(\xi) + 2\nu_{eq}(0)} \exp(-\xi^2)d\xi, \tag{6.61}$$

and the parameter r can be determined from Eq. (6.47). Typical parameters for inverse power-law potential are shown in Table 6.1.

6.5.1.1 Lennard-Jones Potential

Since the collision kernel (4.7) for the Lennard-Jones potential is designed as the linear combination of that of power-law potentials, in the ν-model the velocity-dependent collision frequency can also be expressed as linear combination of Eq. (6.60):

$$\nu_{LJ}(v) = A_{LJ} \sum_{j=1}^{3} b_j \left(\frac{k_B T}{\epsilon}\right)^{0.5-\omega_j} \times \nu_{\omega_j}(\xi), \tag{6.62}$$

where A_{LJ} can be determined in a similar way as Eq. (6.61).

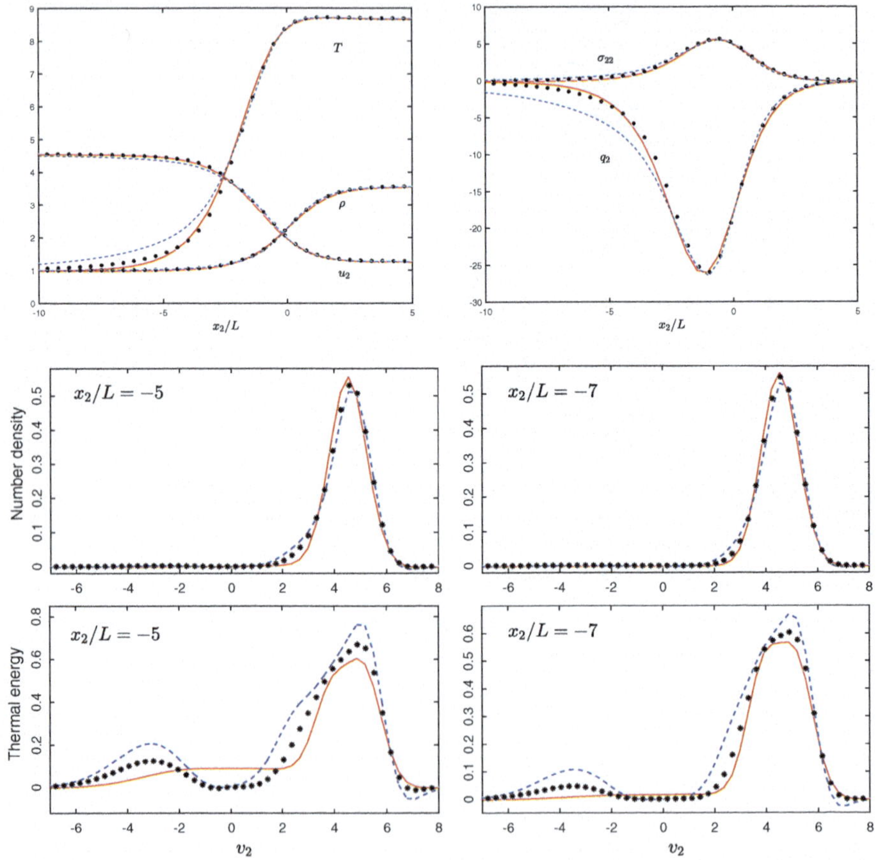

Fig. 6.2 The Ma = 5 argon shock structure using the Lennard-Jones potential [25]. Solid lines: Boltzmann solutions. Asterisks: ν-model with the collision frequency (6.62). Blue dashed lines: Shakhov model

Figure 6.2 shows that the ν-model yields consistent shock structure with those from the Boltzmann equation and MD simulation (see Fig. 4.8). To further interpret the numerical results, the marginal velocity distributions, especially the thermal energy distribution

$$F_{thermal}(x_2, v_2) = \iint_{-\infty}^{\infty} c^2 f(x_2, \boldsymbol{v}) dv_1 dv_3, \tag{6.63}$$

are displayed at upstream locations $x_2 = -5L$ and $-7L$. While the number density distributions $\iint_{-\infty}^{\infty} f dv_1 v_3$ are nearly the same, the thermal energy distributions exhibit large discrepancy among different kinetic models. At $x_2 = -7L$, molecules with $v_2 < 0$ take up 4.20%, 7.62%, 14.85% of the total thermal energy in the Boltzmann equation, ν-model, and Shakhov model, respectively, while those at $x_2 = -5L$ are 15.66%, 16.26%, and 21.94%. That the Shakhov model has larger $F_{thermal}(v_2 < 0)$ is because an extra bump around $v_2 = -3$ emerges in the thermal energy curve, which comes from the molecules with large negative velocities in the high temperature post-shock. To fix this problem, the collision frequency for large-speed molecules should be increased to prevent these molecules traveling too far to the upstream. That's why our ν-model outperforms the Shakhov model.

6.5.2 *Thermal Transpiration*

The 1D thermal transpiration is considered first, see the configuration in Sect. 5.3. In this problem, the viscosity index does not affect the solution of the kinetic models with velocity-independent collision frequency, such as the Shakhov and ESBGK models. This is because, after the linearization, these kinetic models are only characterized by the rarefaction parameter δ_{rp}, irrespective of ω. The ν-model is different, where the collision frequency (6.60) changes with ω.

The Maxwell and HS gases are used to show whether the ν-model can capture the influence of intermolecular potential or not. Figure 6.3 shows that the ν-model predicts the velocity

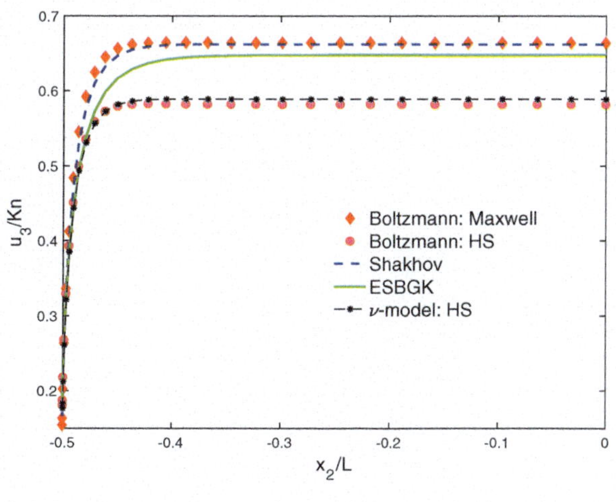

Fig. 6.3 Thermal transpiration between two parallel plates located at $x_2 = 0$ and 1 [25]. Abscissas are for the spatial coordinate x_2 which is perpendicular to the two plates and normalized by the wall distance. Kn = 0.01

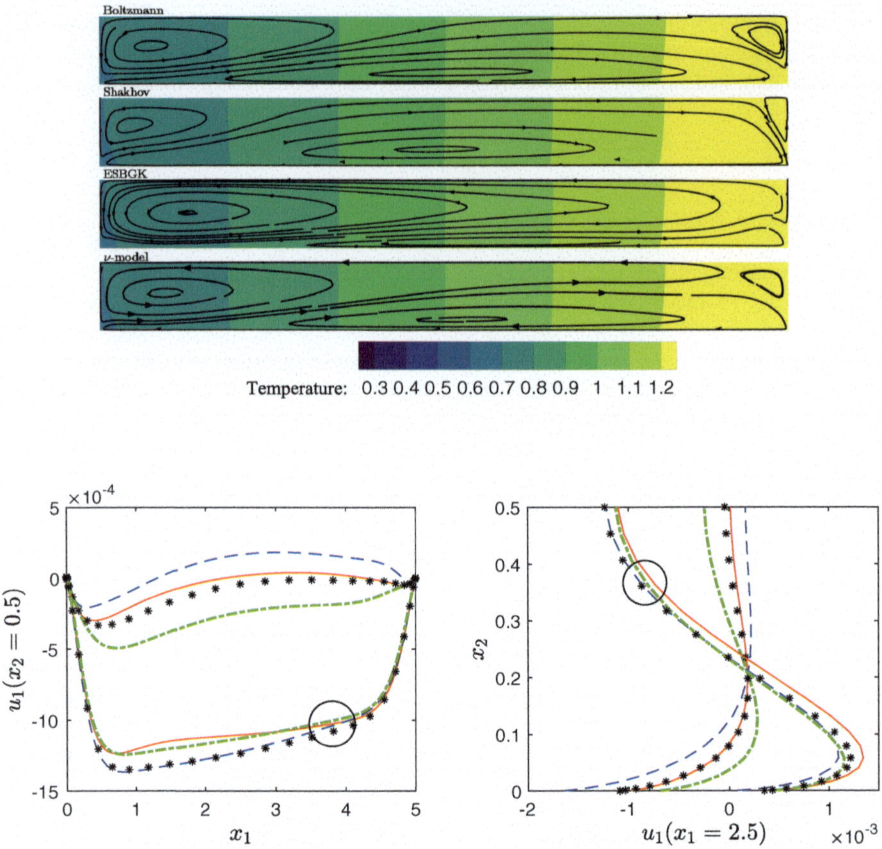

Fig. 6.4 Thermal transpiration of HS gas in a cavity [25]. (Top) temperature fields and streamlines (in half of the channel) calculated by Boltzmann equation and different kinetic models, when the Knudsen number is 0.5. (Bottom) Velocity. The Knudsen numbers are 0.1 (results passing through circles) and 0.5, respectively. Solid lines: Boltzmann equation. Dashed lines: Shakhov model. Dash-dotted lines: ESBGK. Asterisks: ν-model

profiles of Maxwell and HS gases well, while the Shakhov and ESBGK models fail to predict the velocity of HS gas. At large Knudsen numbers, the ν-model continues to perform well than the model with velocity-independent collision frequency.

We then revisit the thermal transpiration in Sect. 4.5.3. The HS gas is used. Figure 6.4 shows that, while all kinetic models predict similar temperature field with that of the Boltzmann solution, the flow patterns are different. Overall, the ν-model is the best kinetic model, while the ESBGK model is the worst.

References

1. Struchtrup H (2005) Macroscopic transport equations for rarefied gas fows: approximation methods in kinetic theory. Springer, Heidelberg, Germany
2. Bhatnagar PL, Gross EP, Krook M (1954) A model for collision processes in gases. I. Small amplitude processes in charged and neutral one-component systems. Phys Rev 94:511–525
3. Gross EP, Jackson EA (1959) Kinetic models and the linearized Boltzmann equation. Phys Fluids 2(4):432–441
4. Holway LH (1966) New statistical models for kinetic theory: methods of construction. Phys Fluids 9:1658–1673
5. Shakhov EM (1968) Approximate kinetic equations in rarefied gas theory. Fluid Dyn 3:112–115
6. Shakhov EM (1968) Generalization of the Krook kinetic relaxation equation. Fluid Dyn 3(5):95–96
7. Liu GJ (1990) A method for constructing a model form for the Boltzmann equation. Phys Fluids A 2:277
8. Cercignani C (1966) The method of elementary solutions for kinetic models with velocity-dependent collision frequency. Ann Phys 40:469
9. Loyalka SK, Ferziger JH (1967) Model dependence of the slip coefficient. Phys Fluids 10:1833–1839
10. Loyalka SK, Ferziger JH (1968) Model dependence of the temperature slip coefficient. Phys Fluids 11:1168–1671
11. Struchtrup H (1997) The BGK-model with velocity-dependent collision frequency. Cont Mech Theromodyn 9:23–32
12. Mieussens L, Struchtrup H (2004) Numerical comparison of Bhatnagar-Gross-Krook models with proper Prandtl number. Phys Fluids 16(8):2297–2813
13. Zheng YS, Struchtrup H (2005) Ellipsoidal statistical Bhatnagar-Gross-Krook model with velocity-dependent collision frequency. Phys Fluids 17(12):127103
14. Welander P (1954) On the temperature jump in a rarefied gas. Ark Fys 7:507–553
15. Andries P, Le Tallec P, Perlat J, Perthame B (2000) The Gaussian-BGK model of Boltzmann equation with small Prandtl number. Eur J Mech B Fluids 19:813–830
16. Chen SZ, Xu K, Cai QD (2015) A comparison and unification of ellipsoidal statistical and Shakhov BGK models. Adv Appl Math Mech 7:245–266
17. Wu L (2013) Deterministic numerical simulation of the boltzmann and kinetic model equations for classical and quantum dilute gases. PhD thesis, University of Strathclyde, Glasgow, 9 2013
18. Cercignani C (1988) The Boltzmann equation and its applications. Springer, New York
19. Gorji MH, Torrilhon M, Jenny P (2011) Fokker-Planck model for computational studies of monatomic rarefied gas flows. J Fluid Mech 680:574–601
20. Mathiaud J, Mieussens L (2016) A Fokker-Planck model of the Boltzmann equation with correct Prandtl number. J Stat Phys 162:397–414
21. Jenny P, Torrilhon M, Heinz S (2010) A solution algorithm for the fluid dynamic equations based on a stochastic model for molecular motion. J Comput Phys 229:1077–1098
22. Gorji MH, Jenny P (2013) A Fokker-Planck based kinetic model for diatomic rarefied gas flows. Phys Fluids 25:062002
23. Liu S, Yuan RF, Javid U, Zhong CW (2019) Conservative discrete-velocity method for the ellipsoidal Fokker-Planck equation in gas-kinetic theory. Phys Rev E 100:033310
24. Fei F, Liu HL, Liu ZH, Zhang J (2020) A benchmark study of kinetic models for shock waves. AIAA J 58:2596–2608
25. Yuan RF, Wu L (2022) Capturing the influence of intermolecular potential in rarefied gas flows by a kinetic model with velocity-dependent collision frequency. J Fluid Mech 942:A13

Chapter 7
Kinetic Modeling of Molecular Gas Flow

This chapter is dedicated to the kinetic modeling of molecular gas. The essential difference between the molecular and monatomic gases is that the molecules exhibit internal relaxation that exchanges the translational and internal energies, which lead to the bulk viscosity and internal thermal conductivity; they also reduce the translational thermal conductivity. As introduced in Sect. 6.2.4 that while the transport coefficients are the appearance valid in continuum flow regime, the corresponding relaxation processes are the essence. Therefore, a kinetic model should reflect these internal relaxation processes properly.

7.1 Bulk Viscosity

The bulk viscosity expresses a type of internal friction that resists the shearless compression or expansion of a fluid. For dilute gas, such a resistance arises from the exchange of translational and internal energies: "in an expansion or contraction the work done by the pressure alters the translational energy immediately, but affects the internal energy (through inelastic collisions) only after a certain time-lag" [1]. The bulk viscosity is strongly related to the time lag.

For simplicity we consider the case where only the rotational modes are activated, and the rotational degrees of freedom d_r is a constant. The relaxation of rotational temperature T_r is described by the Jeans-Landau equation:

$$\frac{\partial T_r}{\partial t} = \frac{T_t - T_r}{\tau_r}, \tag{7.1}$$

where τ_r is the rotational relaxation time and T_t and T_r are the translational and rotational temperature, respectively. The equilibrium temperature T is expressed as

L. Wu, *Rarefied Gas Dynamics*,
https://doi.org/10.1007/978-981-19-2872-7_7

$$T = \frac{3T_t + d_r T_r}{3 + d_r}. \tag{7.2}$$

Ignoring the effect of shear viscosity and heat conduction, from the energy conservation we have

$$\begin{aligned} \frac{3nk_B}{2}\frac{DT_t}{Dt} &= -p_t \nabla \cdot \boldsymbol{u} - \frac{d_r}{2} nk_B \frac{T_t - T_r}{\tau_r}, \\ \frac{d_r}{2} nk_B \frac{DT_r}{Dt} &= \frac{d_r}{2} nk_B \frac{T_t - T_r}{\tau_r}, \end{aligned} \tag{7.3}$$

where $p_t = nk_B T_t$ is the translational pressure. Thus we know

$$\frac{D}{Dt}(T_t - T_r) = -\frac{2}{3} T_t \nabla \cdot \boldsymbol{u} - \frac{3 + d_r}{3} \frac{T_t - T_r}{\tau_r}. \tag{7.4}$$

Applying a similar technique as the Chapman-Enskog expansion to expand $T_t - T_r$ in terms of the power series of the small parameter τ_r, we have

$$T_t - T_r \approx -\frac{2\tau_r}{3 + d_r} T_t \nabla \cdot \boldsymbol{u} \approx -\frac{2\tau_r}{3 + d_r} T \nabla \cdot \boldsymbol{u}. \tag{7.5}$$

Thus, the total pressure $p = nk_B T$ is expressed as

$$p = p_t \left[1 + \frac{2d_r}{(3 + d_r)^2} \tau_r \nabla \cdot \boldsymbol{u} \right]. \tag{7.6}$$

Therefore, the bulk viscosity is $\mu_b = p_t \tau_r (2d_r)/(3 + d_r)^2$. If we introduce the dimensionless rotational collision number Z as

$$Z = \frac{3\tau_r}{(d_r + 3)\tau}, \tag{7.7}$$

with $\tau = \mu / n_0 k_B T_t$ being the mean collision time of gas molecules due to the translational motion, the bulk viscosity μ_b can be expressed in terms of the shear viscosity as

$$\frac{\mu_b}{\mu} = \frac{2d_r Z}{3(d_r + 3)}. \tag{7.8}$$

From Eqs. (7.7) and (7.8) we know that the bulk viscosity is proportional to the energy exchange rate τ_r. It should be emphasized that, however, the derivation is based on the assumption that τ_r^{-1} is much larger than the characteristic frequency ϖ, e.g., the oscillation frequency of a sound wave. When the characteristic frequency is comparable to or even larger than τ_r^{-1}, the bulk viscosity is approximated by a frequency-dependent function [2–6]:

$$\mu_b = 2n_0 k_B T_t \frac{d_r \tau_r/(1+\varpi^2\tau_r^2)}{[3+d_r/(1+\varpi^2\tau_r^2)]^2}. \tag{7.9}$$

That is to say, in the extreme condition $\varpi\tau_r \to \infty$ the bulk viscosity approaches zero rather than infinity, because the energy exchange is so slow that the gas system cannot feel. This makes the bulk viscosity an elusive parameter. Similar situation applies to the effective heat capacity and sound speed, where their effective value depends on the sound frequency [7].

It should be emphasized that, while the bulk viscosity is just one of the transport coefficients obtained from the Chapman-Enskog expansion so only valid in the continuum flow regime, the relaxation process is intrinsic and valid in all flow regimes. This should be taken into account in the gas kinetic modeling.

7.2 Thermal Conductivity

The thermal conductivity of molecular gas is more complicated than the shear and bulk viscosities. Experimentally, although the total thermal conductivity can be measured straightforwardly, the determination of its translational and internal parts is rather difficult. While in continuum flow regime the total thermal conductivity determines the gas dynamics in addition to viscosity, in rarefied regimes both the translational and internal thermal conductivities are important. For example, the mass flow rate in the thermal transpiration is related to the translational thermal conductivity rather than the total thermal conductivity [8–11].

It will be interesting to briefly introduce how the theory of thermal conductivity is developed in history. By analogy with noble gases and simple gas kinetic theory in Sect. 1.3, Eucken expressed the thermal conductivity of a molecular gas as [12]:

$$\frac{m}{\mu k_B}\kappa = \frac{m}{\mu k_B}(\kappa_{tr}+\kappa_{rot}) = \frac{3}{2}f_{tr} + \frac{d_r}{2}f_{rot} \equiv \frac{3+d}{2}f_u, \tag{7.10}$$

where the translational and internal Eucken factors f_{tr} and f_{rot} are

$$f_{tr} = \frac{5}{2}, \quad f_{rot} = 1. \tag{7.11}$$

Later, it was realized that the transport of internal energy is due to the diffusion of gas molecules, f_{rot} is expressed in terms of the self-diffusion coefficient D and gas density ρ (or the Schmidt number $\text{Sc} = \mu/\rho D$):

$$f_{tr} = \frac{5}{2}, \quad f_{rot} = \frac{\rho D}{\mu} \equiv \frac{1}{\text{Sc}}. \tag{7.12}$$

Based on the Wang-Chang and Uhlenbeck Eq. (2.52), Manson and Manchick found via the Chapman-Enskog expansion that, like the energy relaxation (7.1), the translational and rotational heat fluxes ($\boldsymbol{q}_t$ and $\boldsymbol{q}_r$, respectively) for Maxwellian molecules satisfies the following relaxation [13]:

$$\frac{\partial}{\partial t}\begin{bmatrix}\boldsymbol{q}_t\\ \boldsymbol{q}_r\end{bmatrix} = -\frac{p_t}{\mu}\begin{bmatrix}A_{tt} & A_{tr}\\ A_{rt} & A_{rr}\end{bmatrix}\begin{bmatrix}\boldsymbol{q}_t\\ \boldsymbol{q}_r\end{bmatrix}, \tag{7.13}$$

where the four dimensionless thermal relaxation rates in the matrix $\mathbf{A} = [A_{ij}]$ with $i, j = t, r$ are [14]

$$\begin{aligned} A_{tt} &= \frac{2}{3} + \frac{5}{6Z}\frac{d_r}{3+d_r}, \\ A_{rr} &= \frac{\mu}{\rho D'} + \frac{3}{2(3+d_r)Z}, \\ A_{tr} &= -\frac{5}{2(3+d_r)Z}, \\ A_{rt} &= -\frac{d_r}{2(3+d_r)Z} \end{aligned} \tag{7.14}$$

with D' the average diffusion coefficient, which can be markedly different from the self-diffusion coefficient D if strong resonant collision occurs.

The translational and rotational thermal conductivities can be exactly solved according to the Chapman-Enskog expansion:

$$\begin{bmatrix}\kappa_{tr}\\ \kappa_{rot}\end{bmatrix} = \frac{k_B\mu}{2m}\begin{bmatrix}A_{tt} & A_{tr}\\ A_{rt} & A_{rr}\end{bmatrix}^{-1}\begin{bmatrix}5\\ d_r\end{bmatrix}, \tag{7.15}$$

and the final expressions for the translational and rotational Eucken factors are (terms higher than $1/Z$ are ignored) [13]:

$$\begin{aligned} f_{tr} &= \frac{5}{2}\left[1 - \frac{5d_r}{4(d_r+3)Z}\left(1 - \frac{2\rho D'}{5\mu}\right)\right], \\ f_{rot} &= \frac{\rho D'}{\mu}\left[1 + \frac{15}{4(d_r+3)Z}\left(1 - \frac{2\rho D'}{5\mu}\right)\right]. \end{aligned} \tag{7.16}$$

The fact that f_{tr} decreases along with the rotational collision number Z is explained as follows: the translational thermal conductivity decreases because part of the translational energy is converted to the internal energy during the transport of gas molecules; the intensity of this energy exchange becomes faster and stronger when Z decreases and when D' increases. The internal Eucken factor f_{rot} changes with respect to Z in opposite direction. This shows that Eq. (7.16) is physically reasonable. When there is no translational-internal energy exchange ($Z \to \infty$), the translational Eucken factor is reduced to that of the monatomic gas, i.e., $f_{tr} = 5/2$.

One problem remains as one does not know whether the dimensionless thermal relaxation parameter $\boldsymbol{A}$ described in Eq. (7.14) is accurate or not. Mason and

Monchick determined the translational and internal Eucken factors by comparing Eq. (7.16) to experiments through varying Z arbitrarily [13]. Therefore, although the thermal conductivities might be correct, it is very likely that the bulk viscosity is not. On the other hand, when the bulk viscosity is known, the relaxation coefficient A_{rr} in Eq. (7.14) is modified to $-J_{011}$ in Eq. (7.26) below to recover the total thermal conductivity, but there is no justification.

7.3 Thermal Relaxation Rates in DSMC

For monatomic gas, the DSMC is equivalent to the Boltzmann equation [15, 16]. This is because in DSMC the post-collision velocities are updated as that of the Boltzmann equation, and for monatomic gas the structure of BCO ensures that the translational Eucken factor is very close to 2.5, see Eq. (3.30). Therefore, as long as the collision probability in various DSMC models is carefully chosen so that the shear viscosity is recovered, the recovery of thermal conductivity is guaranteed.

However, whether DSMC produces the exact translational and internal thermal conductivities of molecular gases or not has been overlooked. The variable-soft-sphere model has been developed to recover the experimental values of shear viscosity and self-diffusion coefficients, while the phenomenological Larsen-Borgnakke collision model [17] has been developed to recover the energy exchange rate [18–20] so the bulk viscosity is obtained exactly; but whether the total thermal conductivity, f_{tr} and f_{rot} have been accurately modeled in DSMC or not has not been investigated in detail [21]: actually, there is no mechanism to recover the thermal relaxation rates in Eq. (7.13) correctly in DSMC.

We extract the thermal relaxation rates $\boldsymbol{A}$ in DSMC and assess its capability in recovering the thermal conductivities [22]. Both the non-polar gas N_2 and polar gas HCL are studied, at different values of rotational collision number. Note that in DSMC the inelastic collision probability Λ is used which, according to Eqs. (5.67) and (4.50) in Bird's book [15], is related to the rotational collision number Z through the Landau-Jeans equation (7.1) as

$$Z = \frac{1}{\Lambda}\frac{\alpha'(5-2\omega)(7-2\omega)}{5(\alpha'+1)(\alpha'+2)}, \tag{7.17}$$

where α' is the exponent in the variable-soft-sphere model (4.4) that is used to recover the diffusion coefficient. Therefore, it is related to the Schmidt number Sc as

$$\text{Sc} = \frac{\mu}{\rho D} = \frac{5(2+\alpha')}{3(7-2\omega)\alpha'}. \tag{7.18}$$

Here we choose $\text{Sc} = 1/1.33$, $d_r = 2$, and the viscosity index $\omega = 0.74$ for N_2 and $\omega = 1.0$ for HCL.

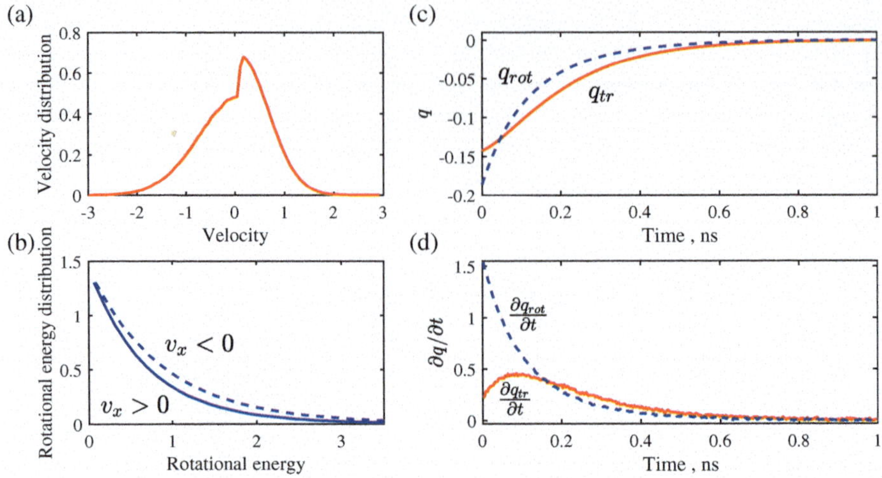

Fig. 7.1 Homogeneous relaxation of N_2 [23]. **a** The VDF used to generate an initial translational heat flux; the abscissa is normalized by the most probable speed v_m. **b** The distribution of rotational energy used to generate an initial rotational heat flux; the abscissa is normalized by $k_B T_0$. **c**, **d** The relaxation of the translational and rotational heat fluxes. Parameters are $Z_{\text{DSMC}} = 4.0$, $\omega_{N_2} = 0.74$, and $\text{Sc} = 1/1.33$

The homogeneous relaxation problem is considered: 10^6 particles are uniformly distributed over a cubic cell of the side length 10 nm, and the periodic condition is implemented at all boundaries. The gas density is $n_0 = 2.69 \times 10^{25}\ \text{m}^{-3}$ and the temperature is $T_0 = 300$ K. Initially, particles with $v_x > 0$ ($v_x < 0$) are generated from the Maxwell distribution of $T_0 = 200$ K ($T_0 = 400$ K), see Fig. 7.1a. Similarly, the rotational energy assigned to the particles with $v_x > 0$ ($v_x < 0$) satisfies the Maxwell distribution of $T_0 = 200$ K ($T_0 = 400$ K), see Fig. 7.1b. Then, the translational and rotational heat fluxes are monitored until the thermal equilibrium is reached. A hundred independent runs are conducted to get smooth results in Fig. 7.1c, d.

The least square method is adopted to solve the linear regression problem (7.13), and the results in Fig. 7.2 show that the thermal relaxation rates **A** are nearly same for N_2 and HCL, which, roughly speaking, vary linearly with $1/Z$.

Additionally, in the limit of $Z \to \infty$, we have $A_{tt} = 2/3$, $A_{rr} = \text{Sc}$, and $A_{tr} = A_{rt} = 0$, which results in Eq. (7.11). The results of Mason and Manchick (7.14) are also plotted for comparison, and the difference with DSMC is visible.

Given the thermal relaxation rates $\boldsymbol{A}$, the Eucken factors are calculated according to Eqs. (7.10) and (7.15), and shown in Fig. 7.3. For N_2, it is seen that the total thermal conductivity measured from experiment might be recovered in DSMC by choosing $\text{Sc} = 1/1.34$, although we do not know whether the translational and rotational thermal conductivities are recovered or not. Note that we use the word "might" here because we do not know exactly the value of bulk viscosity, or the rotational collision number Z: if $1/\Lambda \not\approx 3.5$ the DSMC has some error in recovering the total thermal conductivity. However, for the polar gas HCL, the total Eucken factor measured

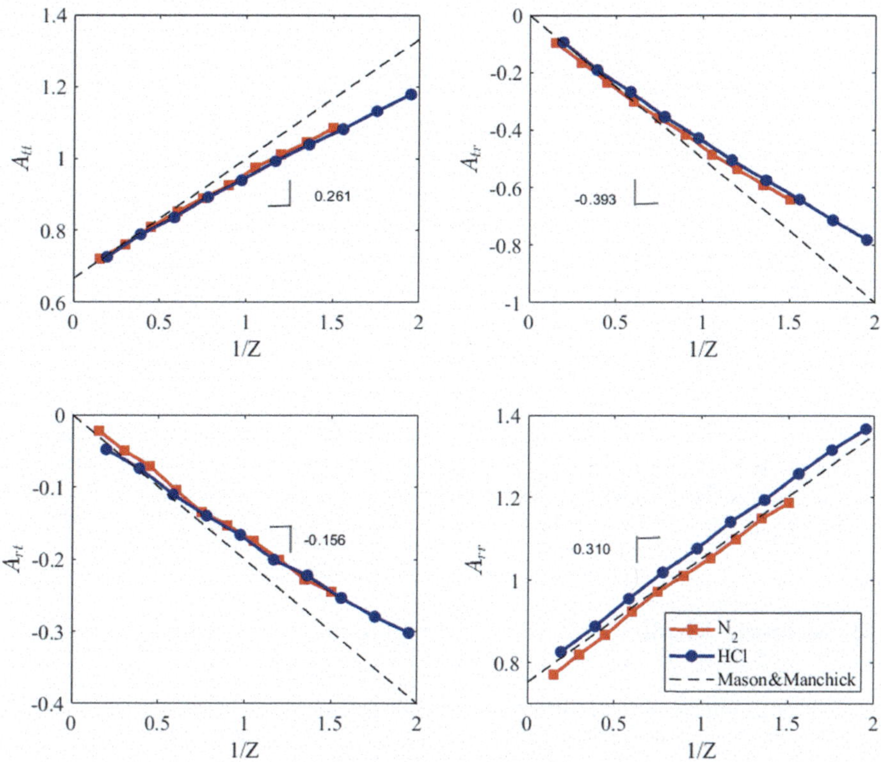

Fig. 7.2 Extracted relaxation rates of heat fluxes as per Eq. (7.13) from the DSMC simulations, as a function of the inverse rotational collision number. Analytical predictions (7.14) from Mason and Manchick are also plotted for comparison

from experiments at room temperature is about 1.7, but from Fig. 7.3a we see that the minimum total Eucken factor the DSMC can reach is about 1.8 when $\text{Sc} = 1/1.34$. Mason and Monchick reported that, for hydrogen chloride, the resonant interaction leads to a significant reduction of $\rho D'/\mu$ from 1.34 to 0.89 [13]; in order to recover the total thermal conductivity measured from experiments, Figure 7.3(b) shows that the inelastic collision probability Λ in DSMC should approach unity, which seems not physical as the maximum possible rate of exchange between translational and internal energy is reached. If one wants to use a reasonable value of Λ (or the bulk viscosity), D' should be much smaller than the self-diffusion coefficient. This will pose another difficulty: in gas mixtures where both diffusion and thermal conductivity are important, how to recover them in DSMC with the Larsen-Borgnakke model?

Although the phenomenological Larsen-Borgnakke model is widely used, we believe that more effort should be done to validate or even improve it. On the other hand, in order to study the effect of relaxation rates $\mathbf{A}$, we need a kinetic model that is able to describe the relaxation phenomena in the above sections.

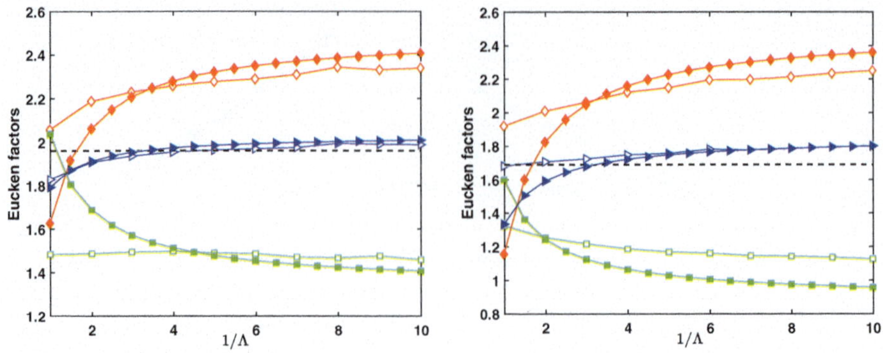

(a) Nitrogen: $\frac{\rho D'}{\mu} = 1.33$ (b) Hydrogen chloride: $\frac{\rho D'}{\mu} = 0.89$

Fig. 7.3 Eucken factors obtained from the analytical solution (7.16) and the DSMC at room temperature [22]. The variable-soft-sphere model is used and the self-diffusion coefficient D in DSMC takes the value of D'. Filled diamonds, squares, and triangles represent the translational, internal, and total Eucken factors from Eq. (7.16), respectively, while the open symbols are the corresponding results from DSMC. Dashed lines: total Eucken factor obtained from experiments at 300 K

7.4 Kinetic Models

To develop a kinetic model which is not only accurate but also permits fast calculations, let us recall that the collision operator (2.52) proposed by Wang-Chang and Uhlenbeck can be decomposed into the elastic and inelastic parts, depending on whether the total kinetic energy is conserved or not during binary collision:

$$
\begin{aligned}
Q_i =& \underbrace{\sum_j \iint P_{ij}^{ij} B\left[f_j(\boldsymbol{v}'_*)f_i(\boldsymbol{v}') - f_j(\boldsymbol{v}_*)f_i(\boldsymbol{v})\right] d\Omega d\boldsymbol{v}_*}_{\text{elastic}} \\
&+ \underbrace{\sum_{jkl}(1-\delta_{jk}\delta_{il}) \iint P_{ij}^{kl} B\left[f_k(\boldsymbol{v}'_*)f_l(\boldsymbol{v}') - f_j(\boldsymbol{v}_*)f_i(\boldsymbol{v})\right] d\Omega d\boldsymbol{v}_*}_{\text{inelastic}}.
\end{aligned} \tag{7.19}
$$

Therefore, on introducing the two reduced VDFs $G(t, \boldsymbol{x}, \boldsymbol{v}) = \sum_i o_i f_i$ and $R(t, \boldsymbol{x}, \boldsymbol{v}) = \sum_i o_i f_i e_i$, where o_i is the occupation fraction (2.58) of the internal energy level i, it is natural to model the rarefied gas flow as

$$
\begin{aligned}
\frac{\partial G}{\partial t} + \boldsymbol{v}\cdot\frac{\partial G}{\partial \boldsymbol{x}} + \boldsymbol{a}\cdot\frac{\partial G}{\partial \boldsymbol{v}} &= \underbrace{\frac{G_t - G}{\tau}}_{elastic} + \underbrace{\frac{G_r - G_t}{Z\tau}}_{inelastic}, \\
\frac{\partial R}{\partial t} + \boldsymbol{v}\cdot\frac{\partial R}{\partial \boldsymbol{x}} + \boldsymbol{a}\cdot\frac{\partial R}{\partial \boldsymbol{v}} &= \underbrace{\frac{R_t - R}{\tau}}_{elastic} + \underbrace{\frac{R_r - R_t}{Z\tau}}_{inelastic},
\end{aligned} \tag{7.20}
$$

where the elastic collisions conserve the kinetic energy, while the inelastic collisions exchange translational and rotational energies. The four reference VDFs $G_t(t, \boldsymbol{x}, \boldsymbol{v})$, $G_r(t, \boldsymbol{x}, \boldsymbol{v})$, $R_t(t, \boldsymbol{x}, \boldsymbol{v})$, and $R_r(t, \boldsymbol{x}, \boldsymbol{v})$ are the distributions where the gas system will relax to, so they must be carefully designed to recover Eqs. (7.1) and (7.13); also in the equilibrium state they should become the Maxwellian distribution.

Macroscopic flow quantities are calculated as the velocity moments of the two VDFs:

$$\left[n, n\boldsymbol{u}, \frac{3}{2}nk_BT_t, p_{ij}, \boldsymbol{q}_t\right] = \int \left[1, \boldsymbol{v}, \frac{m}{2}c^2, mc_ic_j, \frac{m}{2}c^2\boldsymbol{c}\right] G d\boldsymbol{v},$$
$$T_r = \frac{2}{d_r n k_B}\int R d\boldsymbol{v}, \quad \boldsymbol{q}_r = \int R\boldsymbol{c} d\boldsymbol{v}, \tag{7.21}$$

the total heat flux is $\boldsymbol{q} = \boldsymbol{q}_t + \boldsymbol{q}_r$ and the pressures related to the translational, rotational, and total degrees of freedom are $p_t = nkT_t$, $p_r = nkT_r$ and $p = nkT$.

7.4.1 Hanson-Morse Model

We first introduce the Hanson-Morse model [24] which approximates the collision operator of Wang-Chang and Uhlenbeck equation in the linear case. The method of construction is similar to that of Gross-Jackson model [25] for the Boltzmann equation.

By writing the VDF as $f_i(t, \boldsymbol{x}, \boldsymbol{v}) = o_i f_{eq}\,[1 + \phi_i(t, \boldsymbol{x}, \boldsymbol{v})]$, where ϕ_i is the perturbed distribution function for the energy level i, the Wang-Chang and Uhlenbeck equation is linearized as

$$\frac{\partial \phi_i}{\partial t} + \boldsymbol{v}\cdot\frac{\partial \phi_i}{\partial \boldsymbol{x}} - 2\boldsymbol{a}\cdot\boldsymbol{v} = \frac{L}{v_m}\mathcal{J}_i, \tag{7.22}$$

where the linearized collision operator is approximated as

$$\begin{aligned}
\mathcal{J}_i =& J_{030}\phi_i - J_{030}\varrho_i - 2J_{030}\boldsymbol{v}\cdot\boldsymbol{u} + 2\sigma_{\alpha\beta}(J_{020} - J_{030})v_{\langle\alpha}v_{\beta\rangle} \\
&+ T_{tr}\left[(-J_{030} + J_{100})\left(v^2 - \frac{3}{2}\right) - \frac{3}{d_r}J_{100}\,(\epsilon_i - \langle\epsilon\rangle)\right] \\
&+ T_r\left[-J_{100}\left(v^2 - \frac{3}{2}\right) + \left(\frac{3}{d_r}J_{100} - J_{030}\right)(\epsilon_i - \langle\epsilon\rangle)\right] \\
&+ 2\boldsymbol{v}\cdot\boldsymbol{q}_t\left[(J_{110} - J_{030})\frac{2}{5}\left(v^2 - \frac{5}{2}\right) - J_{011}^{110}\sqrt{\frac{4}{5d_r}}\,(\epsilon_i - \langle\epsilon\rangle)\right] \\
&+ 2\boldsymbol{v}\cdot\boldsymbol{q}_r\left[-J_{011}^{110}\sqrt{\frac{4}{5d_r}}\left(v^2 - \frac{5}{2}\right) + 2(J_{011} - J_{030})\frac{\epsilon_i - \langle\epsilon\rangle}{d_r}\right],
\end{aligned} \tag{7.23}$$

with the following relaxation rates (which are closely related to the relaxation matrix **A**):

$$
\begin{aligned}
J_{020} &= -\tau^{-1} = -\frac{p_t}{\mu}, \\
J_{030} &= -\frac{3}{2}\tau^{-1}, \\
J_{100} &= -\frac{1}{Z\tau}\frac{d_r}{3+d_r}, \\
J_{110} &= -\frac{2}{3\tau} - \frac{5}{6Z\tau}\frac{d_r}{3+d_r}, \\
J_{011}^{110} &= -\sqrt{\frac{5}{4d_r}\frac{1}{Z\tau}\frac{d_r}{3+d_r}}, \\
J_{011} &= -\frac{\mu}{\rho D'} - \frac{3}{2(3+d_r)Z}.
\end{aligned}
\tag{7.24}
$$

Note that in Eqs. (7.22) and (7.23), $\boldsymbol{v}$ and $\boldsymbol{x}$ have been normalized by the most probable speed v_m and the characteristic length L, respectively, the acceleration has been normalized by v_m^2/L, while the internal energy has been normalized by $k_B T_0$. Macroscopic quantities are

$$
\varrho = \int f_{eq}\phi_i d\boldsymbol{v}, \quad \boldsymbol{u} = \sum_i o_i \int f_{eq}\phi_i \boldsymbol{v} d\boldsymbol{v}, \quad \langle e\rangle = \sum_i o_i e_i,
$$

$$
T_t = \frac{2}{3}\sum_i o_i \int f_{eq}\phi_i \left(v^2 - \frac{3}{2}\right) d\boldsymbol{v}, \quad T_r = \frac{2}{d_r}\sum_i o_i \int f_{eq}\phi_i (\epsilon_i - \langle\epsilon\rangle) d\boldsymbol{v},
$$

$$
\boldsymbol{q}_t = \sum_i o_i \int f_{eq}\phi_i \left(v^2 - \frac{5}{2}\right) \boldsymbol{v} d\boldsymbol{v}, \quad \boldsymbol{q}_r = \sum_i o_i \int f_{eq}\phi_i (\epsilon_i - \langle\epsilon\rangle) \boldsymbol{v} d\boldsymbol{v},
$$

$$
\sigma_{\alpha\beta} = \sum_i o_i \int f_{eq}\phi_i v_{\langle\alpha} v_{\beta\rangle} d\boldsymbol{v},
$$

and the total perturbation density is $\varrho = \sum_i \varrho_i o_i$.

In the spontaneous Rayleigh-Brillouin scattering, if the traceless pressure tensor $\sigma_{\alpha\beta}$ is contained, the model is called Tenti-S7 model [26, 27], which cannot give reasonable light scattering spectrum; however, if $\sigma_{\alpha\beta}$ is removed (and J_{030} in Eq. (7.24) is replaced by $-1/\tau$), it becomes the Tenti-S6 model, see Sect. 15.3. One may check that the relaxation of stress tensor (6.28), energy (7.1), and heat flux (7.13) are properly recovered. Nevertheless, this does not guarantee the exact recovery of thermal conductivity as compared to experimental measurement. For this reason, the coefficient J_{011} is modified into the following form [26, 28]:

$$
J_{011} = -\frac{2d_r}{3(3+d_r)\tau}\frac{\frac{1}{5}(3+d_r) + \frac{6+d_r}{4Z} + \frac{9}{16Z^2}f_u}{\frac{2}{15}f_u(3+d_r) + \frac{d_r}{6Z}f_u - 1},
\tag{7.25}
$$

which is equivalent to the change of A_{rr} in Eq. (7.14). However, this is not justified.

If we introduce the reduced VDFs $G = \sum_i o_i f_{eq}\phi_i$ and $R = \sum_i o_i\phi_i f_{eq}(\epsilon_i - \langle\epsilon\rangle)$, and let $\phi = \sum_i o_i\phi_i$, we have

$$\frac{\partial G}{\partial t} + \boldsymbol{v}\cdot\frac{\partial G}{\partial \boldsymbol{x}} - 2\boldsymbol{a}\cdot\boldsymbol{v} = \frac{L}{v_m}\mathcal{J}_G,$$
$$\frac{\partial R}{\partial t} + \boldsymbol{v}\cdot\frac{\partial R}{\partial \boldsymbol{x}} - 2\boldsymbol{a}\cdot\boldsymbol{v} = \frac{L}{v_m}\mathcal{J}_R, \tag{7.26}$$

with

$$\begin{aligned}\mathcal{J}_G =& -J_{030}\left[\varrho + 2\boldsymbol{v}\cdot\boldsymbol{u} + T_{tr}\left(v^2-\frac{3}{2}\right) + \underbrace{\frac{4}{5}\boldsymbol{v}\cdot\boldsymbol{q}_t\left(v^2-\frac{5}{2}\right)}_{\text{relaxation of heat flux}} - G\right] \\ &+ 2\sigma_{\alpha\beta}(J_{020}-J_{030})v_{\langle\alpha}v_{\beta\rangle}\underbrace{-(T_r-T_{tr})J_{100}\left(v^2-\frac{3}{2}\right)}_{\text{relaxation of energy}} \\ &\underbrace{+\boldsymbol{v}\cdot\boldsymbol{q}_tJ_{110}\frac{4}{5}\left(v^2-\frac{5}{2}\right) - 2\boldsymbol{v}\cdot\boldsymbol{q}_rJ_{011}^{110}\sqrt{\frac{4}{5d_r}}\left(v^2-\frac{5}{2}\right)}_{\text{relaxation of heat fluxes}},\end{aligned} \tag{7.27}$$

and

$$\begin{aligned}\mathcal{J}_R =& -J_{030}\left(\frac{d_r}{2}T_r + \underbrace{2\boldsymbol{v}\cdot\boldsymbol{q}_r}_{\text{relaxation of heat flux}} - R\right)\underbrace{-\frac{3}{2}J_{100}(T_{tr}-T_r)}_{\text{relaxation of energy}} \\ &\underbrace{-2\boldsymbol{v}\cdot\boldsymbol{q}_tJ_{011}^{110}\sqrt{\frac{d_r}{5}} + 2\boldsymbol{v}\cdot\boldsymbol{q}_rJ_{011}}_{\text{relaxation of heat fluxes}}.\end{aligned} \tag{7.28}$$

It is clear that the underbraced terms describe the relaxation of energy and heat fluxes. Therefore, the Hanson-Morse model can be fitted to the general form (7.20).

We have shown in Sect. 6.2.5 that for monatomic gas, the BGK, Shakhov, and ESBGK models can be derived from the Gross-Jackson kinetic model via the nonlinearization procedure. Likewise, one may design nonlinear kinetic models for molecular gas from the Hanson-Morse model. But it seems that history did not take this road; instead different kinetic models are proposed from different perspectives. Typical models are introduced below.

7.4.2 Rykov Model

The original Rykov model was developed for non-vibrating diatomic gases, but the method of construction can be extended straightforwardly to polyatomic gases [29, 30]. The kinetic equation is given by Eq. (7.20), where the four reference VDFs are given by

$$\begin{aligned} G_t &= F_{eq}(T_t)\left[1+\frac{2m\boldsymbol{q}_t\cdot\boldsymbol{c}}{15k_BT_tp_t}\left(\frac{mc^2}{2k_BT_t}-\frac{5}{2}\right)\right], \\ G_r &= F_{eq}(T)\left[1+\omega_0\frac{2m\boldsymbol{q}_t\cdot\boldsymbol{c}}{15k_BTp}\left(\frac{mc^2}{2k_BT}-\frac{5}{2}\right)\right], \\ R_t &= \frac{d_rk_BT_r}{2}G_t+F_{eq}(T_t)(1-\text{Sc})\frac{m\boldsymbol{q}_r\cdot\boldsymbol{c}}{nk_BT_t}, \\ R_r &= \frac{d_rk_BT}{2}G_r+F_{eq}(T)\omega_1(1-\text{Sc})\frac{m\boldsymbol{q}_r\cdot\boldsymbol{c}}{nk_BT}. \end{aligned} \tag{7.29}$$

Here, the equilibrium VDF F_{eq} is defined in Eq. (2.31), ω_0 and ω_1 are two constants used to recover the thermal conductivities of molecular gases. The elastic relaxation time τ in Eq. (7.20) should be choose as

$$\tau=\frac{\mu}{p_t}, \tag{7.30}$$

so the shear viscosity and bulk viscosity are properly recovered. Also, we find

$$\begin{aligned} A_{tr} &= A_{rt}=0, \\ A_{tt} &= \frac{2}{3}\left(1+\frac{1-\omega_0}{2Z}\right), \\ A_{rr} &= \text{Sc}+\frac{(1-\text{Sc})(1-\omega_1)}{Z}, \end{aligned} \tag{7.31}$$

which means that the relaxation of translational (internal) heat flux is related to the translational (internal) heat flux only; later in the flow driven by Maxwell's demon we will show that this results in errors. The translational and rotational Eucken factors are

$$\begin{aligned} f_{tr} &= \frac{5}{2}\left(1+\frac{1-\omega_0}{2Z}\right)^{-1}, \\ f_{rot} &= \left(\text{Sc}+\frac{(1-\text{Sc})(1-\omega_1)}{Z}\right)^{-1}. \end{aligned} \tag{7.32}$$

The Eucken factors can be modified by the free parameters ω_0 and ω_1.

7.4.3 ESBGK Model

To obtain the proper Prandtl number, Holway proposed the ESBGK model through the maximal entropy principle [31]. This model retains much of the mathematical simplicity of the standard BGK model and fulfills the H-theorem [32]. In this model, the evolution of two reduced VDFs is governed by

$$\begin{aligned}\frac{\partial G}{\partial t}+\boldsymbol{v}\cdot\frac{\partial G}{\partial \boldsymbol{x}}+\boldsymbol{a}\cdot\frac{\partial G}{\partial \boldsymbol{v}}&=\frac{1}{\tau}\left(f_r^{ES}-G\right),\\ \frac{\partial R}{\partial t}+\boldsymbol{v}\cdot\frac{\partial R}{\partial \boldsymbol{x}}+\boldsymbol{a}\cdot\frac{\partial R}{\partial \boldsymbol{v}}&=\frac{1}{\tau}\left(\frac{d_r}{2}T_{rel}f_r^{ES}-R\right),\end{aligned} \tag{7.33}$$

where $T_{rel}=(1-\theta)T_r+\theta T$; the reference distribution function f_r^{ES} is given in Eq. (6.13), but with

$$\lambda_{ij}=(1-\theta)\left[(1-b)\frac{k_BT_t}{m}\delta_{ij}+b\frac{p_{ij}}{nm}\right]+\theta\frac{k_BT}{m}\delta_{ij}. \tag{7.34}$$

The relaxation time in the ESBGK model is

$$\tau=\frac{1}{\Pr}\frac{\mu}{p_t}\equiv(1-b+b\theta)\frac{\mu}{p_t}, \tag{7.35}$$

while the bulk viscosity is

$$\mu_b=\frac{1}{\theta\Pr}\frac{2d_r}{3(d_r+3)}\mu. \tag{7.36}$$

The ESBGK model satisfies the H-Theorem for all $-1/2\le b\le 1$ and $0\le\theta\le 1$, which lead to $2/3\le\Pr\le+\infty$. For monatomic gas, a Prandtl number of 2/3 is obtained with $\theta=0$ and $b=-1/2$. The relaxation coefficients in Eq. (7.13) are $A_{tt}=A_{rr}=\Pr$, $A_{tr}=A_{rt}=0$. Therefore, like the Rykov model, the relaxation of translational (internal) heat flux is related to the translational (internal) heat flux only. However, the translational and rotational Eucken factors are

$$f_{tr}=\frac{5}{3\Pr},\quad f_{rot}=\frac{1}{\Pr}, \tag{7.37}$$

which cannot be adjusted independently when the Prandtl number is fixed.

7.4.4 Wu Model

To properly describe the effect of intermolecular potentials, a kinetic model should not only recover the proper relaxation rates of energy and heat fluxes, but also be reduced to the Boltzmann equation for monatomic gas when the energy and heat flux relaxations between translational and internal modes are absent. This brings us the following model:

$$\begin{aligned}\frac{\partial G}{\partial t}+\boldsymbol{v}\cdot\frac{\partial G}{\partial \boldsymbol{x}}&=Q(G)+\frac{G_r-G_t}{Z\tau},\\ \frac{\partial R}{\partial t}+\boldsymbol{v}\cdot\frac{\partial R}{\partial \boldsymbol{x}}&=\frac{R_t'-R}{\tau}+\frac{R_r-R_t}{Z\tau},\end{aligned}\tag{7.38}$$

where the relaxation time is $\tau=\mu/p_t$, Q is the BCO that will give the correct value of shear viscosity if the inelastic collision term in the first equation of Eq. (7.38) is removed, the four reference VDFs are

$$\begin{aligned}G_t&=F_{eq}(T_t)\left[1+\frac{2m\boldsymbol{q}_t\cdot\boldsymbol{c}}{15k_BT_tp_t}\left(\frac{mc^2}{2k_BT_t}-\frac{5}{2}\right)\right],\\ G_r&=F_{eq}(T)\left[1+\frac{2m\boldsymbol{q}'\cdot\boldsymbol{c}}{15k_BTp}\left(\frac{mc^2}{2k_BT}-\frac{5}{2}\right)\right],\\ R_t=\frac{d_rk_BT_r}{2}G_t,\quad R_r&=\frac{d_rk_BT}{2}G_r+F_{eq}(T)\frac{m\boldsymbol{q}''\cdot\boldsymbol{c}}{nk_BT},\end{aligned}\tag{7.39}$$

and

$$R_t'=\frac{d_rk_BT_r}{2}(\tau Q+G)+\left(\frac{m}{2\pi k_BT_t}\right)^{3/2}\exp\left(-\frac{mc^2}{2k_BT_t}\right)\frac{m\boldsymbol{q}_1\cdot\boldsymbol{c}}{k_BT_t}.\tag{7.40}$$

To recover the correct relaxation of heat fluxes in Eq. (7.13), $\boldsymbol{q}'$ and $\boldsymbol{q}''$ are chosen as

$$\begin{aligned}\boldsymbol{q}'&=\left[-3Z\left(A_{tt}-\frac{2}{3}\right)+1\right]\boldsymbol{q}_t-3ZA_{tr}\boldsymbol{q}_r,\\ \boldsymbol{q}''&=-Z\left[A_{rt}q_t+(A_{rr}-1)\boldsymbol{q}_r\right].\end{aligned}\tag{7.41}$$

Three key features of the Wu model should be emphasized. First, it is flexible to adjust not only the two Eucken factors independently but also the relaxation rates of heat fluxes. Second, when the BCO is replaced by that of the Shakhov mode, then after linearization this model is equivalent to the Hanson-Morse model (7.27) and (7.28) when the stress tensor is not included. Third, when there is no translational-internal energy exchange ($Z\to\infty$), the Wu model reduces to the Boltzmann equation for monatomic gas.

7.5 Accuracy of Kinetic Models

The DSMC is used to assess the Rykov, ESBGK, and Wu models in both hypersonic flows and microflows. We take $Z=2.667$, $\omega=0.74$ and $\alpha'=1.36$ for N_2; this set of parameters recovers the experimental value of total thermal conductivity. In the Wu model, the same values of relaxation coefficient can be chosen as those in the DSMC

Table 7.1 Eucken factors and heat flux relaxation rates used in different gas kinetic models and DSMC [33]. The total Eucken factor of the Nitrogen at room temperature is $f_{eu} = 1.993$

	ESBGK	Rykov	Wu	DSMC
f_{tr}	2.373	2.365	2.365	2.365
f_{rot}	1.424	1.436	1.436	1.436
A_{tt}	0.702	0.705	0.786	0.786
A_{tr}	0	0	−0.201	−0.201
A_{rt}	0	0	−0.059	−0.059
A_{rr}	0.702	0.696	0.842	0.842

simulations. However, the Rykov and ESBGK models can only have the same values of total Eucken factor and bulk viscosity, but the relaxation rates $\boldsymbol{A}$ of heat fluxes are different [33], see the parameters in Table 7.1.

7.5.1 Normal Shock Wave

Figure 7.4 compares the structures of normal shock wave when $\text{Ma} = 7$. When the Rykov and ESBGK models are used, large discrepancy in the profiles of temperature and heat flux can be seen. This is mostly caused by the velocity-independent collision frequency in both models. On the contrary, in the Wu model the BCO is used for the translational motion where the collision frequency is naturally velocity-dependent; although the rotational motion is stilled modeled by the velocity-independent collision frequency, it gives satisfactory agreement with DSMC.

7.5.2 Couette Flow

Figure 7.5 shows that the Wu model agrees well with the DSMC, the Rykov model has some discrepancy in the temperature and heat flux, while the ESBGK model has errors in density, temperature, and heat flux.

7.5.3 Maxwell's Demon

In the thermal transpiration of monatomic gas along long channel, the gas flow is driven by the wall temperature gradient, which is equivalent to applying an external acceleration that is a function of the molecular velocity, see the last term in Eq. (5.19).

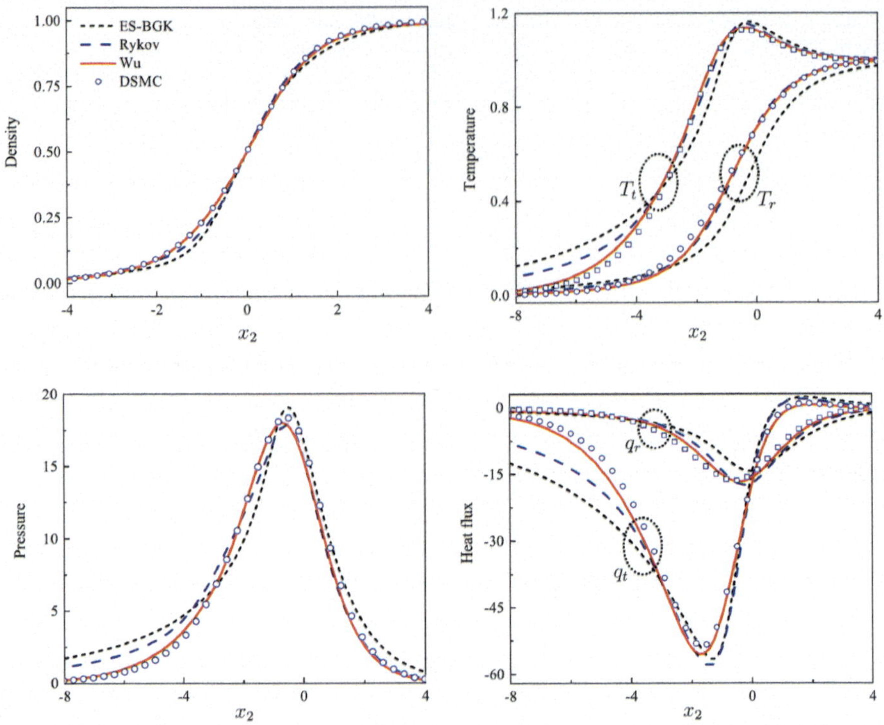

Fig. 7.4 Comparisons in the density, temperature, stress, and heat flux between different kinetic models for a normal shock in nitrogen gas with Mach number 7. The spatial variable is normalized by the upstream MFP [23, 33]

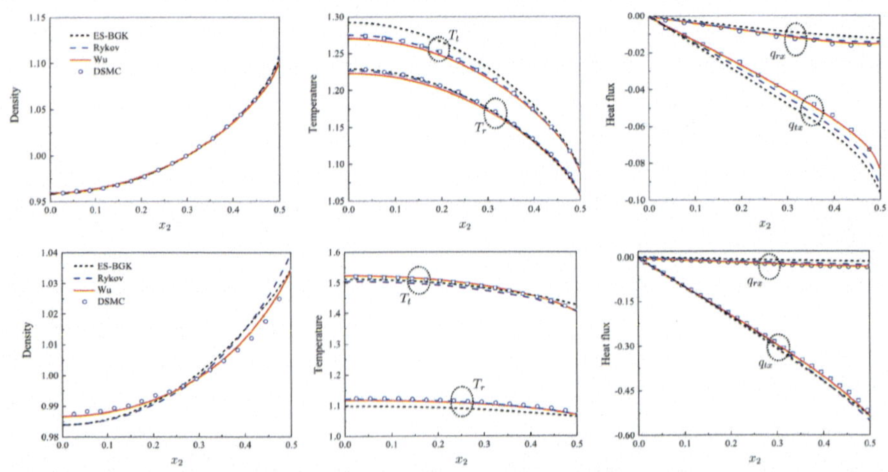

Fig. 7.5 Density, temperature, and heat flux (perpendicular to the plates) in the planar Couette flow of the Knudsen number Kn $= 0.1$ (first row) and 1 (second row). Due to symmetry, the other half region $0 \leq x_1 \leq 0.5$ is not shown [33]

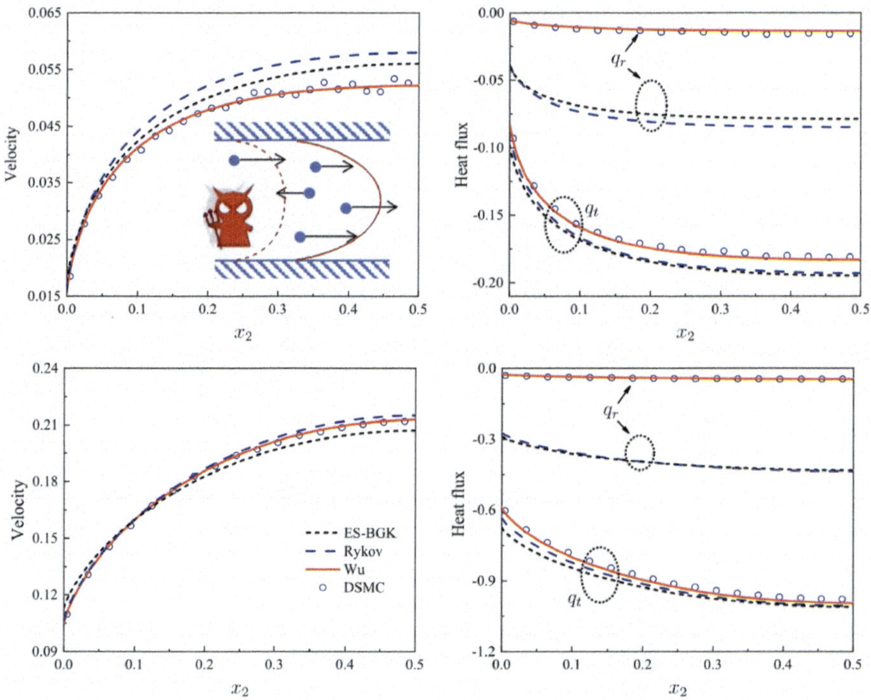

Fig. 7.6 Rarefiefd gas flow driven by Maxwell's demon [23, 33]. The Knudsen number in the first and second rows is Kn = 0.1 and 1, respectively. Due to symmetry, the other half region $0.5 \leq x_2 \leq 1$ is not shown. Results are further normalized by the dimensionless acceleration

Here, as a thought experiment, we consider the creep flow where each individual molecule is subject to an external acceleration based on its kinetic energy:

$$a_x = a_0 \left(\frac{v^2}{v_m^2} - \frac{3}{2} \right), \tag{7.42}$$

where a_0 is a small parameter to guarantee the gas flow is slightly deviated from equilibrium; here it is $2a_0L/v_m^2 = 0.0718$.

The Wu model is linearized and solved by the GSIS [34, 35] at Kn = 0.1 and Kn = 1.0. Figure 7.6 shows that the velocity and heat flux in the horizontal direction obtained from the Wu model match well with the DSMC data. Besides, the rotational heat flux is smaller than the translational one. It implies that the translational thermal conductivity plays the dominant role in determining the velocity and hence the mass flux in this problem. The Rykov and ESBGK models, due to the lack of cross-relaxation raters A_{tr} and A_{rt}, predict wrong rotational heat flux.

7.6 Uncertainty Quantification

Although in DSMC and other kinetic models the effect of temperature relaxation (7.1) that is related to the bulk viscosity has been extensively studied [36, 37], the role of thermal relaxation rates reflected in the matrix **A** has never been investigated. In experiments, since the total thermal conductivity can be measured straightforwardly, and in some cases the translational thermal conductivity [8–10, 22] is measured, according to Eq. (7.15) we know that at least two elements in **A** are not determined. Therefore, it is necessary to quantify these uncertainties.

To this end, different values of A_{ij} are chosen in the Wu model (7.38) to see the effect of thermal relaxation rates, as the DSMC does not have this capability when the bulk viscosity is fixed. To quantify the uncertainties in molecular gas flow, two scenarios are considered:

- Test 1: we choose $Z = 2.667$, $f_{tr} = 2.365$, and $f_{rot} = 1.435$, but vary A_{ij}. Specifically, A_{rt} and A_{tr} are selected uniformly in the region $[-5/6Z, 0]$ and $[-1/3Z, 0]$, respectively, while A_{tt} and A_{rr} are determined according to Eq. (7.15).
- Test 2: we choose $f_u = 1.993$, $Z = 2.667$, $A_{tr} = -5/6Z$, and $A_{rt} = -1/3Z$, but vary f_{tr} and f_{rot}. Specifically, we choose $f_{tr} = 1.5, 2.0,$ and 2.5, while A_{tt} and A_{rr} are determined according to Eq. (7.15). The small value of f_{tr} is possible, especially in the polar gas, e.g., $f_{tr} = 1.78$ for H_2O and $f_{tr} = 0.41$ for CH_3OH [13].

7.6.1 *Normal Shock Wave*

Numerical simulation reveals that Test 1 has little influence on the shock structure. However, for Test 2, significant variations of macroscopic quantities are observed in Fig. 7.7, especially in the temperature and heat flux.

7.6.2 *Flow Driven by Maxwell's Demon*

The results of Tests 1 and 2 with Kn = 0.2 are shown in Fig. 7.8. For Test 1, different relaxation rates **A** induce the maximum relative uncertainty of about 17% in the velocity and translational heat flux. Meanwhile, the uncertainty occurs in the channel center, while the velocity slip and heat flux on the walls rarely change.

For Test 2, both the velocity and translational heat flux vary significantly with f_{tr}. Contrast to the results of Test 1 where the velocity slip and heat flux on the walls do not change with fixed f_{tr}, in Test 2 both velocity and heat flux on the walls increase with f_{tr}. Thus it can be concluded that the translational Eucken factor f_{tr} plays a dominant role.

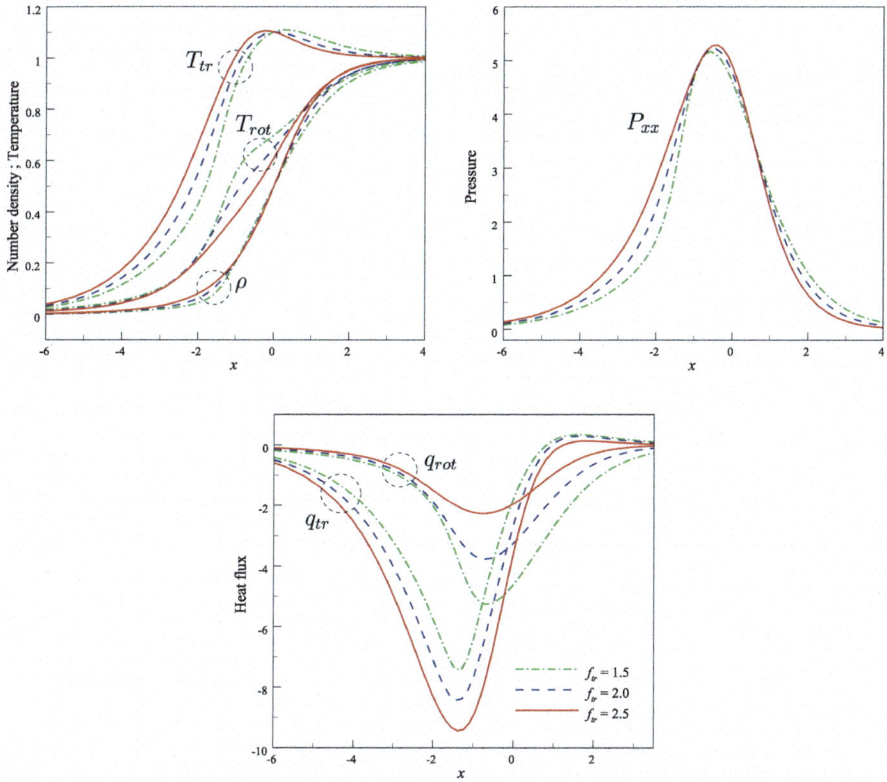

Fig. 7.7 Influence of the Eucken factors in the Ma = 4 normal shock wave [23]. All the cases have the same f_u. P_{xx} is the normal stress

7.6.3 Thermal Transpiration in Cavity

The configuration is the same as that in Sect. 4.5.3, but here the diatomic gas is used. The flow velocity along several horizontal lines parallel to the wall, from wall surface to the central are plotted in Fig. 7.9. The results show that the Wu model gives good agreement with DSMC.

Similar to 1D creep flow driven by Maxwell's demon, the flow field is determined by f_{tr} rather than f_u, where both the normal pressure and the velocity magnitude increase with increase of f_{tr}. Therefore, the mass flow rate has the same trend. For the situations with small value of f_{tr} which may occur in some polar gases, the flow pattern and flow rate could be very different from those for non-polar molecular and monatomic gases. When f_{tr} and f_u are fixed, the variation of $\boldsymbol{A}$ has little impact on the flow fields.

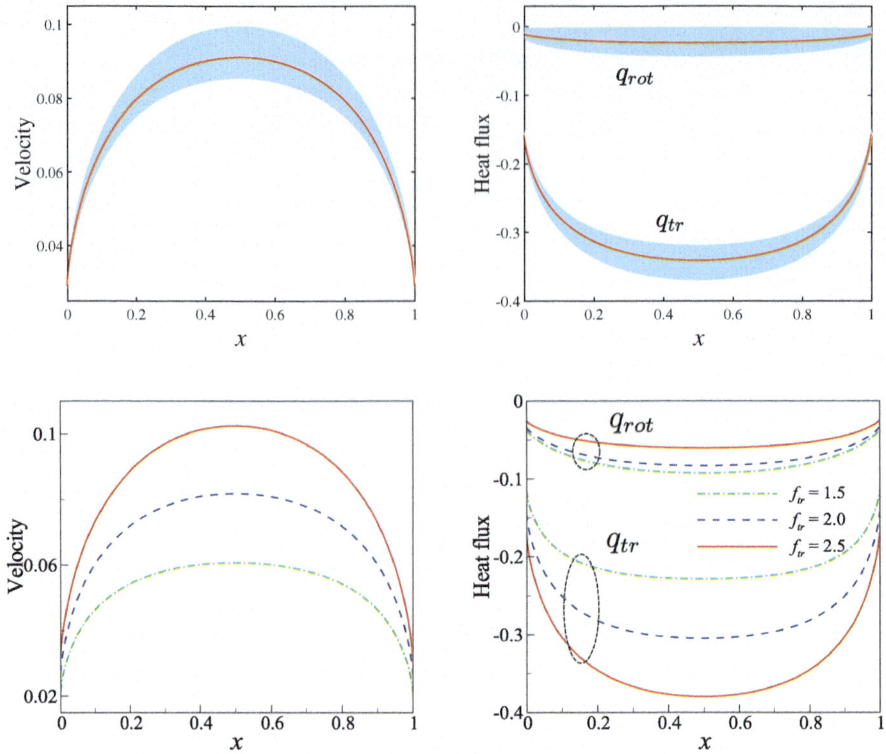

Fig. 7.8 Influence of the thermal relaxation rates in the creep flow driven by Maxwell's demon [23, 33]. (First row) Solid lines: results with **A** extracted from DSMC. Shade region: results for the parameters in Test 1. (Second row) Results from Test 2. The Knudsen number is 0.2

7.7 Conclusions and Discussions

The origins of bulk viscosity and translational/rotational thermal conductivities of molecular gas are introduced, and the deficiency of DSMC with the Larsen-Borgnakke phenomenological model in recovering the thermal conductivities is analyzed. We note that, however, if the quantum kinetics are considered, and the collision kernel B, transfer probability P_{ij}, and the degeneracy are all properly modeled in Eq. (2.57), then the DSMC may be able to describe the rarefied dynamics of molecular gas [38]. However, this is a very difficult task, since even for the simple N_2 molecule, there are more than 70 rotational energy levels. Thus, a coarse-grained kinetic modeling has important practical meaning.

The Wu model is proposed and validated by the DSMC when the same values of energy and heat flux relaxation rates are chosen. Since the translation motion is described by the BCO for monatomic gas, the Wu model has much better performance than the Rykov and ESBGK models where the collision frequency is independent

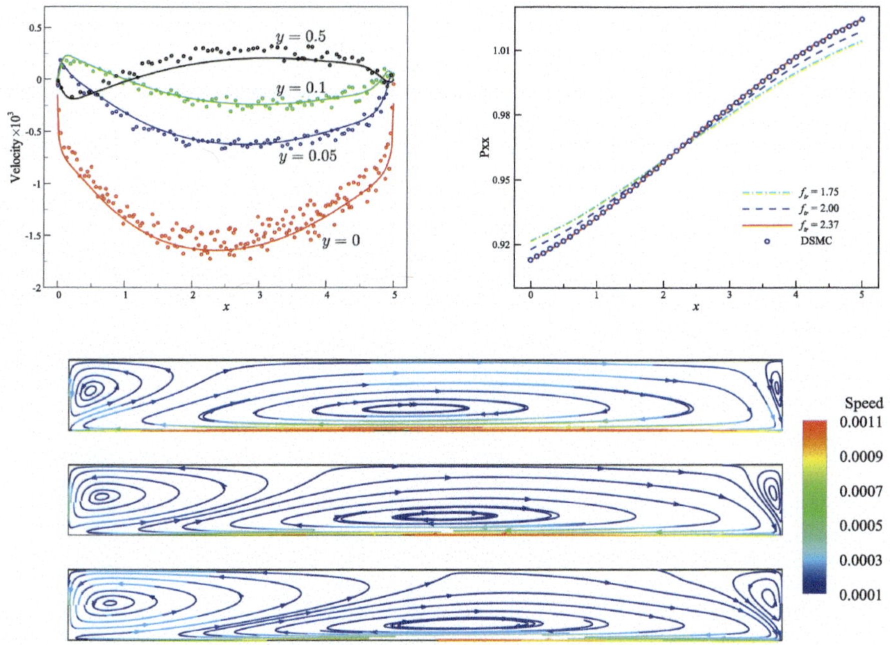

Fig. 7.9 Thermal transpiration of nitrogen gas in a cavity [23]. (Left) Comparison in the velocity between the DSMC (dots) and Wu model (lines), when the DSMC and Wu model use the same relaxation parameters. (Right) Influence of Eucken factors in the normal pressure along $y = 0.5$. (Bottom) Flow fields normalized by v_m in the lower half part of the cavity with $f_{tr} = 2.5, 2.0, 1.75$ from top to bottom. Calculated from the Wu model

of the molecular velocity. Also, in terms of the modeling capability, the Wu model is better than the Larsen-Borgnakke model, since it has the flexibility to adjust the bulk viscosity and translational/rotational thermal conductivities.

Since in experiment normally only the total thermal conductivity is measured, the Wu model has uncertainty in the translational and rotational thermal conductivities; and this uncertainty has big impact on the rarefied gas flows. To eliminate this uncertainty, we may use the Green-Kubo formulas and MD simulation in Sect. 1.5, or use the Rayleigh-Brillouin scattering experiment in Chap. 15.

One may extend the Wu model to molecular gas with vibrational modes excited [39]. Since the energy exchange between the vibrational and translational motions is much more slowly than the rotational-translational exchange, the vibrational collision number is large and it is safe to ignore the cross-relaxation rates A_{vt} and A_{tv}, where the subscript v stands for vibrational mode. Representative vibrational energy levels may be selected and each is assigned a VDF so that the population inversion is built into the kinetic model and the radiation is modeled.

References

1. Chapman S, Cowling TG (1970) The mathematical theory of non-uniform gases. Cambridge University Press
2. Jaeger F, Matar OK, Muller EA (2018) Bulk viscosity of molecular fluids. J Chem Phys 148:174504
3. Wang Y, Ubachs W, van de Water W (2019) Bulk viscosity of CO_2 from Rayleigh-Brillouin light scattering spectroscopy at 532 nm. J Chem Phys 150:154502
4. Bruno D, Frezzotti A, Ghiroldi GP (2015) Oxygen transport properties estimation by classical trajectory-direct simulation Monte Carlo. Phys Fluids 27:057101
5. Meador WE, Miner GA, Townsend LW (1996) Bulk viscosity as a relaxation parameter: fact or fiction? Phys Fluids 8:258
6. Meijer AS, de Wijn AS, Peters MFE, Dam NJ, van de Water W (2010) Coherent Rayleigh-Brillouin scattering measurements of bulk viscosity of polar and nonpolar gases, and kinetic theory. J Chem Phys 133:164315
7. Lambert L (1977) Vibrational and rotational relaxation in gases. Clarendon
8. Mason EA (1963) Molecular relaxation times from thermal transpiration measurements. J Chem Phys 39:522–526
9. Gupta AD, Storvick TS (1970) Analysis of the heat conductivity data for polar and nonpolar gases using thermal transpiration measurements. J Chem Phys 52(2):742–749
10. Porodnov BT, Kulev AN, Tuchvetov FT (1978) Thermal transpiration in a circular capillary with a small temperature difference. J Fluid Mech 88:609–622
11. Loyalka SK, Storvick TS (1979) Kinetic theory of thermal transpiration and mechanocaloric effect. III. Flow of a polyatomic gas between parallel plates. J Chem Phys 71:339–350
12. Eucken A (1913) Über das Wärmeleitvermögen, die spezifische Wärme und die innere Reibung der Gase. Phys Z 14:324
13. Mason EA, Monchick L (1962) Heat conductivity of polyatomic and polar gases. J Chem Phys 36:1622
14. McCormack FJ (1968) Kinetic equations for polyatomic gases: the 17-moment approximation. Phys Fluids 11:2533
15. Bird GA (1994) Molecular gas dynamics and the direct simulation of gas flows. Oxford Science Publications, Oxford University Press Inc, New York
16. Wagner W (1992) A convergence proof for Bird's direct simulation Monte Carlo method for the Boltzmann equation. J Stat Phys 66:1011–1044
17. Borgnakke C, Larsen P (1975) Statistical collision model for Monte Carlo simulation of polyatomic gas mixture. J Comput Phys 18(4):405–420
18. Boyd ID (1990) Rotational-translational energy transfer in rarefied nonequilibrium flows. Phys Fluids A 2:447
19. Haas BL, Hash DB, Bird GA, Lumpkin FE III, Hassan HA (1994) Rates of thermal relaxation in direct simulation Monte Carlo methods. Phys Fluids 6:2191
20. Gimelshein NE, Gimelshein SF, Lavin DA (2002) Vibrational relaxation rates in the direct simulation Monte Carlo method. Phys Fluids 14:4452
21. Gallis MA, Rader DJ, Torczynski JR (2004) A generalized approximation for the thermophoretic force on a free-molecular particle. Aerosol Sci Technol 38:692–706
22. Wu L, Li Q, Liu H, Ubachs W (2020) Extraction of the translational Eucken factor from light scattering by molecular gas. J Fluid Mech 901:A23
23. Li Q, Zeng JN, Su W, Wu L (2021) Uncertainty quantification in rarefied dynamics of molecular gas: rate effect of thermal relaxation. J Fluid Mech 917:338
24. Hanson FB, Morse TF (1967) Kinetic models for a gas with internal structure. Phys Fluids 10:345
25. Gross EP, Jackson EA (1959) Kinetic models and the linearized Boltzmann equation. Phys Fluids 2(4):432–441
26. Boley CD, Desai RC, Tenti G (1972) Kinetic models and Brillouin scattering in a molecular gas. Can J Phys 50:2158–2172

27. Tenti G, Boley C, Desai R (1974) On the kinetic model description of Rayleigh-Brillouin scattering from molecular gases. Can J Phys 52:285
28. Pan X, Shneider MN, Miles RB (2004) Coherent Rayleigh-Brillouin scattering in molecular gases. Phys Rev A 69:033814
29. Rykov V (1975) A model kinetic equation for a gas with rotational degrees of freedom. Fluid Dyn 10:959–966
30. Wu L, White C, Scanlon TJ, Reese JM, Zhang YH (2015) A kinetic model of the Boltzmann equation for non-vibrating polyatomic gases. J Fluid Mech 763:24–50
31. Holway LH (1966) New statistical models for kinetic theory: methods of construction. Phys Fluids 9:1658–1673
32. Andries P, Le Tallec P, Perlat J, Perthame B (2000) The Gaussian-BGK model of Boltzmann equation with small Prandtl number. Eur J Mech B Fluids 19:813–830
33. Zeng JN, Li Q, Wu L (2022) Kinetic modeling of rarefied molecular gas dynamics. Acta Aerodynamica Sinica 40:1–30 (in Chinese)
34. Wang P, Su W, Wu L (2020) Thermal transpiration in molecular gas. Phys Fluids 32:082005
35. Su W, Zhang YH, Wu L (2021) Multiscale simulation of molecular gas flows by the general synthetic iterative scheme. Comput Methods Appl Mech Engrg 373:113548
36. Frezzotti A, Ytrehus T (2006) Kinetic theory study of steady condensation of a polyatomic gas. Phys Fluids 18:027101
37. Kosuge S, Aoki K (2018) Shock-wave structure for a polyatomic gas with large bulk viscosity. Phys Rev Fluids 3:023401
38. Bird GA (2011) The Q-K model for gas-phase chemical reaction rates. Phys Fluids 23(10):106101
39. Li Q, Wu L (2022) A kinetic model for rarefied flows of molecular gas with vibrational modes arXiv:2201.06855

Chapter 8
General Synthetic Iterative Scheme

The conventional iterative scheme used in previous chapters works well in finding the steady-state solution of the Boltzmann equation when the Knudsen number is large. However, it becomes inefficient and inaccurate when the Knudsen number is small. The general synthetic iterative scheme is proposed to fix these problems, with the properties of fast convergence, asymptotic preserving, and universality.

8.1 Problems of CIS

We take the following linearized Shakhov model to analyze the mathematical properties of CIS in the search of steady-state solutions:

$$\boldsymbol{v} \cdot \nabla h^{k+1} = \mathcal{L}_s^{+,k} - \delta_{rp} h^{k+1} + \text{source term}, \tag{8.1}$$

where the gain part of the linearized collision operator is

$$\mathcal{L}_s^{+} = \delta_{rp} \left[\varrho + 2\boldsymbol{u} \cdot \boldsymbol{v} + T \left(v^2 - \frac{3}{2} \right) + \frac{4(1-\text{Pr})}{5} \boldsymbol{q} \cdot \boldsymbol{v} \left(v^2 - \frac{5}{2} \right) \right] f_{eq}, \tag{8.2}$$

and the source term is due to, e.g., the presence of pressure and temperature gradients.

Figure 8.1 shows the iteration numbers needed to get the steady-state solution in the Poiseuille flow between two parallel plates. When the Knudsen number Kn is large, stationary solution can be found in about 10 iterations, which means that CIS is very efficient. However, when Kn is small, about one million iterations are needed, and yet the solution is contaminated by large numerical dissipation when the spatial cell size is not small enough.

L. Wu, *Rarefied Gas Dynamics*,
https://doi.org/10.1007/978-981-19-2872-7_8

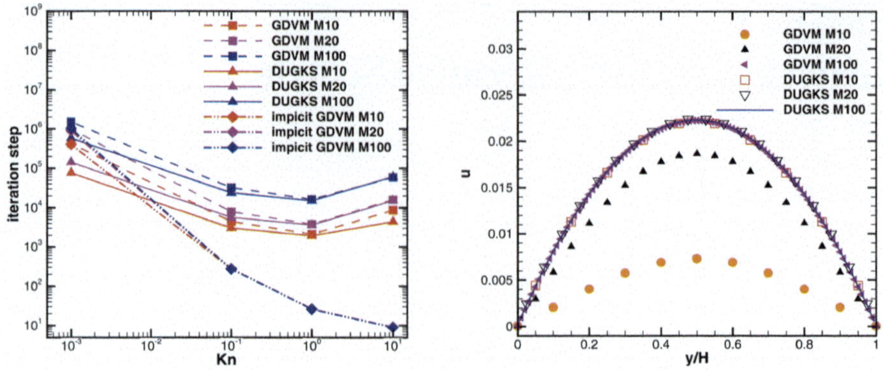

Fig. 8.1 (Left) The iteration number needed to find the steady-state solution of force-driven Poiseuille flow, for both the CIS and discrete UGKS [1]. (Right) Velocity profiles obtained at different numbers of spatial cells (M) when Kn = 0.001, which demonstrates that the CIS is highly dissipative when the spatial resolution is not adequate [2]

8.1.1 *Slow Convergence*

We adopt the Fourier stability analysis to rigorously investigate the efficiency of CIS. We define the error functions between VDFs at two consecutive iterations as

$$Y^{k+1}(\boldsymbol{x},\boldsymbol{v}) = h^{k+1}(\boldsymbol{x},\boldsymbol{v}) - h^{k}(\boldsymbol{x},\boldsymbol{v}), \tag{8.3}$$

and the corresponding error functions for perturbed macroscopic quantities $M = [\varrho, \boldsymbol{u}, T, \boldsymbol{q}]$ as

$$\Phi^{k+1}(\boldsymbol{x}) = M^{k+1}(\boldsymbol{x}) - M^{k}(\boldsymbol{x}) = \int Y^{k+1}(\boldsymbol{x},\boldsymbol{v})\phi(\boldsymbol{v})d\boldsymbol{v}, \tag{8.4}$$

where

$$\phi(\boldsymbol{v}) = \left[1, v_1, v_2, v_3, \frac{2}{3}v^2 - 1, v_1\left(v^2 - \frac{5}{2}\right), v_2\left(v^2 - \frac{5}{2}\right), v_3\left(v^2 - \frac{5}{2}\right)\right]. \tag{8.5}$$

To determine how fast the error decays, we seek the eigenfunctions $\bar{Y}(\boldsymbol{v})$ and $\alpha = [\alpha_\varrho, \boldsymbol{\alpha}_u, \alpha_T, \boldsymbol{\alpha}_q]$ of the following forms:

$$\begin{aligned} Y^{k+1}(\boldsymbol{x},\boldsymbol{v}) &= e^{k}\bar{Y}(\boldsymbol{v})\exp(i\boldsymbol{\theta}\cdot\boldsymbol{x}), \\ \Phi^{k+1}(\boldsymbol{x}) &= e^{k+1}\alpha\exp(i\boldsymbol{\theta}\cdot\boldsymbol{x}), \end{aligned} \tag{8.6}$$

where $\boldsymbol{\theta} = (\theta_1, \theta_2, \theta_3)$ is the wavevector of perturbance. Note that the two exponents in the right-hand side are different, due to the fact that in CIS we first need macroscopic quantities to start the iteration. The iteration is unstable when the error decay rate e is larger than unity, and slow (fast) convergence occurs when the error decay rate $|e|$ approaches one (zero).

From Eqs. (8.4) and (8.6) we have

$$e\alpha = \int \bar{Y}(\boldsymbol{v})\phi(\boldsymbol{v})d\boldsymbol{v}, \tag{8.7}$$

and from Eqs. (8.1), (8.3), and (8.6), we have

$$\bar{Y}(\boldsymbol{v}) = \frac{\alpha_\varrho + 2\boldsymbol{\alpha}_u \cdot \boldsymbol{v} + \alpha_T\left(v^2 - \frac{3}{2}\right) + \frac{4(1-\text{Pr})}{5}\boldsymbol{\alpha}_q \cdot \boldsymbol{v}\left(v^2 - \frac{5}{2}\right)}{1 + i\delta_{rp}^{-1}\boldsymbol{\theta} \cdot \boldsymbol{v}} f_{\text{eq}}. \tag{8.8}$$

On multiplying Eq. (8.8) with $\phi(\boldsymbol{v})$ and integrating the resultant equations with respect to $\boldsymbol{v}$, we obtain eight linear equations for eight unknown elements in α with the help of Eq. (8.7):

$$C_8\alpha^\top = e\alpha^\top, \tag{8.9}$$

where the superscript $\top$ is the transpose operator. The error decay rate is the maximum eigenvalue in magnitude of matrix C_8; the result[1] is shown in Fig. 8.2. Specifically, when Kn $\to$ 0, the error decay rate is calculated analytically [3]:

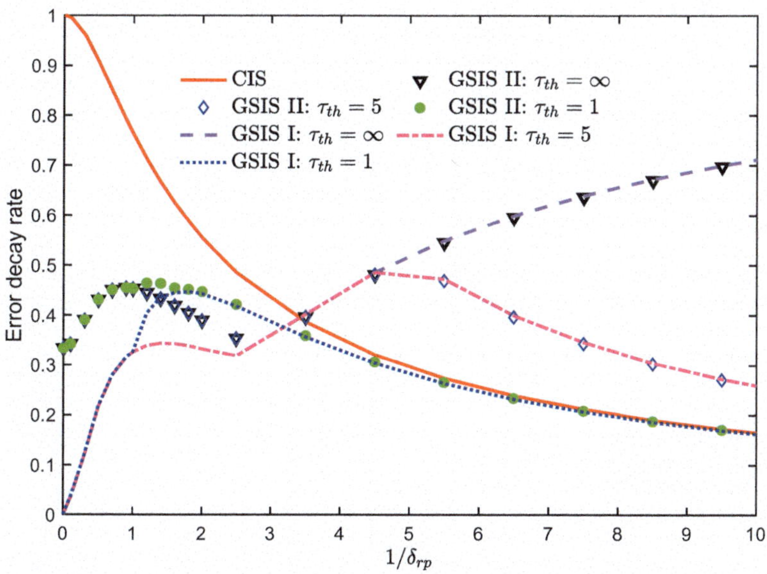

Fig. 8.2 The error decay rate calculated from the linearized Shakhov model by the Fourier stability analysis [4]. In GSIS, different threshold rarefaction parameters δ_{rp}^{th} are considered, see Eq. (8.33)

[1] If without specification, the perturbation wavevector is chosen as $|\boldsymbol{\theta}| = 1$; Behaviors of the error decay rate are similar for other values of $|\boldsymbol{\theta}|$.

$$e_{CIS} = 1 - \frac{1}{2\delta_{rp}^2}. \tag{8.10}$$

The results show that CIS is efficient in the free-molecular regime, as $e \to 0$ so that the error decays quickly. On the other hand, the CIS is extremely slow in the near-continuum flow regime, as $e \to 1$ when $\mathrm{Kn} \to 0$.

8.1.2 *False Convergence*

In addition to the slow convergence when $\mathrm{Kn} \to 0$, CIS suffers the problem of "false convergence". This can be analyzed following the work of Adam and Larsen for radiation transfer equation [5]. We rewrite the iterative scheme (8.1) as

$$h^k = \mathcal{I}h^{k-1}, \quad \text{with } \mathcal{I} = \frac{\mathcal{L}_s^+}{\delta_{rp} + \boldsymbol{v} \cdot \nabla}. \tag{8.11}$$

The exact solution h (the corresponding macroscopic quantities are denoted by Φ_M) satisfies $h = \mathcal{I}h$. Therefore, at the k-th step, the difference from the exact solution is $h - h^k = \mathcal{I}(h - h^{k-1}) = \mathcal{I}(h - h^k) + \mathcal{I}(h^k - h^{k-1})$, so

$$h - h^k = \frac{\mathcal{I}}{1 - \mathcal{I}}(h^k - h^{k-1}). \tag{8.12}$$

This yields an estimation:

$$|h - h^k| \le \left|\frac{\mathcal{I}}{1 - \mathcal{I}}\right| |h^k - h^{k-1}| \approx \frac{e}{1 - e}|h^k - h^{k-1}|, \tag{8.13}$$

which implies that, if the iteration is terminated with the convergence criterion $|\Phi_M^{k+1} - \Phi_M^k| < \epsilon$, the relative difference from the true steady-state solution Φ_M can be estimated as

$$|\Phi_M^{k+1} - \Phi_M| < \frac{e_{CIS}}{1 - e_{CIS}}\epsilon. \tag{8.14}$$

With Eq. (8.10), it is found that the error in the final step of iteration is much larger than the preassigned value ϵ:

$$|\Phi_M^{k+1} - \Phi_M| \to \frac{\epsilon}{\mathrm{Kn}^2}, \quad \text{when } \mathrm{Kn} \to 0. \tag{8.15}$$

Therefore, false convergence occurs if ϵ is not set small enough.

8.2 General Synthetic Iterative Scheme

To eliminate the deficiencies in CIS, the synthetic iterative scheme, which is initially designed for radiation transport problem [5], is extended to solve rarefied gas flows. Success examples include the Poiseuille flow, Couette flow, thermal transpiration, and flows driven by concentration gradient [6–14], where the flow velocity is perpendicular to the computational domain.

In synthetic iterative scheme, macroscopic synthetic equation is exactly derived from the kinetic equation, which, in the limit of Kn $\to 0$, becomes the diffusion equation. When the Knudsen number is not negligible, however, the macroscopic equation has an additional source term, or high-order term, which describes the rarefaction effects. The gas kinetic equation and macroscopic synthetic equation are solved on the same spatial grids in the entire domain: at each iteration, the kinetic equation provides high-order terms to the macroscopic equation, which, when solved to the steady state, is used to correct the VDF and macroscopic quantities. Since the diffusion equation is efficient in exchange the flow information, faster convergence is achieved when Kn $\to 0$.

Synthetic equations should be carefully designed when the general rarefied gas flow is concerned. For generality we consider the LBE. We multiply Eq. (5.3) by the five fundamental collision invariants and integrate the resultant equations with respect to the molecular velocity $\boldsymbol{v}$, yielding (suppose the external acceleration is absent):

$$\begin{aligned} \frac{\partial \varrho}{\partial t}+\frac{\partial u_i}{\partial x_i}&=0,\\ 2\frac{\partial u_i}{\partial t}+\frac{\partial \varrho}{\partial x_i}+\frac{\partial T}{\partial x_i}+\frac{\partial \sigma_{ij}}{\partial x_j}&=0,\\ \frac{3}{2}\frac{\partial T}{\partial t}+\frac{\partial q_j}{\partial x_j}+\frac{\partial u_j}{\partial x_j}&=0. \end{aligned} \tag{8.16}$$

To develop a general fast-converging scheme, it is beneficial to construct macroscopic diffusion-type equations that contain Newton's law for stress and Fourier's law for heat conduction explicitly to recover the macroscopic transport mechanism; we express the shear stress and heat flux as follows:

$$\sigma_{ij} = -2\delta_{rp}^{-1}\frac{\partial u_{<i}}{\partial x_{j>}}+\text{HoT}_{\sigma_{ij}}, \tag{8.17}$$

$$q_i = -\frac{5}{4\text{Pr}}\delta_{rp}^{-1}\frac{\partial T}{\partial x_i}+\text{HoT}_{q_i}, \tag{8.18}$$

where $\text{HoT}_{\sigma_{ij}}$ and HoT_{q_i} are the high-order terms containing contributions from all the orders $O(\text{Kn}^\alpha)$ with $\alpha=2,3,\cdots,\infty$. These high-order terms cannot be obtained from Burnett, super-Burnett equations, nor from the Grad moment equations, but have to be determined from the kinetic equation that are valid in the whole range of gas rarefaction.

8.2.1 Scheme-I GSIS

To obtain Eq. (8.17), we multiply Eq. (5.3) by $2v_{<i}v_{j>}$ and integrate the resultant equation with respect to $\boldsymbol{v}$:

$$\frac{\partial \sigma_{ij}}{\partial t}+2\int v_{<i}v_{j>}\boldsymbol{v}\cdot\frac{\partial h}{\partial \boldsymbol{x}}d\boldsymbol{v}=2\int \mathcal{L}v_iv_j\mathrm{d}\boldsymbol{v}, \tag{8.19}$$

which is rewritten as

$$\underbrace{\frac{\partial \sigma_{ij}}{\partial t}+2\int v_{<i}v_{j>}\boldsymbol{v}\cdot\frac{\partial h}{\partial \boldsymbol{x}}d\boldsymbol{v}-2\frac{\partial u_{<i}}{\partial x_{j>}}}_{\mathrm{HoT}_{\sigma_{ij}}} \\ +\underbrace{2\frac{\partial u_{<i}}{\partial x_{j>}}=-\delta_{rp}\sigma_{ij}}_{\text{Newton's law}}+2\int(\mathcal{L}-\mathcal{L}_s)v_iv_j\mathrm{d}\boldsymbol{v}. \tag{8.20}$$

Note that the purpose of introducing $\mathcal{L}_s$ is only to produce the term $\delta_{rp}\sigma_{ij}$, so that the Newton's law appears in the synthetic equation naturally; this turns out to be crucial in boosting convergence. Also note that, for the linearized BCO, comparing to $\delta_{rp}\sigma_{ij}$, the term $2\int(\mathcal{L}-\mathcal{L}_s)v_iv_jd\boldsymbol{v}$ is negligible. For instances, this term is zero for Maxwell gas, while for the HS gas this term is only about 2% of $\delta_{rp}\sigma_{ij}$, see Eq. (3.32). It will be shown later that the high-order terms are evaluated at the k-th iteration step, while Newton's law for stress is calculated at the $(k+1)$-th step.

Similarly, to obtain Eq. (8.18), we multiply Eq. (5.3) by $v_i(v^2-5/2)$ and integrate the resultant equation with respect to $\boldsymbol{v}$:

$$\underbrace{\frac{\partial q_i}{\partial t}+\int\left(v^2-\frac{5}{2}\right)v_i\boldsymbol{v}\cdot\frac{\partial h}{\partial \boldsymbol{x}}d\boldsymbol{v}-\frac{3C_q}{2}\frac{\partial T}{\partial x_i}}_{\mathrm{HoT}_{q_i}} \\ +\underbrace{\frac{3C_q}{2}\frac{\partial T}{\partial x_i}=-\delta_{rp}\mathrm{Pr}q_i}_{\text{Fourier's law}}+\int(\mathcal{L}-\mathcal{L}_s)v_iv^2\mathrm{d}\boldsymbol{v}. \tag{8.21}$$

For the linearized BCO, the term $\int(\mathcal{L}-\mathcal{L}_s)v_iv^2d\boldsymbol{v}$ is negligibly small when compared to $\delta_{rp}q_i$. If we choose $C_q=5/6$, Fourier's heat conduction law appears naturally.

The GSIS is designed to find steady-state solutions of the LBE through the following steps (5.3):

- Step 1. When the VDF h^k and the corresponding macroscopic quantities are known, calculate $2\int(\mathcal{L}-\mathcal{L}_s)v_iv_j\mathrm{d}\boldsymbol{v}$ and $\int(\mathcal{L}-\mathcal{L}_s)v_iv^2\mathrm{d}\boldsymbol{v}$. Also calculate the VDF $h^{k+1/2}$ at the intermediate iterative step as

$$\nu_{eq}(\boldsymbol{v})h^{k+1/2} + \boldsymbol{v} \cdot \frac{\partial h^{k+1/2}}{\partial \boldsymbol{x}} = \mathcal{L}^+(h^k, f_{eq}). \tag{8.22}$$

- Step 2. From $h^{k+1/2}$, calculate the macroscopic quantities at the intermediate step, and the high-order terms.
- Step 3. Calculate the macroscopic quantities, in the bulk region, at the $(k+1)$-th step by solving synthetic equations (8.16), (8.20) and (8.21) to the steady state, where the boundary conditions in the vicinity of walls are obtained from Step 2.
- Step 4. The VDF is corrected in the following manner so that its corresponding macroscopic quantities is the same as those obtained in Step 3:

$$\begin{aligned} h^{k+1} =& h^{k+1/2} + \left[\lambda_\rho + 2\lambda_{\boldsymbol{u}} \cdot \boldsymbol{v} + \lambda_T\left(v^2 - \frac{3}{2}\right)\right] f_{eq} \\ &+ \left[\lambda_{\sigma_{ij}}\left(v_i v_j - \frac{v^2}{3}\delta_{ij}\right) + \frac{4}{5}\lambda_q \cdot \boldsymbol{v}\left(v^2 - \frac{5}{2}\right)\right] f_{eq}, \end{aligned} \tag{8.23}$$

  where $\lambda_{\boldsymbol{u}} = \boldsymbol{u}^{k+1} - \boldsymbol{u}^{k+1/2}, \lambda_{\boldsymbol{q}} = \boldsymbol{q}^{k+1} - \boldsymbol{q}^{k+1/2}, \lambda_\rho = \rho^{k+1} - \rho^{k+1/2}, \lambda_T = T^{k+1} - T^{k+1/2}$, and $\lambda_{\sigma_{ij}} = \sigma_{ij}^{k+1} - \sigma_{ij}^{k+1/2}$.
- Step 5. The above steps are repeated until convergence.

8.2.2 *Scheme-II GSIS*

High-order terms in Eqs. (8.20) and (8.21) are constructed by considering the evolution equation of stress and heat flux, which involves the calculation of complicated collision operator when the Boltzmann equation is considered. This is not a problem for deterministic numerical methods, but will be not easy for DSMC. As we aim to develop a generalized scheme, not only for deterministic methods, but also for stochastic methods such as DSMC, we propose the following easy-to-use constitutive relations [4]:

$$\begin{aligned} \mathrm{HoT}_{\sigma_{ij}} &= \sigma_{ij}^{k+1/2} + 2\delta_{rp}^{-1}\frac{\partial u_{<i}^{k+1/2}}{\partial x_{j>}}, \\ \mathrm{HoT}_{q_i} &= q_i^{k+1/2} + \frac{5}{4\mathrm{Pr}}\delta_{rp}^{-1}\frac{\partial T^{k+1/2}}{\partial x_i}. \end{aligned} \tag{8.24}$$

8.3 Properties of GSIS

Two important questions in the multiscale simulation of rarefied gas flows are: can we find the steady-state efficiently and can we get accurate results even at coarse spatial grid? This section is dedicated to proving both.

8.3.1 Super Convergence

The linearized Shakhov model is used to analyze the error decay rate of GSIS. That of the Boltzmann equation can only be shown in numerical simulation, due to the complicated structure of the BCO.

The error functions in Eqs. (8.3), (8.4), and (8.6) are redefined as

$$\begin{aligned} Y^{k+1/2}(\boldsymbol{x}, \boldsymbol{v}) = h^{k+1/2}(\boldsymbol{x}, \boldsymbol{v}) - h^k(\boldsymbol{x}, \boldsymbol{v}) = e^k \bar{Y}(\boldsymbol{v}) \exp(i\boldsymbol{\theta} \cdot \boldsymbol{x}), \\ \Phi^{k+1}(\boldsymbol{x}) = M^{k+1}(\boldsymbol{x}) - M^k(\boldsymbol{x}) = e^{k+1}\alpha \exp(i\boldsymbol{\theta} \cdot \boldsymbol{x}), \end{aligned} \tag{8.25}$$

where $\bar{Y}(\boldsymbol{v})$ is still given by Eq. (8.8). Note that now Φ are calculated from the synthetic equations, rather than from the VDF. That is, macroscopic quantities at the $(k+1)$-th iteration step are obtained by solving the following synthetic equations:

$$\frac{\partial u_i^{k+1}}{\partial x_i} = 0, \quad \frac{\partial \varrho^{k+1}}{\partial x_i} + \frac{\partial T^{k+1}}{\partial x_i} + \frac{\partial \sigma_{ij}^{k+1}}{\partial x_j} = 0, \quad \frac{\partial q_i^{k+1}}{\partial x_i} = 0. \tag{8.26}$$

With Eqs. (8.8), (8.20), (8.21), (8.26), and (8.25), we obtain the following eight linear equations:

$$\begin{aligned} e(i\theta_1\alpha_{u_1} + i\theta_2\alpha_{u_2} + i\theta_3\alpha_{u_3}) = 0, \\ e[i\theta_j(\alpha_\varrho + \alpha_T) + |\boldsymbol{\theta}|^2\delta_{rp}^{-1}\alpha_{u_j}] = S_{j+1}, \\ e(i\theta_1\alpha_{q_1} + i\theta_2\alpha_{q_2} + i\theta_3\alpha_{q_3}) = 0, \\ e\left(\frac{5i}{4\mathrm{Pr}}\theta_j\delta_{rp}^{-1}\alpha_T + \alpha_{q_j}\right) = S_{j+5}, \end{aligned} \tag{8.27}$$

where $j = 1, 2, 3$, and the source terms in the scheme I, due to the HoTs in Eqs. (8.20) and (8.21), are also linear functions of α_M:

$$\begin{aligned} S_{j+1} = \delta_{rp}^{-1} \int \left(|\boldsymbol{\theta}|^2 v_j - 2\Theta\theta_k v_{<j} v_{k>}\right) \bar{Y}(\boldsymbol{v}) \mathrm{d}^3\boldsymbol{v}, \\ S_{j+5} = \frac{i}{\delta_{rp}\mathrm{Pr}} \int \left[\frac{5}{4\mathrm{Pr}}\theta_j\left(\frac{2}{3}v^2 - 1\right) - \Theta v_j\left(v^2 - \frac{5}{2}\right)\right] \bar{Y}(\boldsymbol{v}) \mathrm{d}^3\boldsymbol{v}, \end{aligned} \tag{8.28}$$

where $\Theta = \theta_1 v_1 + \theta_2 v_2 + \theta_3 v_3$, $k = 1, 2, 3$ is the dummy index, while that in the scheme II, with Eq. (8.24), are

$$\begin{aligned} S_{j+1} = \int \left(\delta_{rp}^{-1}|\boldsymbol{\theta}|^2 v_j - 2i\theta_k v_{<j} v_{k>}\right) \bar{Y}(\boldsymbol{v}) \mathrm{d}^3\boldsymbol{v}, \\ S_{j+5} = \int \left[\frac{5i}{4\mathrm{Pr}}\theta_j\delta_{rp}^{-1}\left(\frac{2}{3}v^2 - 1\right) + v_j\left(v^2 - \frac{5}{2}\right)\right] \bar{Y}(\boldsymbol{v}) \mathrm{d}^3\boldsymbol{v}. \end{aligned} \tag{8.29}$$

For the scheme I, the error decay rate is obtained by solving Eqs. (8.27) and (8.28), which are rewritten in the matrix form as

$$L_8 e\alpha^\top = R_8\alpha^\top, \tag{8.30}$$

where L_8 and R_8 are two 8×8 matrices. By introducing $G_1 = L_8^{-1}R_8$ and numerically computing its eigenvalues, we obtain the error decay rate e in Fig. 8.2. When $\text{Kn} \to 0$, $e \propto \text{Kn}^2$, so GSIS can boost convergence in near-continuum flows. As a matter of fact, compared to the false convergence of CIS described by Eq. (8.15), GSIS possesses the properties of super convergence, since according to the following equation

$$|\Phi_M^{k+1} - \Phi_M| \to \epsilon\text{Kn}^2, \quad \text{when Kn} \to 0, \tag{8.31}$$

the convergence criterion ϵ can be set at a relative larger value.

When $\text{Kn} \to \infty$, however, $e \to 1$. To fix this problem, macroscopic quantities are updated as

$$M^{k+1}(\boldsymbol{x}) = \beta M_{\text{syn}}(\boldsymbol{x}) + (1-\beta)M^{k+1/2}(\boldsymbol{x}), \tag{8.32}$$

where the parameter β is chosen as

$$\beta = \frac{\delta_{rp}}{\max(\delta_{rp}, \delta_{rp}^{\text{th}})} \tag{8.33}$$

with δ_{rp}^{th} being the threshold rarefaction parameter. That is, $\beta = 1$ when the rarefaction parameter is larger than δ_{rp}^{th}; when $\delta_{rp} < \delta_{rp}^{\text{th}}$, β gradually decreases to zero as the Knudsen number approaches infinity. The error decay rate is obtained by computing the eigenvalues of the matrix $G = \beta L_8^{-1}R_8 + (1-\beta)C_8$.

Results in Fig. 8.2 show that the maximum error decay rate can be restrained to be less than 0.5 for all Knudsen numbers, by choosing approximate value of β. Thus, theoretically, GSIS can achieve fast convergence in the whole range of Knudsen number, because the error is reduced by three orders of magnitude in 10 iterations.

Likewise, for the scheme II, the error decay rate is obtained by solving Eqs. (8.27) and (8.29). It is seen from Fig. 8.2 that when $\text{Kn} \to 0$, the error decay rate of the scheme II is reduced from one in CIS but does not go to zero.[2] Nevertheless, like the scheme I, by choosing appropriate value of δ_{rp}^{th}, the error decay rate can also be controlled within 0.5 in the whole range of rarefaction parameter.

[2] This is because the heat flux enters the collision term in the Shakhov model, and in the Scheme-II the evolution equation (8.21) of heat flux is not considered in the macroscopic synthetic equation. If we use the linearized BGK model, we find that the error decay rate approaches zero when $\text{Kn} \to 0$.

8.3.2 *Asymptotic Preserving*

In GSIS the mesoscopic kinetic equation and macroscopic synthetic equations can be solved by different numerical methods with different order of accuracy. It is important to investigate the influence of spatial discretization of the gas kinetic solver on the accuracy of GSIS, based on the assumptions that (i) synthetic equations can be solved exactly and (ii) the spatial cell size Δx is adequate to capture the physical solution of NSF equations. This is equivalent to check whether the NSF equations can be exactly derived or not, through the Chapman-Enskog expansion [15], from the discretized gas kinetic equation

$$\boldsymbol{v} \cdot \frac{\partial h}{\partial \boldsymbol{x}} + O(\Delta x^n)\delta(h) = \mathcal{L}_s - \delta_{rp} h, \tag{8.34}$$

with the following scaling (Δx has been normalized by the characteristic flow length L):

$$\Delta x \sim \text{Kn}^{1/\alpha}, \tag{8.35}$$

where n is the order of approximation for the spatial derivative, $\delta(h)$ is the $(n+1)$-th order derivative of h, and α is the order of accuracy in the asymptotic preserving of NSF equations.

8.3.2.1 Scheme-I

By substituting the expansion $h = h_0 + \text{Kn}h_1 + \cdots$ into Eq. (8.34) and collecting terms with the order of Kn^{-1}, we have $h_0 = f_{eq}$, when the following largest scaling is chosen

$$\Delta x \sim \text{Kn}^{1/\infty} = O(1). \tag{8.36}$$

Under this scaling, at the order Kn^0, we obtain

$$h_1 = -\boldsymbol{v} \cdot \frac{\partial h_{eq}}{\partial \boldsymbol{x}} - \delta(h_{eq}). \tag{8.37}$$

Thus, according to Eqs. (8.20) and (8.21), the linear constitutive relations are recovered in synthetic equations with accuracy $O(\text{Kn}^2)$:

$$\begin{aligned} \sigma_{ij} &= -2\delta_{rp}^{-1} \frac{\partial u_{<i}}{\partial x_{j>}} + O(\text{Kn}^2), \\ q_i &= -\frac{5}{4\text{Pr}} \delta_{rp}^{-1} \frac{\partial T}{\partial x_i} + O(\text{Kn}^2). \end{aligned} \tag{8.38}$$

Thus, GSIS asymptotically preserves the NSF equations, provided that the spatial resolution $\Delta x = O(1)$ is able to capture the physical solution of NSF equations.

In practice, however, $\Delta x = O(1)$ cannot be used in regions where the physical solutions require a resolution of $O(\text{Kn})$, e.g., the Knudsen layer and shock structure. Fortunately, these kinetic layers only occupy a small fraction of the computational domain, which can be captured by implicit schemes with non-uniform spatial discretization.

8.3.2.2 Scheme-II

In order to recover the NSF constitutive relations (8.38), the VDF to the first order of Knudsen number must be exactly recovered as

$$h_1 = -\boldsymbol{v} \cdot \frac{\partial h_{eq}}{\partial \boldsymbol{x}}. \tag{8.39}$$

And this requires the following largest scaling:

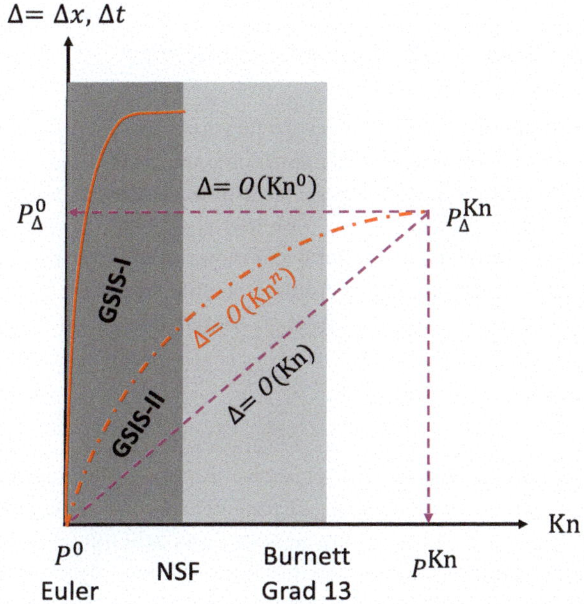

Fig. 8.3 Schematic of the asymptotic path to the limiting hydrodynamic flow regimes. The NSF equations are valid in the dark gray region. The region below the line $\Delta = O(\text{Kn})$ suggests resolved kinetic scale. The red solid line stands for the maximum spatial grid size/time step to solve the NSF equations accurately using some discretization method (certainly Δ depends on the numerical method for NSF equations and specific flow problems; Δ could be larger than $O(1)$ because the time step can be proportional to Kn^{-1} in some flows), below which GSIS-I can capture the hydrodynamic behavior. GSIS-II works at a smaller value of Δ, say, beneath the dash-dotted line $\Delta = O(\text{Kn}^n)$ with $n \geq 1/2$

$$\Delta x \sim \text{Kn}^{1/2}. \tag{8.40}$$

The asymptotic path to the limiting hydrodynamic flow regimes is summarized in Fig. 8.3. Note that both GSIS schemes can be extended to time-dependent problems [16], and the property of asymptotic preserving in the temporal domain is the same as the spatial domain.

8.4 Numerical Tests

Several numerical simulations are carried out to demonstrate the major properties of GSIS, including the 1D coherent Rayleigh-Brillouin scattering problem where the influence of gas-surface interaction is absent, the planar Fourier flow, and the two-dimensional (2D) Couette flow between two non-coaxial cylinders. More challenging numerical examples can be found in Refs. [3, 4, 17–19].

8.4.1 Coherent Rayleigh-Brillouin Scattering

This problem is ideal to show the fast-converging and asymptotic-preserving properties of the GSIS, as well as the flexibility, importance, and advantage of GSIS that the kinetic and synthetic equations are solved by different methods with different order of accuracy.

The theory of coherent RBS is elaborated in Chap. 15. Here we directly write the evolution equation for the perturbed VDF as

$$2\pi i f_s h + v_2 \frac{\partial h}{\partial x_2} = \mathcal{L}_s^{+,k} - \delta_{rp} h + 2 v_2 \cos(2\pi x_2), \tag{8.41}$$

where the normalized scattering frequency is chosen as $f_s = \sqrt{5/6}$ and the last term presents the external acceleration generated by the optical lattice. The problem is spatially periodic, without the influence of gas-surface interaction. The presence of the first term in Eq. (8.41) is due to the temporal periodicity. Correspondingly, the time derivative $\partial/\partial t$ in the synthetic equations should be replaced by $2\pi i f_s$.

We first consider the case where the kinetic and synthetic equations are solved by the second-order upwind finite difference and Fourier spectral methods, respectively. Figure 8.4a shows the convergence history of both CIS and GSIS, when the rarefaction parameter is $\delta_{rp} = 200$. It is clearly seen that, GSIS produces the converged solution after only 3 iterations, and when the convergence criterion

$$\epsilon = \left| \frac{\int |\varrho^{(k+1)} dx_2|}{\int |\varrho^{(k)}| dx_2} - 1 \right| \tag{8.42}$$

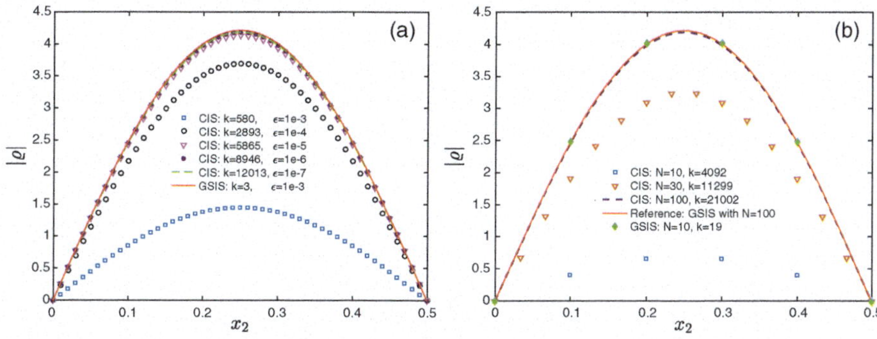

Fig. 8.4 Density profile in the coherent Rayleigh-Brillouin scattering [3], when $\delta_{rp} = 200$. **a** The convergence history of CIS. The spatial region $x_2 \in [0, 1]$ is divided into $N = 100$ uniform cells. Due to symmetry only half of the density profile is plotted. **b** Density profiles and iteration numbers, where the iteration is terminated when $\epsilon < 10^{-10}$.

is 10^{-3}. This is due to the super-converging property given in Eq. (8.31). However, due to the false convergence proven in Eq. (8.15), CIS needs huge number of iterations and the convergence criterion has to be set very small, i.e., $\epsilon = 10^{-10}$ in this case.

Figure 8.4b shows that GSIS-I asymptotically preserves the NSF limit, when only 10 uniform spatial cells are used. This is in agreement with the theoretical proof (8.36). When the same number of spatial cells are used in CIS, the density amplitude is about 5 times smaller than the true solution, which indicates a strong numerical dissipation. Only when the number of cells is increased to 100 can the CIS capture the density profile at $\delta_{rp} = 200$. Taking into account both the spatial discretization and iteration number, GSIS is about 6,000 times faster than CIS.

We further consider an extreme case of $\delta_{rp} = 10^8$. In this case, normally the Euler equation is used to describe the gas dynamics, and many kinetic schemes are proposed to asymptotically preserve the Euler limit. However, it should be emphasized that the Euler equation cannot be used here, otherwise the amplitude of perturbation will go to infinity due to the omission of viscosity. The extremely large value of δ_{rp} poses a grand challenge to the numerical method for kinetic equations, as any small value of numerical dissipation could easily contaminate the final solution, which leads to a significantly smaller amplitude of density perturbation. Even at such a small Knudsen number, if the macroscopic equation is solved exactly (by the Fourier spectral method) and $\Delta x \sim O(1)$ is able to capture the spatial variation, GSIS has infinite order of accuracy, see the circles in Fig. 8.5. However, when the second-order finite difference scheme is used to solve the synthetic equations, we need more than 2000 spatial points to capture the density perturbation; this is understandable since the numerical dissipation (or numerical viscosity) $(\Delta x)^2 = 2.5 \times 10^{-7}$ is larger than the physical viscosity (here it is reflected by the inverse rarefaction parameter). When the synthetic equations are solved by the fourth-order finite difference, we see that $N = 100$ leads to wrong solutions, while $N = 300$ yields accurate solution;

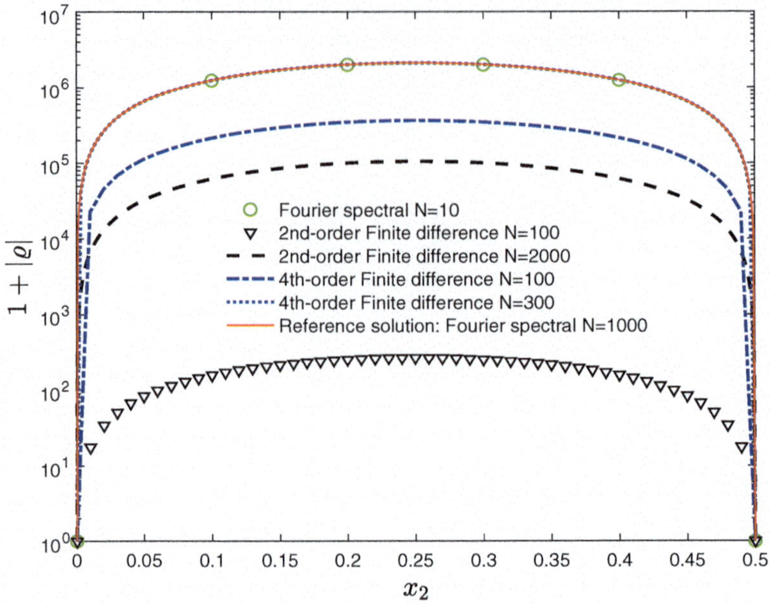

Fig. 8.5 Comparisons of the density profile in coherent Rayleigh-Brillouin scattering [3], where the kinetic equation is solved by the second-order upwind finite difference, while macroscopic equations are solved by various schemes with different number of spatial points (N). The rarefaction parameter is $\delta_{rp} = 10^8$. The density amplitude is shifted by one in order to show it in the log scale.

This is because the corresponding numerical dissipation is about $(\Delta x)^4 = 10^{-8}$ and 1.2×10^{-10}, respectively, so that the former is too dissipative while the latter is accurate.

The last numerical example in Fig. 8.5 illustrates the flexibility, importance, and advantage of GSIS that the kinetic and synthetic equations are solved by different methods with different order of accuracy. Since many sophisticated high-order methods have been developed for NSF equations, they can be directly incorporated in the GSIS framework to boost convergence and reduce the use of spatial cells and hence the computational memory.

8.4.2 Planar Fourier Flow

We consider the heat conduction in gas between two plates located at $x_2 = 0$ and 1, see Sect. 5.5.4, to see the influence of solid walls in the fast-converging properties of GSIS.

From the synthetic equations (8.16), (8.20) and (8.21), as well as the symmetry condition, we know $\boldsymbol{u} = 0, \sigma_{ij} = 0$ when $i \neq j$, and $q_1, q_3 = 0$; the heat flux q_2 is a constant, and the perturbed temperature is governed by the following ordinary differential equation:

$$\frac{\partial T}{\partial x_2} = -\frac{4\delta_{rp}}{9C_q}q_2 + \underbrace{\frac{2}{3C_q}\int v_2 v^2(\mathcal{L}-\mathcal{L}_s)\mathrm{d}\boldsymbol{v}}_{H_1^k(x_2)}$$
$$-\underbrace{\frac{2}{3C_q}\frac{\partial}{\partial x_2}\int (v_2^2 - C_q)\left(v^2-\frac{3}{2}\right)h\mathrm{d}\boldsymbol{v}}_{H_2^{k+1/2}(x_2)}, \tag{8.43}$$

whose solution at the $(k+1)$-th iteration step is given by

$$T^{k+1}(x_2) = -\frac{4\delta_{rp}q_2}{9C_q}\left(x_2 - \frac{1}{2}\right) + \int_{1/2}^{x_2} H_1^k(x_2)\mathrm{d}x_2 - H_2^{k+1/2}(x_2), \tag{8.44}$$

where the constant heat flux q_2 is

$$q_2 = \frac{9C_q}{2\delta_{rp}}\left[T^{k+1/2}(0) + H_2^{k+1/2}(0) - H_1^k(0)\right]. \tag{8.45}$$

The density can be easily obtained by $\varrho + T + \sigma_{22} = 0$, where from Eq. (8.20) the stress σ_{22} is calculated as

$$\sigma_{22} = -\frac{\frac{\partial}{\partial x_2}\int 2v_{<2}v_{2>}v_2 h\mathrm{d}\boldsymbol{v}}{\delta_{rp}} + \frac{2}{\delta_{rp}}\int (\mathcal{L}-\mathcal{L}_s)v_2^2\mathrm{d}\boldsymbol{v}. \tag{8.46}$$

The efficiency of GSIS and CIS is compared in Fig. 8.6. Starting from the initial condition $h(x_2, \boldsymbol{v}) = 0$, the perturbance from the solid surface quickly changes the density and temperature nearby in the CIS. However, the perturbance can hardly penetrates into the bulk region. Such a slow convergence is completely changed in GSIS. As can be seen from Fig. 8.6c and d, after the first iteration, although the gas density and temperature near the wall are the same as that in CIS, the bulk temperature and density in GSIS are adjusted to be nearly linear, by the synthetic equation whose dominant parts are

$$\frac{\partial T}{\partial x_2} = -\frac{4\mathrm{Pr}\delta_{rp}}{5}q_2, \quad \text{and} \quad \varrho = -T. \tag{8.47}$$

Then, due to the correction of VDF as per Eq. (8.23), at the second iterative step, the boundary condition (5.35) in GSIS is more close to the final steady state than that in CIS. Hence after the second iteration, the GSIS results are closer to the final solution both at the boundary and in the bulk, i.e., fast convergence is achieved.

Figure 8.7 shows the history of error decay of both CIS and GSIS, from the continuum to transition flow regimes. In CIS, the number of iteration steps increases drastically with δ_{rp}, while in GSIS converged solutions are found within 20 iterations for $\delta_{rp} = 1$, 10, 30, and 50. Theoretically, according to Eq. (8.31), the number of iterations should be very small when δ_{rp} is large. This discrepancy is in fact caused by the presence of solid wall, where the local Knudsen number inside the Knudsen layer

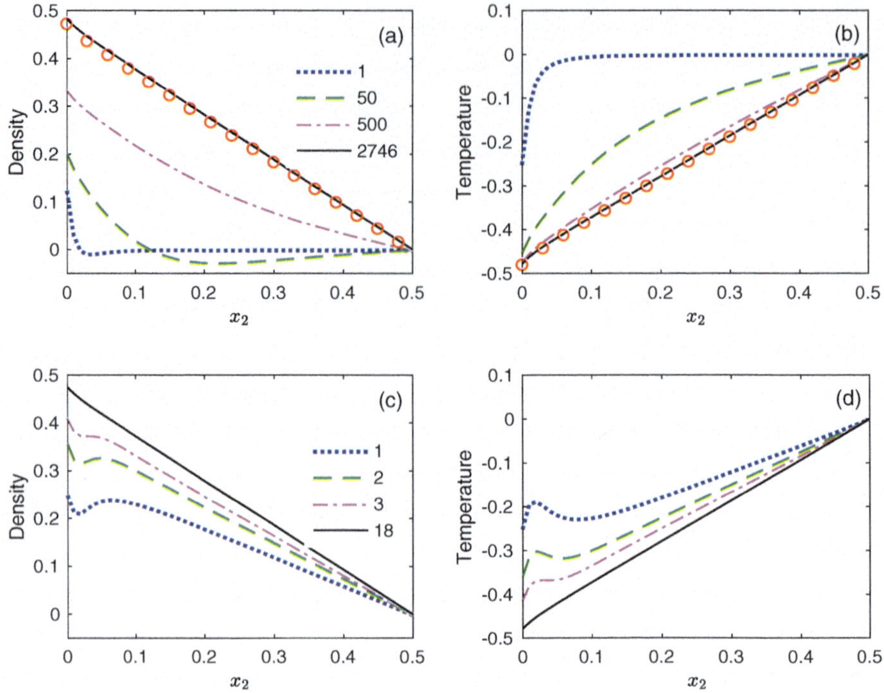

Fig. 8.6 Convergence history in planar Fourier flow at $\delta_{rp} = 50$ [17]. (Top) CIS, (Bottom) GSIS. Circles: converged solution from GSIS. The linearized Shakhov model is used with the initial condition $h(x_2, \boldsymbol{v}) = 0$. Iteration steps are shown in the legends

is about 1, therefore, the effective error decay rate of GSIS, as shown in Fig. 8.2, is about 0.5. This means that, for wall-bounded rarefied gas flows, the error is reduced by 3 orders of magnitude after 10 iterations in GSIS, no matter what the Knudsen number is. This explanation is in strong agreement with the results in Fig. 8.7. Therefore, in terms of fast-converging, GSIS-I and GSIS-II have similar performance.

8.4.3 Couette Flow Between Eccentric Cylinders

Consider a Couette flow between two non-coaxial cylinders in Fig. 8.8. The inner cylinder is stationary, while the outer cylinder rotates clockwise at a constant speed of u_w. The Shakhov model is linearized when β in Eq. (5.1) takes the value of u_w/v_m. The high-order discontinuous (DG) methods are employed to solve the linearized Shakhov model and synthetic equations, on structured triangular mesh [17].

Figure 8.8 shows the velocity contours and streamlines. When $\delta_{rp} = 1000$, the flow is in the continuum regime, and the GSIS results overlap with the ones from the NSF equations even when the maximum cell size is about 260 times of MFP, which

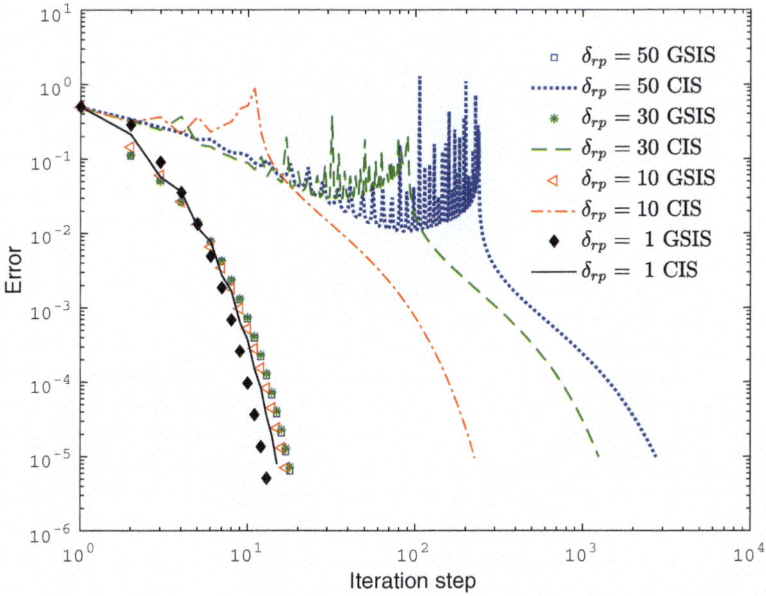

Fig. 8.7 The decay of error ϵ as a function of the iteration step, for the Fourier flow between two parallel plates [17]. The spatial region is discretized by 51 equidistant points

proves the asymptotically preserving property of GSIS. However, the CIS fails to predict the velocity profile due to its numerical dissipation on such a coarse mesh. When $\delta_{rp} = 10$, CIS produces close solutions to GSIS, because the mesh resolution is adequate.

Table 8.1 shows the iteration numbers and CPU time. While GSIS obtains the converged solution within dozens of iterations, CIS needs 49454 and 296 iteration steps when $\delta_{rp} = 1000$ and 10, respectively; as a result, GSIS is about 1300 and 5 times faster.

8.5 Concluding Remarks and Outlooks

Various numerical results have demonstrated that GSIS is able to find the converged solution within dozens of iterations at any Knudsen number, even when the spatial cell size is much larger than the MFP in the bulk flow region. Moreover, the fast-converging property of GSIS enables the numerical error to decay fast and the convergence criterion can be set at a much higher value than the CIS.

In addition to the fast-converging and asymptotic-persevering properties that have been proven rigorously, GSIS has the following advantages:

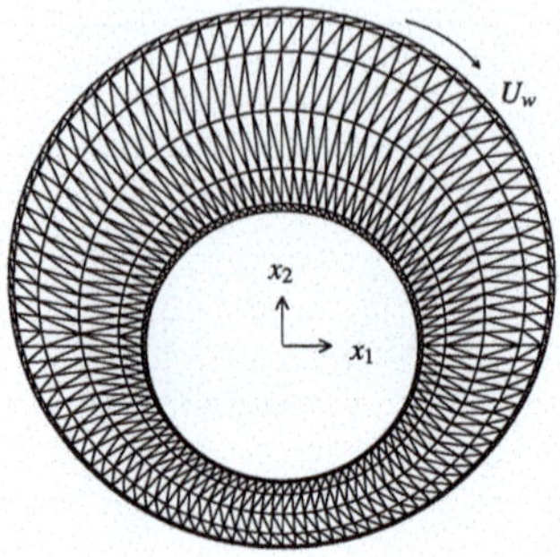

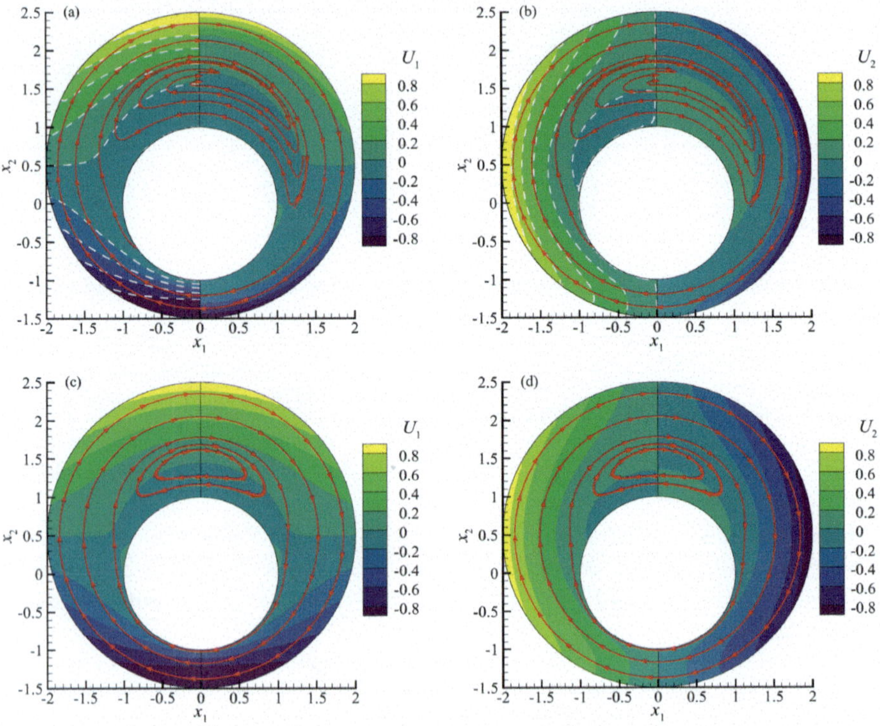

Fig. 8.8 Contours and streamlines in the Couette flow between two eccentric cylinders [17]. **a**, **b** $\delta_{rp} = 1000$. (c, d) $\delta_{rp} = 10$. GSIS (CIS) results are plotted in the left (right) half domain. Dashed lines in (**a**) and (**b**): velocity contours from the Navier-Stokes equations with non-slip velocity boundary

Table 8.1 Number of iteration steps and CPU time to reach convergence for the Couette flow between two eccentric cylinders [17]. Iteration terminates when the relative error in flow velocity between two consecutive iteration steps is less than 10^{-5}. Simulations are run on 12 processors using OpenMP, on double precision Intel Xeon-E5-2680 processors

δ_{rp}	Number of Triangles	Discrete Velocities	Iteration steps		CPU time (s)	
			CIS	GSIS	CIS	GSIS
1000	2400	$8 \times 8 \times 12$	49454	26	33861.2	26.3
10	1600	$32 \times 32 \times 24$	296	26	2849.8	580.3

1. **Compatibility with CFD techniques.** The gas kinetic equation and synthetic equations can be solved by sophisticated methods in computational fluid dynamics, and the kinetic and macroscopic solvers can have different orders of accuracy. For highly rarefied gas flows, the spatial cell size is usually smaller than the MFP and both methods yield high accuracy. For continuum/or near-continuum flows, as long as the macroscopic solver for synthetic equations (which are essentially the NSF equations) is able to capture the continuum flow behaviors, the accuracy of GSIS is guaranteed, that is, the numerical cell size can be much larger than the MFP, but should be smaller than the variation scale (such as the sound wavelength) of the flow.
2. **Universality.** GSIS can be easily extended to nonlinear flows, time-dependent systems, and rarefied flow of polyatomic gas [4, 18]. Extension to DSMC (to remove the limitation on cell size and boost convergence) is also possible, since the GSIS relies on no specific collision operator so that it can be extended naturally to multi-species flows and even flows involving chemical reactions.

With these development implemented, it is hoped that the multiscale simulation of rarefied gas flows is solved completely in the near future. Also, the essential idea is ready to be applied to other kinetic equations such as the Enskog equation for dense gases with applications to gas extraction in unconventional reservoirs and non-equilibrium evaporation and condensation at the nanoscale [20, 21], as well as the Callaway model on non-equilibrium phonon transport with dual relaxation times [22].

References

1. Guo ZL, Xu K, Wang RJ (2013) Discrete unified gas kinetic scheme for all Knudsen number flows: low-speed isothermal case. Phys Rev E 88:033305
2. Wang P, Ho MT, Wu L, Guo ZL, Zhang YH (2018) A comparative study of discrete velocity methods for low-speed rarefied gas flows. Comput Fluids 161:33–46
3. Su W, Zhu LH, Wu L (2020) Fast convergence and asymptotic preserving of the general synthetic iterative scheme. SIAM J Sci Comput 42:B1517–B1540

4. Zhu LH, Pi XC, Su W, Li ZH, Zhang YH, Wu L (2021) General synthetic iterative scheme for nonlinear gas kinetic simulation of multi-scale rarefied gas flows. J Comput Phys 430:110091
5. Adams ML, Larsen EW (2002) Fast iterative methods for discrete-ordinates particle transport calculations. Progr Nucl Energy 40:3–159
6. Valougeorgis D, Naris S (2003) Acceleration schemes of the discrete velocity method: Gaseous flows in rectangular microchannels. SIAM J Sci Comput 25:534–552
7. Lihnaropoulos J, Naris S, Valougeorgis D (2007) Formulation and stability analysis of rapidly convergent iteration schemes for the 2?D linearized BGK equation. Transp Theory Stat Phys 36(4–6):513–528
8. Szalmás L, Valougeorgis D (2010) A fast iterative model for discrete velocity calculations on triangular grids. J Comput Phys 229:4315–4326
9. Naris S, Valougeorgis D, Sharipov F, Kalempa D (2004) Discrete velocity modelling of gaseous mixture flows in MEMS. Superlattices Microstruct 35:629–643
10. Naris S, Valougeorgis D, Kalempa D, Sharipov F (2004) Gaseous mixture flow between two parallel plates in the whole range of the gas rarefaction. Physica A: Statist Mech Appl 336:294–318
11. Naris S, Valougeorgis D, Kalempa D, Sharipov F (2005) Flow of gaseous mixtures through rectangular microchannels driven by pressure, temperature, and concentration gradients. Phys Fluids 17:100607
12. Szalmás L (2016) An accelerated discrete velocity method for flows of rarefied ternary gas mixtures in long rectangular channels. Comput Fluids 128:91–97
13. Wu L, Zhang J, Liu HH, Zhang YH, Reese JM (2017) A fast iterative scheme for the linearized Boltzmann equation. J Comput Phys 338:431–451
14. Su W, Wang P, Liu HH, Wu L (2019) Accurate and efficient computation of the Boltzmann equation for Couette flow: influence of intermolecular potentials on Knudsen layer function and viscous slip coefficient. J Comput Phys 378:573–590
15. Chapman S, Cowling TG (1970) The mathematical theory of non-uniform gases. Cambridge University Press
16. Wu L, Su W, Zeng JN, General synthetic iterative scheme for unsteady rarefied gas flows
17. Su W, Zhu LH, Wang P, Zhang YH, Wu L (2020) Can we find steady-state solutions to multiscale rarefied gas flows within dozens of iterations? J Comput Phys 407:109245
18. Su W, Zhang YH, Wu L (2021) Multiscale simulation of molecular gas flows by the general synthetic iterative scheme. Comput Methods Appl Mech Engrg 373:113548
19. Su W, Ho MT, Zhang YH, Wu L (2020) GSIS: an efficient and accurate numerical method to obtain the apparent gas permeability of porous media. Comput Fluids 206:104576
20. Wu L, Zhang Y, Reese JM (2015) Fast spectral solution of the generalized Enskog equation for dense gases. J Comput Phys 303:66–79
21. Wu L, Liu HH, Reese JM, Zhang YH (2016) Non-equilibrium dynamics of dense gas under tight confinement. J Fluid Mech 794:252–266
22. Liu J, Wu L (2021) A fast-converging scheme for the Phonon Boltzmann equation with dual relaxation times, arXiv:2107.06688

Chapter 9
Acoustics in Rarefied Gas

Oscillatory rarefied gas flows, which are common in microelectromechanical systems subject to external oscillations, are solved in this chapter. Specifically, the resonance and anti-resonance of rarefied gas flows are discussed.

9.1 Formulation of the Problem

When the amplitude of oscillation is small, the problem can be linearized; and the time derivative can be removed when one is interested in the final steady oscillation. This benefits the numerical simulation since the time step is not limited by the oscillation period. To be specific, when the oscillation is fully established, the VDF changes periodically with the external oscillation frequency ϖ:

$$f(t, \boldsymbol{x}, \boldsymbol{v}) = f_{eq}(\boldsymbol{v}) + \beta\Re\left[\exp(i\mathrm{St}t)h(\boldsymbol{x}, \boldsymbol{v})\right], \tag{9.1}$$

where $\Re$ is a real part of a variable, and the Strouhal number St (related to the temporal Knudsen number (1.14) as $\mathrm{St} = \mathrm{Kn}_t\delta_{rp}$) is defined as

$$\mathrm{St} = \frac{\varpi L}{v_m}. \tag{9.2}$$

The perturbation VDF h satisfies the following LBE (the same normalization in Chap. 5 is used):

$$i\mathrm{St}h + \boldsymbol{v}\cdot\frac{\partial h}{\partial \boldsymbol{x}} = \mathcal{L}(h). \tag{9.3}$$

L. Wu, *Rarefied Gas Dynamics*,
https://doi.org/10.1007/978-981-19-2872-7_9

The perturbation VDF h is a complex function, so are the macroscopic quantities. For example, the velocity, which is normalized by $v_m\beta$, is given by $\boldsymbol{u}(\boldsymbol{x},t)=\Re\left[\exp(i\text{St}t)\int \boldsymbol{v}hd\boldsymbol{v}\right]$. For simplicity, the velocity will be referred to $\boldsymbol{u}(\boldsymbol{x})=\int \boldsymbol{v}h(\boldsymbol{x},\boldsymbol{v})d\boldsymbol{v}$. The GSIS can be applied to boost convergence for near-continuum flows, or the flow with small value of Strouhal number, where the synthetic equations are obtained if the time derivative $\partial/\partial t$ is replaced by $i\text{St}$.

9.2 Oscillatory Couette Flow

Consider the gas between two parallel plates located at $x_2=0$ and 1, respectively (the spatial variable is normalized by plate distance L). The plate at $x_2=1$ is stationary, and the one at $x_2=0$ oscillates with the velocity

$$u_{w,1}=\Re\left[\frac{u_0}{v_m}\exp(i\text{St}t)\right], \tag{9.4}$$

where u_0 is the amplitude of oscillation speed. Both plates have a temperature T_0. The Boltzmann equation can be linearized by choosing $\beta=u_0/v_m$ in Eq. (9.1). The diffuse BC is used, so $h(0,\boldsymbol{v})=2v_1 f_{eq}$ when $v_2>0$, and $h(1,\boldsymbol{v})=0$ when $v_2<0$. The synthetic equations (8.16), (8.20), and (8.21) are simplified to

$$\begin{aligned}2i\text{St}(i\text{St}+\delta_{rp})u_1^{k+1} &- \frac{\partial^2 u_1^{k+1}}{\partial x_2^2}\\ &=\frac{\partial}{\partial x_2}\left[2\int(\mathcal{L}^k-\mathcal{L}_s^k)v_1v_2\text{d}\boldsymbol{v}-\text{HoT}_{\sigma_{12}}^k\right],\end{aligned} \tag{9.5}$$

which is solved by the central finite difference scheme with 5 stencils, in a nonuniform spatial grid (5.15). Figure 9.1 confirms the asymptotic preserving and fast-converging property of GSIS, even in oscillatory flows.

When the Knudsen number is small and the Strouhal number is zero, the Navier-Stokes equations predict a stress $4\text{Kn}/\sqrt{\pi}$ at the solid wall, see Eq. (1.13). When either spatial or temporal Knudsen number is large, the collision operator is negligible and hence we have

$$i\text{St}h+v_2\frac{\partial h}{\partial x_2}=0, \tag{9.6}$$

which bears the analytical solution

$$h(x_2,\boldsymbol{v})=\begin{cases}2v_1 f_{eq}\exp\left(-i\text{St}\frac{x_2}{v_2}\right); & v_2>0,\\ 0; & v_2<0.\end{cases} \tag{9.7}$$

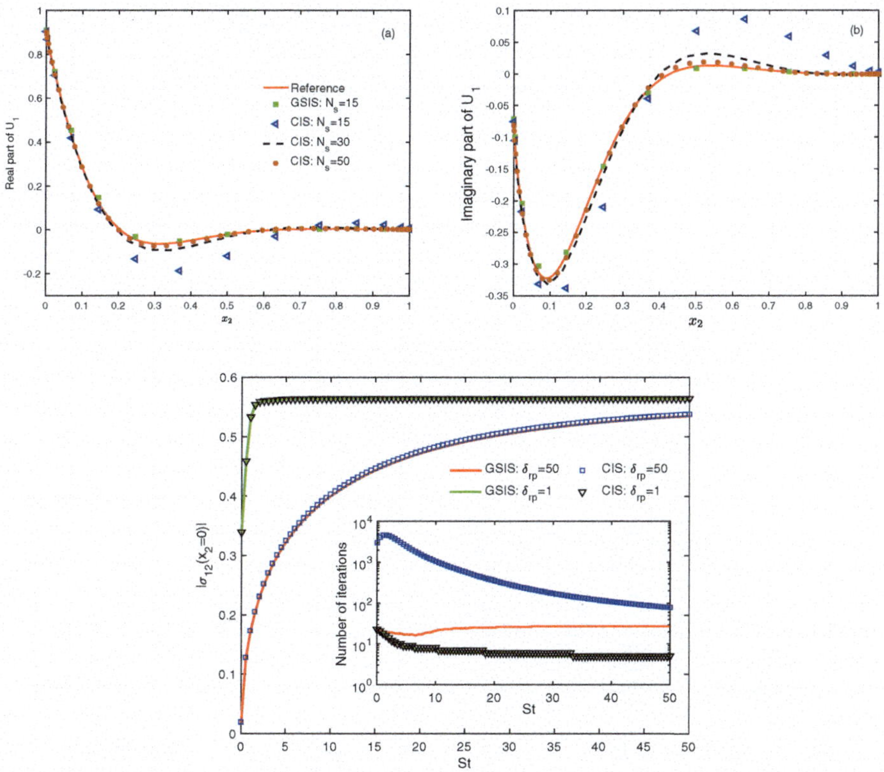

Fig. 9.1 Oscillating Couette flow [1], when $\delta_{rp} = 50$ and St $= 1$. The reference solution is obtained from GSIS, when the spatial domain is discretized by Eq. (5.15) with $N_s = 100$. When $\delta_{rp} = 1$, the iteration numbers in GSIS and CIS are the same. The Shakhov model is solved, where the solution is converged when the relative error in macroscopic quantities is less than 10^{-5}

Hence, the amplitudes of the velocity and shear stress are given by

$$\begin{aligned} u_1 &= \int v_1 h(\boldsymbol{v}, x_2) d\boldsymbol{v} = \frac{1}{\sqrt{\pi}} I_0[i\text{St}x_2], \\ \sigma_{12} &= 2\int v_1 v_2 h(\boldsymbol{v}, x_2) d\boldsymbol{v} = -\frac{2}{\sqrt{\pi}} I_1[i\text{St}x_2], \end{aligned} \tag{9.8}$$

where

$$I_m(z) = \int_0^\infty c^m \exp\left(-c^2 - \frac{z}{c}\right) dc.$$

Thus, at the oscillating plate, we have

$$u_1 = \frac{1}{2}, \quad \sigma_{12} = \frac{1}{\sqrt{\pi}}, \tag{9.9}$$

which is consistent with the numerical results in Fig. 9.1. The fact that, as the Strouhal number increases, the wall stress of $\delta_{rp} = 1$ approaches the collisionless limit faster than that of $\delta_{rp} = 50$, is because the temporal Knudsen number is $\text{Kn}_t = \text{St}/\delta_{rp}$.

9.3 Oscillating Lid-Driven Cavity Flow

Figure 9.2 shows a gas inside a 2D cavity with wall temperature T_0, driven by the top lid with the velocity $u_{w,1} = \Re[(u_0/v_m)\exp(i\text{St}t)]$. The other three walls are stationary. The gas flow inside is characterized by the aspect ratio of the cavity $A = L/H$, St, and Kn if we choose the height of cavity as the characteristic flow length.

The Boltzmann equation is linearized according to Eq. (9.1) with $\beta = u_0/v_m$. The diffuse BC reads

$$h = \begin{cases} \left(2v_x + 2\sqrt{\pi}\int_{v_y>0} v_y h d\boldsymbol{v}\right) f_{eq}; & y = 1, v_y < 0, \\ -2\sqrt{\pi} f_{eq} \int_{v_x<0} v_x h d\boldsymbol{v}; & x = 0, v_x > 0, \\ -2\sqrt{\pi} f_{eq} \int_{v_y<0} v_y h d\boldsymbol{v}; & y = 0, v_y > 0, \end{cases} \tag{9.10}$$

while at the central vertical line, $h(y, -v_x, v_y, v_z) = -h(y, v_x, v_y, v_z)$ due to symmetry.

When the oscillation frequency is high, binary collisions are negligible, and the LBE becomes $i\text{St}h + v_x \partial h/\partial x + v_y \partial h/\partial y = 0$. Integrating this equation along the x direction and introducing the average VDF $g(y, \boldsymbol{v}) = \int_0^A h dx/A$, we have $i\text{St}h + v_y \frac{\partial h}{\partial y} = v_x[h(x = 0) - h(x = A)]/A$. The last term can further be ignored, and the average amplitude of shear stress is also described by Eq. (9.9).

It is interesting to note that average shear stress in the free-molecular flow regime is 0.69 when $A = 1$ and $\text{St} = 0$ [2], which is larger than $1/\sqrt{\pi}$ at high oscillation frequency. This indicates that it is possible to reduce the friction at the lid by changing the oscillation frequency.

9.3.1 *Scaling Law for Anti-resonant Frequency*

The LBE is solved by the CIS with FSM [3] and the results are summarized in Fig. 9.2. While in 1D oscillatory Couette flow, the average amplitude of stress on the lid always increases with the Strouhal number, in 2D oscillatory cavity flow the stress amplitude is a non-monotonic function of St, which exhibits local minimum at the critical Strouhal number.

Figure 9.3 shows that the critical Strouhal number is roughly proportional to the inverse aspect ratio, which can be explained qualitatively by the gas anti-resonance.

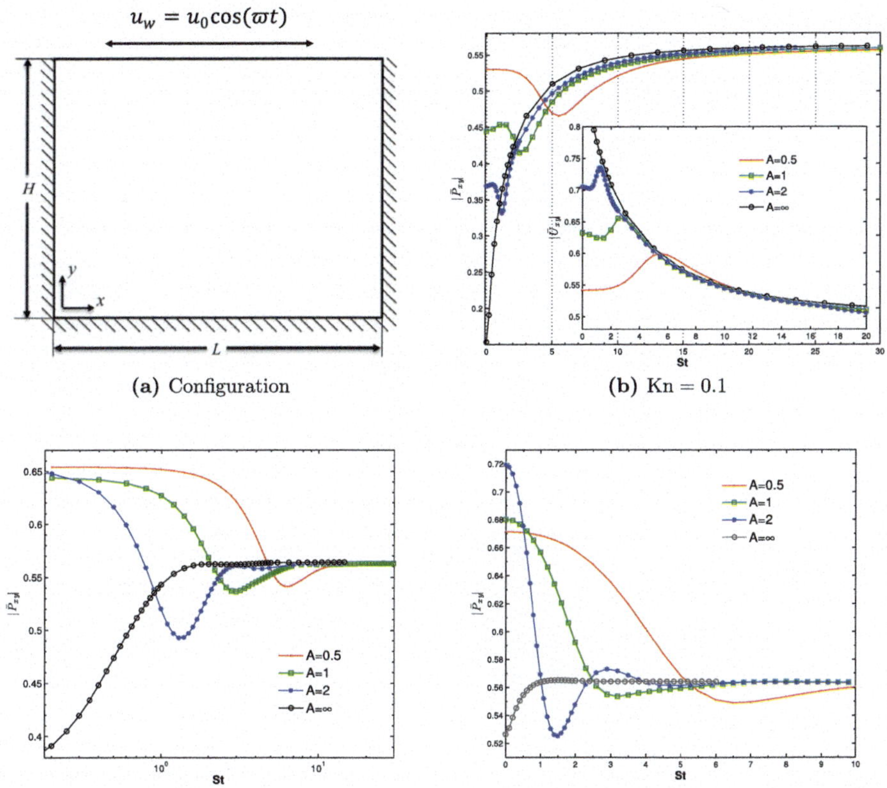

(a) Configuration (b) Kn = 0.1

(c) Kn = 1 (d) Kn = 10

Fig. 9.2 Average amplitudes of the shear stress and horizontal velocity along the lid [3], for different values of St and A. The linearized Boltzmann equation for HS gas is solved

Using the method of characteristics, the collisionless LBE can be rewritten as $i\mathrm{St}h + \mu\frac{\partial h}{\partial s} = 0$, where $\mu = \sqrt{v_x^2 + v_y^2}$ and s is the coordinate along the characteristic line. Choose v_m as the effective velocity, Eq. (9.11) becomes

$$i\mathrm{St}h + \frac{\partial h}{\partial s} = 0, \tag{9.11}$$

which shows that the phase of VDF is recovered after a propagation distance of $2\pi/\mathrm{St}$. Imagine molecules leaving the top lid with velocities nearly parallel to the top lid, hitting the left wall, then being reflected and hitting the right wall, and finally returning to where they left. The distance they have traveled is about $2A$. If

$$A = \frac{\pi}{\mathrm{St}}, \tag{9.12}$$

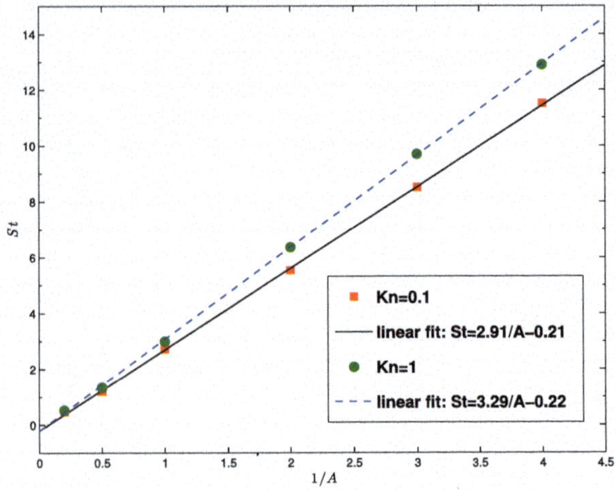

Fig. 9.3 The critical Strouhal number versus the aspect ratio A [3]. The case of Kn = 10 overlaps with Kn = 1

molecules leaving and hitting the top lid have the same phase. Thus, the shear stress is a minimum because the v_y components of the leaving and returning molecules cancel each other out. This explanation agrees with the results in Fig. 9.3 for Kn = 1 and 10. When Kn is small, the anti-resonance Strouhal number is slightly less than π/A, due to the frequent binary collisions such that Eq. (9.11) is less valid.

9.4 Planar Sound Propagation

The configuration is the same as that in 1D oscillatory Couette flow, but the plate at $x_2 = 0$ oscillates with the velocity $u_{w,2} = \Re\left[(u_0/v_m)\exp(i\mathrm{St}t)\right]$. The Boltzmann equation is linearized by choosing $\beta = u_0/v_m$ in Eq. (9.3). VDFs reflected from the boundaries are

$$\begin{aligned} h(0, \boldsymbol{v}) &= \left[\sqrt{\pi} + 2v_2 - 2\sqrt{\pi}\int_{v_2<0} v_2 h(0, \boldsymbol{v})\mathrm{d}\boldsymbol{v}\right] f_{eq}, \\ h(1, \boldsymbol{v}) &= 2\sqrt{\pi} f_{eq}\int_{v_2>0} v_2 h(x_2 = 1, \boldsymbol{v})\mathrm{d}\boldsymbol{v}. \end{aligned} \tag{9.13}$$

When the oscillation frequency is much larger than the mean collision frequency of gas molecules, the collision operator is neglected, and the analytical solution for VDF reads

$$h(x_2, \boldsymbol{v}) = \begin{cases} (2v_2 + \nu_L) f_{eq}(\boldsymbol{v}) \exp\left(-i\text{St}\frac{x_2}{v_2}\right); & v_2 > 0, \\ \nu_R f_{eq}(\boldsymbol{v}) \exp\left[-i\text{St}\frac{(x_2-1)}{v_2}\right]; & v_2 < 0, \end{cases} \tag{9.14}$$

where

$$\nu_L = \frac{\sqrt{\pi} + 8I_1(i\text{S})I_2(i\text{St})}{1 - 4I_1(i\text{St})}, \quad \nu_R = \frac{2\sqrt{\pi}I_1(i\text{S}) + 4I_2(i\text{St})}{1 - 4I_1(i\text{St})}. \tag{9.15}$$

The pressure $p(x_2) = 2\int v_2^2 h(x_2, \boldsymbol{v}) d\boldsymbol{v}$ is then calculated to be

$$p(x_2) = \frac{2\nu_R}{\sqrt{\pi}} I_2[i\text{St}(x_2 - 1)] + \frac{4}{\sqrt{\pi}} I_3(i\text{St}x_2) + \frac{2\nu_L}{\sqrt{\pi}} I_2(i\text{St}x_2). \tag{9.16}$$

When $\text{St} \to \infty$, the average pressure at $x_2 = 0$ (related to the damping force exerting on the oscillating plate) is

$$p(0) \to \frac{2}{\sqrt{\pi}} + \frac{\sqrt{\pi}}{2} \approx 2.01, \tag{9.17}$$

while that on the right plate is $\bar{p}(1) \to 0$.

In GSIS, the synthetic equations (8.16), (8.20), and (8.21) can be simplified to

$$\begin{aligned} i\text{St}\varrho + \frac{\partial u_2}{\partial x_2} &= 0, \\ 2i\text{St}u_2 + \frac{\partial \varrho}{\partial x_2} + \frac{\partial T}{\partial x_2} + \frac{\partial \sigma_{22}}{\partial x_2} &= 0, \\ \frac{3}{2}i\text{St}T + \frac{\partial q_2}{\partial x_2} + \frac{\partial u_2}{\partial x_2} &= 0, \\ i\text{St}\sigma_{22} + \text{HoT}_{\sigma_{22}} + \frac{4}{3}\frac{\partial u_2}{\partial x_2} &= -\delta_{rp}\sigma_{22} + 2\int (\mathcal{L} - \mathcal{L}_s)\left(v_2^2 - \frac{v^2}{3}\right) \mathrm{d}\boldsymbol{v}, \\ i\text{St}q_2 + \text{HoT}_{q_2} + \frac{3C_q}{2}\frac{\partial T}{\partial x_2} &= -\frac{2}{3}\delta_{rp}q_2 + \int (\mathcal{L} - \mathcal{L}_s)v_2 v^2 \mathrm{d}\boldsymbol{v}. \end{aligned} \tag{9.18}$$

Again, Fig. 9.4 shows that the GSIS outperforms CIS.

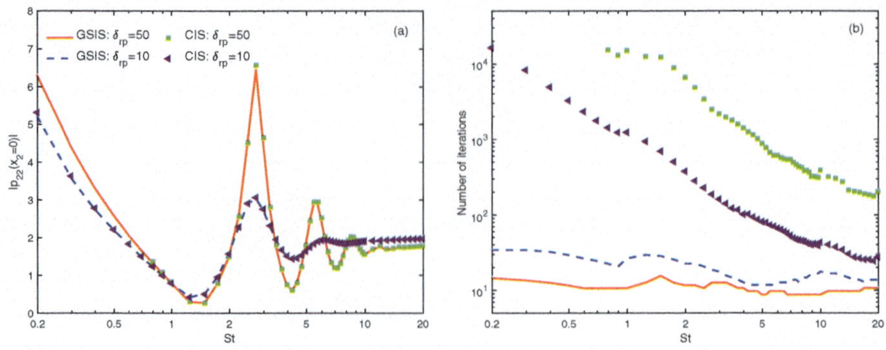

Fig. 9.4 Planar sound propagation [1]. Comparisons between CIS and GSIS, in terms of the **a** amplitude and **b** iteration number. The Shakhov model is solved, where the solution is converged when the relative error in macroscopic quantities is less than 10^{-5}

9.5 Sound Propagation in Cavity

Consider a gas flow inside a 2D cavity, driven by a plate at $x = 0$ that oscillates as $u_{w,x} = \Re[(u_0/v_m)\exp(i\mathrm{St}t)]$, while the other walls are fixed, see Fig. 9.5. All the walls have temperature T_0. The rarefied gas flows inside are characterized by the cavity aspect ratio $A = H/L$ (note the definition is different to that of Couette flow), the Strouhal number, and the Knudsen number when L is the characteristic flow length.

The Boltzmann equation is linearized according to Eq. (9.1) with $\beta = u_0/v_m$. The diffuse BC is

$$\frac{h}{f_{eq}} = \begin{cases} \sqrt{\pi} + 2v_x - 2\sqrt{\pi}\int_{v_x<0} v_x h d\boldsymbol{v}; & x = 0, v_x > 0, \\ -2\sqrt{\pi}\int_{v_y<0} v_y h d\boldsymbol{v}; & y = 0, v_y > 0, \\ 2\sqrt{\pi}\int_{v_x>0} v_x h d\boldsymbol{v}; & x = 1, v_x < 0, \end{cases} \tag{9.19}$$

while at the horizontal line $y = A/2$: $h(x, y, v_x, v_y, v_z) = h(x, y, v_x, -v_y, v_z)$ due to symmetry.

The average gas pressure is defined as

$$\bar{p}(x) = |\bar{p}(x)|e^{i\phi(x)} = \frac{\int_0^A p_{xx}(x, y)dy}{2A} \equiv \frac{\int_0^A \int_{\infty}^{\infty} v_x^2 h(\boldsymbol{v})d\boldsymbol{v}dy}{A}, \tag{9.20}$$

where the amplitude is $|\bar{p}(x)|$, while the phase is ϕ. In the limit of high oscillation frequency, the average amplitude of the pressure can be calculated analytically: it is half the result of Eq. (9.17) since we define the pressure according to Eq. (9.20). Therefore, the average pressure has nothing to do with the cavity aspect ratio.

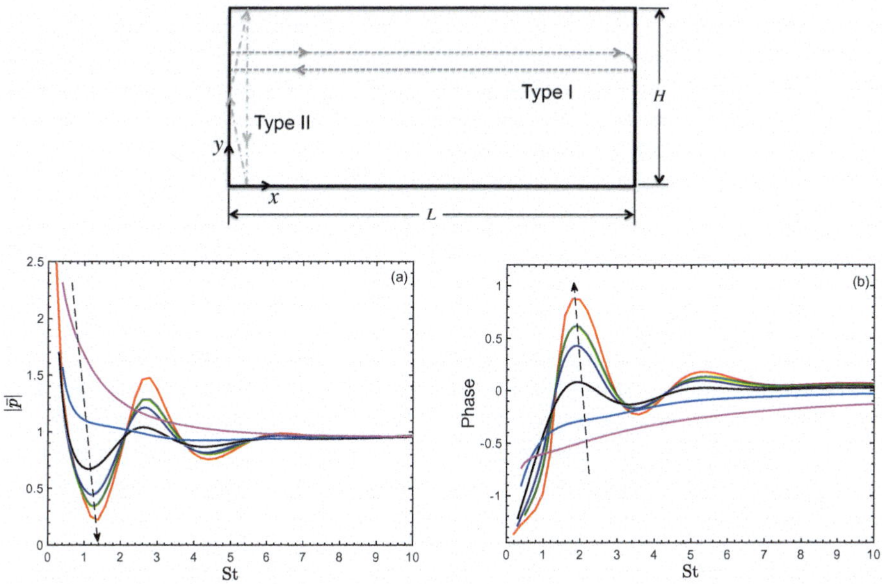

Fig. 9.5 Sound propagation in cavity [4]. The sound source (left plate) oscillates harmonically in the x direction. Type I and II interference occur in the direction parallel and perpendicular to the motion of the oscillating plate, respectively. The amplitude (**a**) and phase (**b**) of the average pressure at the oscillating plate when Kn $= 0.1$. Along the arrow, the cavity aspect ratio $A = H/L$ is 0.125, 0.25, 0.5, 1, 2, and infinity, respectively

9.5.1 *Two Types of Resonances*

In 1D sound propagation, the oscillatory behavior of average pressure as the function of St, as observed in Fig. 9.4a, is explained by the sound interference [5]. In the free-molecular flow regime, according to Eq. (9.11), we know that the distance molecules travel in the type I interference in Fig. 9.5 is 2, hence the phase change is 2St. Therefore, the sound pressure becomes locally maximum at the resonance frequency (molecules leaving and returning to the osculating plate are in phase, and the normal pressure is related to v_x^2, in sharp contrast to that in the oscillating lid-driven cavity flow where the stress is related to v_y)

$$\mathrm{St}_{r,1} \simeq m_1\pi, \quad m_1 = 0, 1, 2, \cdots, \tag{9.21}$$

and locally minimum at the anti-resonance frequency

$$\mathrm{St}_{a,1} \simeq \frac{(2m_2+1)}{2}\pi, \quad m_2 = 0, 1, 2, \cdots. \tag{9.22}$$

However, for the sound propagation in the cavity, a type II interference in Fig. 9.5 might become dominant, when A is small. Imagine molecules leaving the left plate with velocities nearly parallel to the left plate, hitting the bottom plates, then being reflected and hitting the top plate, and finally returning to the point where they left. The distance is about $2A$. Therefore, the pressure on the oscillating plate is maximum (minimum) at the resonance (anti-resonance) frequency:

$$\begin{aligned} \mathrm{St}_{r,2} &\simeq \frac{m_1}{A}\pi, \\ \mathrm{St}_{a,2} &\simeq \frac{(2m_2+1)}{2A}\pi. \end{aligned} \tag{9.23}$$

The average amplitude of pressure exerting on the oscillating plates is shown in Fig. 9.5a as a function St, when Kn $= 0.1$. When A is large, the oscillatory behavior of $\bar{p}$ versus St is seen, like that in 1D sound propagation. This is due to the type I resonance and anti-resonance. This oscillatory behavior becomes weaker when A decreases, which eventually disappears when $A < 0.25$. In this case, $\bar{p}$ decreases when St increases; and the smaller the value of A, the slower the decrease. This is due to the type II anti-resonance. For example, when $A = 0.125$, the first anti-resonance frequency is reached when $\mathrm{St} = 4\pi$, while the first resonance frequency is $\mathrm{St} = 8\pi$. Therefore, theoretically, $\bar{p}$ decreases in the region $0 \le \mathrm{St} \le 4\pi$ and increases in the region $4\pi \le \mathrm{St} \le 8\pi$; but the latter does not occur since $\mathrm{St} = 4\pi$ is close to the high oscillation frequency limit, where $\bar{p} = 1$.

9.5.2 *Sound Speed*

The variation of sound phase as a function of the Strouhal number is opposite to that of sound amplitude. It is interesting to investigate how the cavity aspect ratio and Strouhal number affect the sound speed near the oscillating plate. The sound speed, normalized by the most probable speed v_m, is calculated as

$$v_{ph} = -\frac{\mathrm{St}}{(\partial\phi/\partial x)|_{x=0}}, \tag{9.24}$$

where ϕ is the phase of $\bar{p}$. Equation (9.16) tells us that, in the free-molecular limit, the average sound pressure at $x \to 0$ is $\bar{p}(x) = \frac{1}{\sqrt{\pi}} + \frac{\sqrt{\pi}}{4} - i\mathrm{St}x$. Thus, $\phi \approx \mathrm{St}x$ and $v_{ph} \approx 1$.

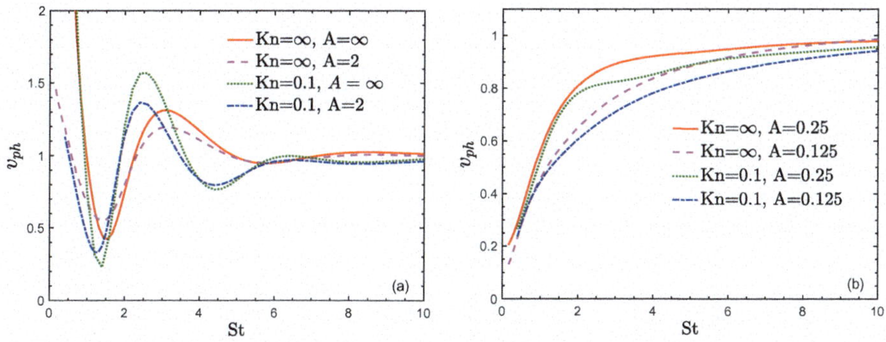

Fig. 9.6 Sound speed near the sound source [4], at large (**a**) and small (**b**) aspect ratios

Figure 9.6 depicts the sound speed at typical values of aspect ratio and Knudsen number. When A is large, the sound speed exhibits the same oscillatory behavior (in-phase) as the amplitude of average pressure. However, when A is small, the sound speed increases monotonically with St, which is out of phase with the amplitude of average pressure.

References

1. Su W, Zhu LH, Wang P, Zhang YH, Wu L (2020) Can we find steady-state solutions to multiscale rarefied gas flows within dozens of iterations? J Comput Phys 407:109245
2. Naris S, Valougeorgis D (2005) The driven cavity flow over the whole range of the Knudsen number. Phys Fluids 17:097106
3. Wu L, Reese JM, Zhang YH (2014) Oscillatory rarefied gas flow inside rectangular cavities. J Fluid Mech 748:350–367
4. Wu L (2016) Sound propagation through a rarefied gas in rectangular channels. Phys Rev E 94:053110
5. Desvillettes L, Lorenzani S (2012) Sound wave resonance in micro-electro-mechanical systems devices vibrating at high frequencies according to the kinetic theory of gases. Phys Fluids 24:092001

Chapter 10
Slip and Jump Coefficients

In wall-bounded gas flow with small Knudsen number, rarefaction effects appear in the Knudsen layer which is within a few mean free path away from the solid surface. Therefore, the Navier-Stokes-Fourier equations can be used to describe the gas flow with the velocity-slip and temperature-jump boundary conditions; the accuracy of fluid-dynamics equations can be further improved when using the effective shear viscosity and thermal conductivity, modified from the Knudsen layer function. This chapter is dedicated to the calculation of slip and jump coefficients, as well as the Knudsen layer functions, based on the linearized Boltzmann equation solved by the fast spectral method and general synthetic iterative scheme.

10.1 State of the Problem

A sketch of the slip velocity and Knudsen layer in the planar Couette and thermal creep flows is depicted in Fig. 10.1. The thickness of the Knudsen layer is just a few MFP. Therefore, in engineering applications with small Knudsen number, its detailed structure is ignored, and the NSF equations are used to describe the gas dynamics. In order to get the correct velocity profiles in the bulk region, the slip velocity Δu_3 at the solid surface is calculated as [1]:

$$\Delta u_3 = \sigma_P \frac{v_m \mu}{p} \frac{\partial u_3}{\partial x_2} + \sigma_T \frac{\mu}{\rho T} \frac{\partial T}{\partial x_3}, \tag{10.1}$$

where σ_P and σ_T are the viscous slip coefficient (VSC) and thermal slip coefficient (TSC), respectively, and the spatial derivatives are evaluated at the solid wall.

Likewise, in order to capture the temperature profile in the bulk region, the Knudsen layer (the schematic is the same as that in the Couette flow, when the velocity is replaced by the temperature) for the temperature defect is dropped, but

L. Wu, *Rarefied Gas Dynamics*,
https://doi.org/10.1007/978-981-19-2872-7_10

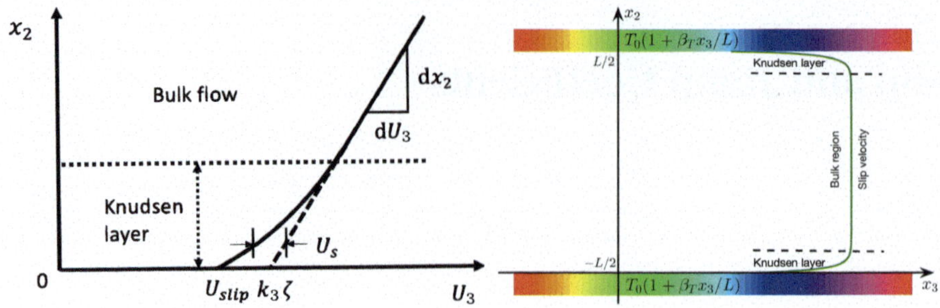

Fig. 10.1 Schematic of the slip velocity and Knudsen layer structure [2, 3]. (Left) In the Couette flow, the velocity profile in the bulk region is linear, while that in the vicinity of solid wall deviates from the linear profile, and this deviation is related to the slip velocity and Knudsen layer function. (Right) In the thermal transpiration, the velocity profile in the bulk region is constant, which is proportional to the slip velocity. Inside the Knudsen layer, the Knudsen layer function describes the difference between the true velocity and the slip velocity

the gas temperature T at the solid wall is calculated as

$$T = T_w + \zeta_T \frac{\mu}{p}\sqrt{\frac{2k_B T_w}{m}}\frac{\partial T}{\partial \boldsymbol{n}}, \tag{10.2}$$

where T_w is the wall temperature, ζ_T is the temperature jump coefficient (TJC), and $\boldsymbol{n}$ is the unit inward normal vector at the wall.

In this following, the slip and jump coefficients, as well as the corresponding Knudsen layer functions (KLFs), will be investigated systematically, based on the LBE for monatomic gas and the Wu model for molecular gas, at different kinetic boundary conditions, and with different intermolecular potentials.

10.2 Viscous Slip

The LBE is solved to find the steady-state solution of the planar Couette flow [2], see the numerical example in Sect. 5.5.5. The LBE is used since the slip velocity and KLF are the same for monatomic and molecular gases. The rarefaction parameter is chosen as $\delta_{rp} = 100$, so that the interference between the Knudsen layers adjacent to the two plates is absent.

The KLF is calculated as

$$u_s\left(\frac{x_2}{\text{Kn}}\right) = \frac{u_{NS}(x_2) - u_3(x_2)}{k_3\text{Kn}}, \tag{10.3}$$

and the VSC is calculated as

$$\sigma_P = -\frac{2-k_3}{2\sigma_{23}}, \tag{10.4}$$

Table 10.1 The viscous slip coefficient for HS, argon, and Maxwellian molecules under Cercignani-Lampis BC with different TMAC α_t and EAC α_n [2]

α_t	ω	$\alpha_n = 0.25$	0.5	0.75	1
0.25	0.5	6.365	6.343	6.324	6.307
	0.81	6.400	6.372	6.348	6.325
	1	6.427	6.395	6.366	6.340
0.5	0.5	2.800	2.785	2.773	2.761
	0.81	2.829	2.811	2.795	2.780
	1	2.852	2.831	2.812	2.795
0.75	0.5	1.598	1.591	1.585	1.579
	0.81	1.623	1.614	1.606	1.599
	1	1.641	1.631	1.622	1.613
1.0	0.5	0.988	0.988	0.988	0.988
	0.81	1.008	1.008	1.008	1.008
	1	1.023	1.023	1.023	1.023
1.25	0.5	0.616	0.622	0.628	0.634
	0.81	0.630	0.638	0.646	0.653
	1	0.641	0.651	0.660	0.668
1.5	0.5	0.361	0.374	0.386	0.397
	0.81	0.371	0.387	0.402	0.417
	1	0.379	0.397	0.415	0.431
1.75	0.5	0.174	0.193	0.211	0.227
	0.81	0.181	0.204	0.226	0.248
	1.0	0.186	0.212	0.238	0.263
2	0.5	0.029	0.054	0.077	0.099
	0.81	0.033	0.063	0.092	0.121
	1.0	0.036	0.069	0.103	0.136

where $u_{NS} = k_3 x_2$ is the linear velocity fitted from the numerical solution in the bulk region $x_2 \in [-0.2, 0.2]$.

10.2.1 Viscous Slip Coefficient

Table 10.1 summarizes the VSC calculated from the Cercignani-Lampis BC. The power-law potentials for the HS, argon, and Maxwell gases are considered, where the viscosity index is $\omega = 0.5, 0.81$, and 1, respectively. When the EAC α_n and ω are fixed, the VSC increases rapidly when the TMAC α_t decreases. When $\alpha_t \lesssim 1.25$ is fixed, the influence of α_n and ω on the VSC is very limited. However, when $\alpha_t \to 2$ (the "backward" scattering), the influence α_n and ω becomes strong. For example,

Table 10.2 Fitting coefficients in Eq. (10.5) for HS, argon, and Maxwellian molecules under the Maxwell and Cercignani-Lampis BCs [2]

BC	α_n	ω	a	$b \cdot 10$	c
Diffuse-specular	n/a	0.5	1.773	1.1660	0.6687
	n/a	0.81	1.773	1.4270	0.6238
	n/a	1	1.773	1.6370	0.5885
Cercignani-Lampis	0.25	0.5	1.774	0.7266	0.7127
		0.81	1.775	0.8889	0.6781
		1	1.776	1.0130	0.6516
Cercignani-Lampis	0.5	0.5	1.773	0.4772	0.7367
		0.81	1.774	0.5867	0.7069
		1	1.774	0.5742	0.6837
Cercignani-Lampis	0.75	0.5	1.772	0.2434	0.7597
		0.81	1.773	0.2915	0.7358
		1	1.773	0.3373	0.7164
Cercignani-Lampis	1.0	0.5	1.772	0.0218	0.7818
		0.81	1.766	0.0543	0.7370
		1	1.772	0.0000	0.7499

when $\alpha_t = 2$ and $\alpha_n = 1$, Maxwellian molecules have a VSC 37% higher than that for HS molecules; the case of $\alpha_n = 1$ is 3.8 times of the case of $\alpha_n = 0.25$.

The VSC is fitted by a general function as used by Lilley and Sader [4] for both Maxwell and Cercignani-Lampis BCs:

$$\sigma_P(\alpha) = \frac{a}{\alpha} - b\alpha - c, \tag{10.5}$$

where the fitting coefficients a, b, and c are shown in Table 10.2 for typical power-law intermolecular potentials. For the Cercignani-Lampis BC, $\alpha = \alpha_t$, while for the Maxwell BC, $\alpha = \alpha_M$.

10.2.2 Knudsen Layer Function

While the VSC determines the gas velocity at solid wall, the KLF can be used to construct effective viscosity [5], such that these two combined effects can be used to capture the velocity profiles in Poiseuille and Couette flows, up to Kn = 0.4.

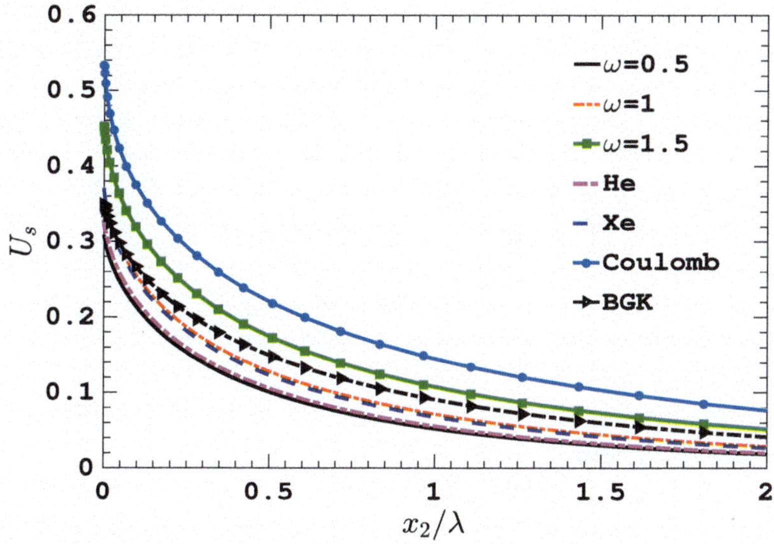

Fig. 10.2 Knudsen layer functions in the planar Couette flow [2], obtained from the LBE solutions for power-law potentials (2.9), Lennard-Jones potentials (2.10) (He and Xe at $T = 300$ K), and shielded Coulomb potential (2.11) with $\epsilon = k_B T_w$. The diffuse BC is used

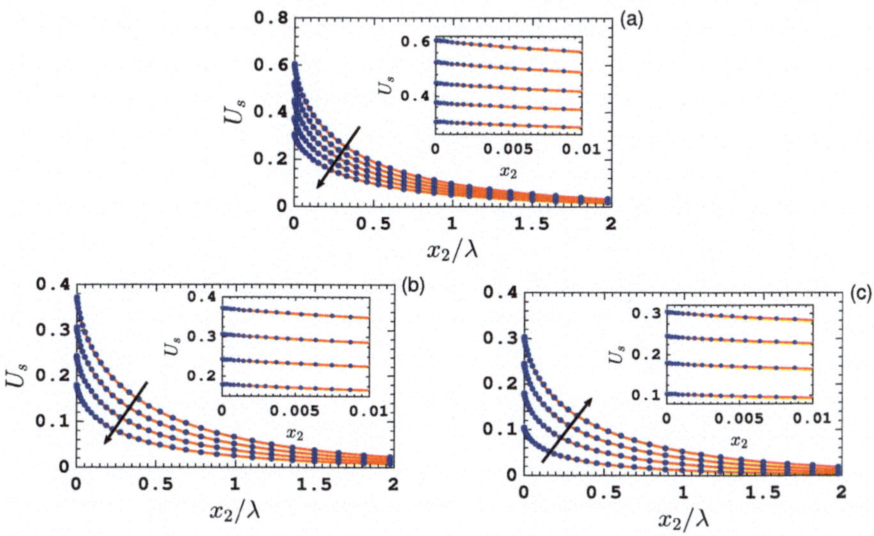

Fig. 10.3 The KLF for HS molecules in the thermal transpiration [2]. **a** The diffuse-specular BC. Along the arrow, $\alpha_M = 0.2, 0.4, 0.6, 0.8, 1$. **b** The Cercignani-Lampis BC with $\alpha_n = 0.25$. Along the arrow, $\alpha_t = 0.5,\ 1,\ 1.5$ and 2. **c** The Cercignani-Lampis BC with $\alpha_t = 2$. Along the arrow, $\alpha_n = 0.25,\ 0.5,\ 0.75$ and 1. Insets: zoomed regions in the vicinity of the solid surface

Typical profiles of the KLF, obtained from the power-law, Lennard-Jones and shielded Coulomb potentials, are compared in Fig. 10.2. The strong influence of the intermolecular potential is clearly seen. Considering the fact that the effective viscosity indexes of helium and xenon at $T = 300$ K are respectively $\omega = 0.66$ and 0.85 [6], and that of the shielded Coulomb potential is close to 2.5 [7], we reach the conclusion that the KLF increases with the viscosity index in the whole Knudsen layer. The BGK model, which does not reflect the influence of intermolecular potential in linear flows (see also the Sect. 6.5.2), predicts a KLF even larger than that from the Maxwellian molecules.

The Maxwell and Cercignani-Lampis BCs are considered in Fig. 10.3, where one finds their influences on the VSC and KFL share the same trend.

10.3 Thermal Slip

The thermal transpiration described in Sect. 5.3 is considered. The KLF is calculated as

$$u_d\left(\frac{x_2}{\text{Kn}}\right) = \frac{1}{\text{Kn}}\left[u_3\left(0\right) - u_3(x_2)\right]. \tag{10.6}$$

Meanwhile, the TSC is calculated as

$$\sigma_T = 2\delta_{rp}u_3\left(0\right). \tag{10.7}$$

10.3.1 Thermal Slip Coefficient

We first consider the Maxwell BC in the framework of LBE for monatomic gas. Figure 10.4 shows that, when the intermolecular potential is fixed, the TSC can be fitted by linear functions of α_M, and the slope of fitting increases with the viscosity index.

Table 10.3 and Fig. 10.5 summarize the TSC computed from the LBE for HS, Maxwellian, and VHS ($\omega = 1.5$) molecules with the Cercignani-Lampis BC. When α_t and α_n are fixed, the TSC increases with the viscosity index. When the viscosity index is fixed, however, the variation of TSC with respect to the effective TMAC α_t and EAC α_n is complicated. We first consider the case when α_t and ω are fixed. From Table 10.3 we see that, when $\alpha_t < 1$, the TSC increases with α_n, where the maximum increment is less than 20% when α_n varies from 0.25 to 1. Furthermore, this kind of variation reduces when α_t approaches 1; when $\alpha_t = 1$, the Cercignani-Lampis BC is reduced to the fully diffuse one in this problem, and the TSC does not vary with α_n; when $\alpha_t > 1$, the variation of TSC on α_n reverses when compared to that of $\alpha_t < 1$. We then fix the value of α_n and the intermolecular potential to see how α_t affects the

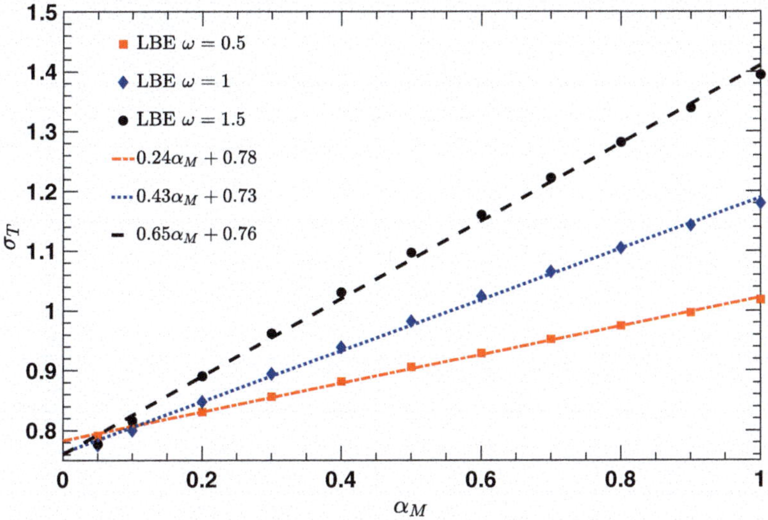

Fig. 10.4 Symbols: TSCs obtained from the LBE for HS, Maxwellian, and variable hard sphere (VHS, $\omega = 1.5$) molecules, when the Maxwell BC is used. Lines: linear fittings of TSCs with respect to the TMAC [3]

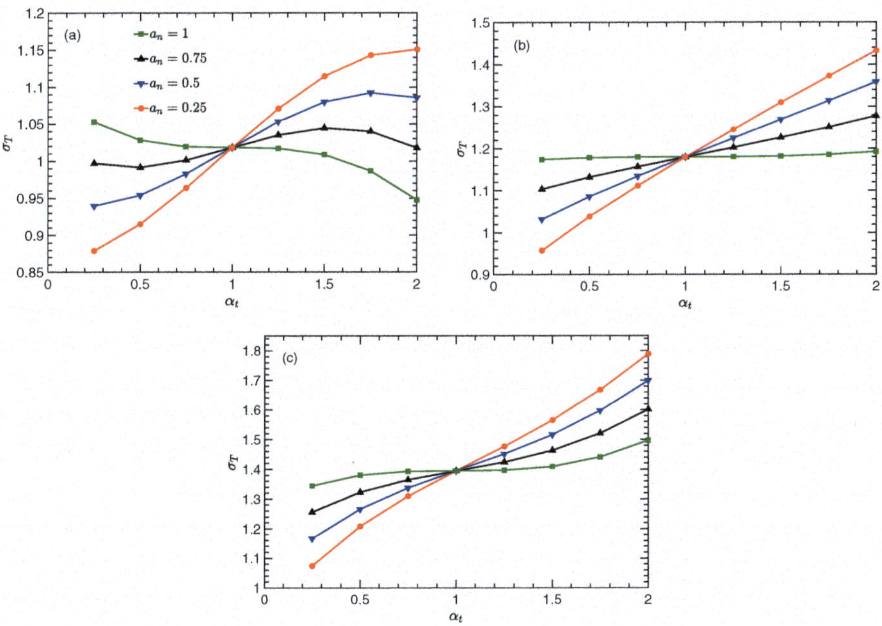

Fig. 10.5 TSCs obtained from LBE for **a** HS, **b** Maxwellian, and **c** VHS ($\omega = 1.5$) molecules, when the Cercignani-Lampis BC is used [3]

Table 10.3 TSCs for HS, Maxwellian and VHS ($\omega = 1.5$) molecules under the Cercignani-Lampis BC with different values of TMAC and EAC [3]

α_t	ω	$\alpha_n = 0.25$	$\alpha_n = 0.5$	$\alpha_n = 0.75$	$\alpha_n = 1$
0.25	0.5	0.879	0.939	0.997	1.053
	1.0	0.958	1.032	1.104	1.174
	1.5	1.073	1.166	1.256	1.343
0.5	0.5	0.915	0.954	0.99	1.028
	1.0	1.038	1.085	1.132	1.178
	1.5	1.208	1.266	1.323	1.380
0.75	0.5	0.964	0.983	1.001	1.020
	1.0	1.111	1.134	1.157	1.180
	1.5	1.309	1.337	1.365	1.393
1.0	0.5	1.018	1.018	1.018	1.018
	1.0	1.180	1.180	1.180	1.180
	1.5	1.395	1.395	1.395	1.395
1.25	0.5	1.071	1.053	1.035	1.017
	1.0	1.246	1.224	1.203	1.180
	1.5	1.476	1.451	1.424	1.396
1.5	0.5	1.115	1.080	1.045	1.009
	1.0	1.310	1.269	1.226	1.181
	1.5	1.565	1.515	1.463	1.409
1.75	0.5	1.143	1.092	1.040	0.987
	1.0	1.373	1.314	1.251	1.185
	1.5	1.668	1.597	1.521	1.440
2	0.5	1.151	1.086	1.018	0.948
	1.0	1.433	1.358	1.278	1.192
	1.5	1.789	1.699	1.601	1.497

TSC. Unlike the Maxwell BC where the TSC is nearly a linear function of TMAC, here the change of TSC with respect to α_t is nonlinear. Also, it is observed from Fig. 10.5 that the smaller the value of α_n, the stronger the variation in TSC with α_t. For instance, when $\omega = 0.5$ and $\alpha_n = 0.75$, the TSC first decreases to a minimum value at $\alpha_t \approx 0.5$, then increases to a maximum value at $\alpha_t \approx 1.5$.

10.3.2 Knudsen Layer Function

Figure 10.6a shows the KLF obtained from the LBE for HS, Maxwellian, and VHS ($\omega = 1.5$) molecules, when the diffuse BC is applied. Again, the KLF is strongly affected by the intermolecular potential, i.e., its value increases with the viscosity

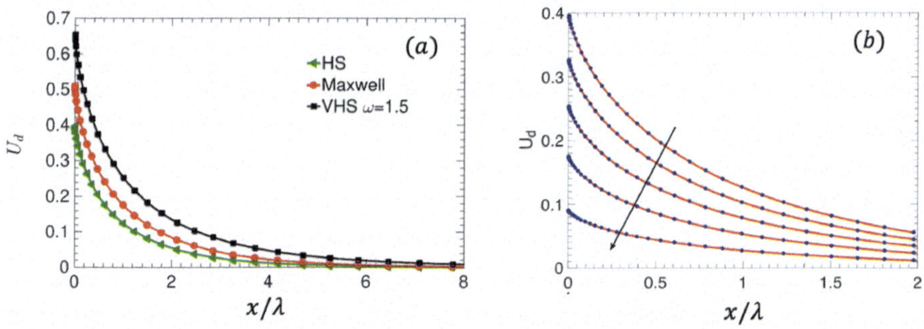

Fig. 10.6 Knudsen layer functions [3]. **a** KLFs obtained from the LBE for HS, Maxwellian, and VHS ($\omega = 1.5$) molecules, when the diffuse BC is applied. **b** KLFs for HS molecules with the Maxwell BC [3]. Along the arrow, the TMAC is $\alpha_M = 1.0, 0.8, 0.6, 0.4$, and 0.2, respectively

index ω. For example, the value of the KLF at solid surface for VHS molecules with $\omega = 1.5$ is larger than that of the HS molecules by approximately 60%.

Figure 10.6b further depicts the KLF when the Maxwell BC is used. It is seen that the KLF decreases with the TMAC; this is in sharp contrast to the KLF in Kramer's problem where its value increases when the TMAC decreases [2].

10.3.3 Molecular Gases

Based on the numerical solution of the Wu model [3, 8, 9], we find the TSC in the Maxwell BC can be fitted by

$$\sigma_T = f_{tr} \left(C_1 \alpha_M + 0.31 \right), \tag{10.8}$$

where the fitting coefficients are $C_1 = 0.095, 0.125$, and 0.165, for $\omega = 0.5, 0.75$, and 1, respectively. If one chooses $f_{tr} = 2.5$, this expression also well approximates the TSC of monatomic gas.

For the Cercignani-Lampis BC, when the TMAC α_t does not deviate too much from 1, say, when $|\alpha_t - 1| < 0.3$, the TSC $\sigma_T(\alpha_n, \alpha_t)$ can be expressed as

$$\sigma_T(\alpha_n, \alpha_t) = \frac{2}{5} f_{tr} \sigma_T^m(\alpha_n, \alpha_t), \tag{10.9}$$

where $\sigma_T^m(\alpha_n, \alpha_t)$ is the TSC for monatomic gas.

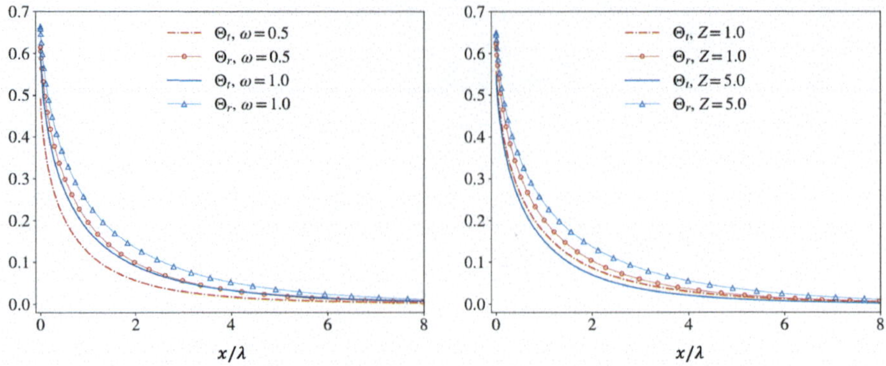

Fig. 10.7 KLFs for the translational and rotational temperatures, when the diffuse BC is used [10]. (Left) Influence of the intermolecular potential. (Right) Influence of the bulk viscosity. The data extracted from the DSMC are used, see Table 7.1, except (left) ω is changed, and (right) Z is changed

10.4 Temperature Jump

After simulating the planar Fourier flow (e.g., see Sect. 5.5.4) with small Knudsen numbers based on the Wu model, the TJC is calculated as

$$\zeta_T = -\delta_{rp}\frac{k+\Delta\tau}{2k}, \tag{10.10}$$

where k is the slope coefficient in the linear fitting $T_{NS}(x_2) = kx_2\Delta\tau$ and $\Delta\tau = \Delta T/T_0$ is the perturbation of wall temperature. Note that here the jump coefficients for the translational and rotational temperatures are the same, while the corresponding KLFs

$$\begin{aligned}\Theta_t(x_2) &= \frac{\delta_{rp}}{k\Delta\tau}(T_{NS}-T_t),\\ \Theta_r(x_2) &= \frac{\delta_{rp}}{k\Delta\tau}(T_{NS}-T_r),\end{aligned} \tag{10.11}$$

are different [10].

Influence of Intermolecular Potential
The translational and rotational KLFs are plotted in Fig. 10.7. When the viscosity index ω increases from 0.5 to 1, the TJC slightly increases.

Influence of Rotational Collision Number
We investigate the influence of the temperature relaxation (7.1) by changing Z but fixing the heat relaxation rates $\boldsymbol{A}$. The viscosity index is $\omega = 0.74$. This has little effect on the TJC, but affects the translational and rotational KLFs in Fig. 10.7. When $Z = 1$, i.e., the exchange of translational and rotational energies is frequent, Θ_t and Θ_r are close; as Z increases to 5, difference between the two KLFs enlarges.

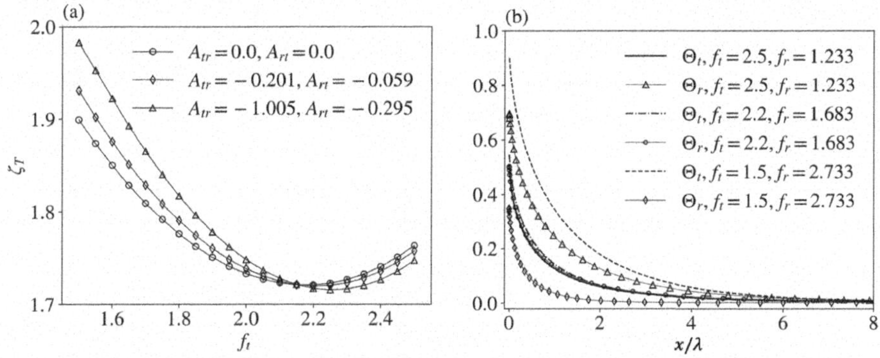

Fig. 10.8 TJC and KLF [10]. **a** The influence of the translational Eucken factor when $Z = 2.667$, $f_u = 1.993$, and the diagonal terms A_{tt} and A_{rr} altering. **b** The influence of the Eucken factors when $Z = 2.667$, $f_u = 1.993$, $A_{tr} = -1.005$ and $A_{rt} = -0.295$

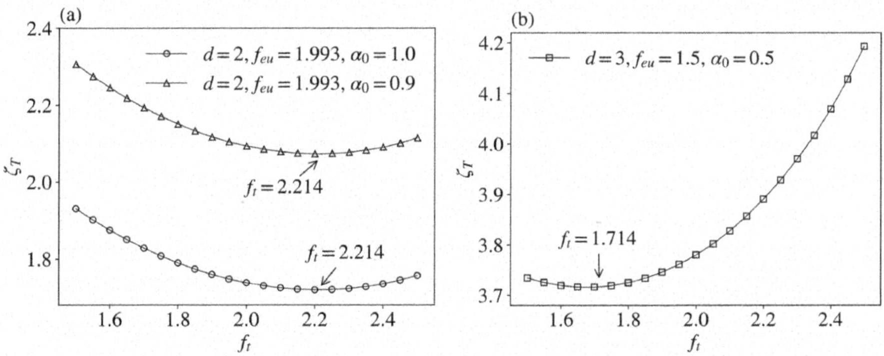

Fig. 10.9 Under the classical Maxwell's BC, a minimum TJC can be found when the translational and internal components f_t and f_r vary but the total Eucken factor f_{eu} is fixed [10]. **a** $d = 2$ and $f_{eu} = 1.993$, $Z = 2.667$, $A_{tr} = -0.201$ and $A_{rt} = -0.059$. **b** $d = 3$ and $f_{eu} = 1.5$, $Z = 3$ and $A_{tr} = A_{rt} = 0.0$. The diffuse BC is used

Influence of Thermal Relaxation

We fix A_{tr} and A_{rt} but alter f_{tr} and f_{rot}. The diagonal terms A_{tt} and A_{rr} will vary correspondingly. Figure 10.8a shows that the TJC first falls and then slightly rises as the translational Eucken factor increases (or the rotational Eucken factor decreases), and the minimum value appears around $f_{tr} = 2.25$. The most significant variation of ζ_T as a function of f_{tr} occurs when the off-diagonal elements in thermal relaxation is intensified ($A_{tr} = -1.005$ and $A_{rt} = -0.295$), where the minimum TJC is about 86% of the maximum one ($\zeta_T = 1.992$ at $f_{tr} = 1.5$).

The KLFs are plotted in Fig. 10.8b. When $f_{tr} = 1.5$ and $f_{rot} = 2.733$, Θ_t is larger than Θ_r; when f_{tr} increases to 2.5 and f_{rot} decreases, the situation reverses. When the two KLFs meet in the middle at $f_{tr} = 2.25$, the minimum TJC emerges.

The Minimum TJC

We find from Fig. 10.9 that the minimum TJC emerges when $f_t = \frac{4}{3} f_r$, independent of the rotational collision number and the accommodation coefficient. Eventually, we propose a new formulation for the TJC in molecular gases:

$$\begin{aligned}\zeta_T =& \frac{2-\alpha_M}{\alpha_M}(1+0.162\alpha_M)\frac{\sqrt{\pi}}{8} \\ &\times \left(\frac{4(3+d)}{4+d} f_{eu} + \left|\frac{4}{4+d} f_{eu} - \frac{3}{3+d} f_t\right|\right),\end{aligned} \tag{10.12}$$

where the minimum TJC

$$\zeta_T = \frac{2-\alpha_M}{\alpha_M}(1+0.162\alpha_M)\frac{\sqrt{\pi}}{2}\frac{(3+d)}{4+d} f_{eu}, \tag{10.13}$$

occurs at

$$f_t = \frac{4(3+d)}{3(4+d)} f_{eu}. \tag{10.14}$$

References

1. Maxwell JC (1879) VII. On stresses in rarified gases arising from inequalities of temperature. Proc Royal Soc Lond 170:231–256
2. Su W, Wang P, Liu HH, Wu L (2019) Accurate and efficient computation of the Boltzmann equation for Couette flow: influence of intermolecular potentials on Knudsen layer function and viscous slip coefficient. J Comput Phys 378:573–590
3. Wang P, Su W, Wu L (2020) Thermal transpiration in molecular gas. Phys Fluids 32:082005
4. Lilley CR, Sader JE (2008) Velocity profile in the Knudsen layer according to the Boltzmann equation. Proc Royal Soc A 464:2015–2035
5. Lockerby DA, Reese JM, Gallis MA (2005) Capturing the Knudsen layer in continuum-fluid models of nonequilibrium gas flows. AIAA J 43:1391–1393
6. Bird GA (1994) Molecular gas dynamics and the direct simulation of gas flows. Oxford Science Publications, Oxford University Press Inc, New York
7. Chapman S, Cowling TG (1970) The mathematical theory of non-uniform gases. Cambridge University Press, Cambridge
8. Wu L, White C, Scanlon TJ, Reese JM, Zhang YH (2015) A kinetic model of the Boltzmann equation for non-vibrating polyatomic gases. J Fluid Mech 763:24–50
9. Su W, Zhang YH, Wu L (2021) Multiscale simulation of molecular gas flows by the general synthetic iterative scheme. Comput Methods Appl Mech Engrg 373:113548
10. Su W, Li Q, Zhang YH, Wu L (2022) Temperature jump and Knudsen layer in rarefied molecular gas. Phys Fluids 34:032010

Chapter 11
Accuracy of Kinetic Boundary Condition

In this chapter, we assess the accuracy of Maxwell, Cercignani-Lampis, and Epstein boundary conditions, by comparing numerical results of the Boltzmann equation to the experimental data in microflows. Especially, in recent experiments of thermal transpiration, the mass flow rate and thermalmolecular pressure difference are measured simultaneously, which is ideal to assess the gas kinetic boundary condition.

11.1 Reynolds Lubrication Equation

The rarefied gas flow manifests at small length scale, however, real engineering applications work at macroscopic scales. Therefore, the upscaling method is needed, e.g., the mass and heat flow rates in Poiseuille flow and thermal transpiration through 2D cross sections in Chap. 5 is upscaled to predict the thermal molecular pressure difference in three-dimensional (3D) channel, where the full 3D kinetic simulation might be extremely time-consuming, if not impossible, in geometries with large aspect ratio. To this end, the Reynolds lubrication equation can be applied [1].

Take the geometry in Fig. 11.1 for an example. Averaging the continuity equation (1.1) along the film thickness, we have

$$\int_0^d \frac{\partial \rho}{\partial t} dy + \int_0^d \frac{\partial(\rho u)}{\partial x} dy + \int_0^d \frac{\partial(\rho v)}{\partial z} dy + \int_0^d \frac{\partial(\rho w)}{\partial y} dy = 0. \tag{11.1}$$

Since $\int_0^d \frac{\partial g}{\partial \xi} dy = \frac{\partial}{\partial \xi} \int_0^d g dy - g \frac{\partial d}{\partial \xi}$, and the normal velocity of the top surface (i.e., $v = \partial d/\partial t + u_{w2} \partial d/\partial t$) equals the temporal change in film thickness plus the spatial change of film thickness due to the lateral motion of the surface, the following Reynolds' equation can be derived

L. Wu, *Rarefied Gas Dynamics*,
https://doi.org/10.1007/978-981-19-2872-7_11

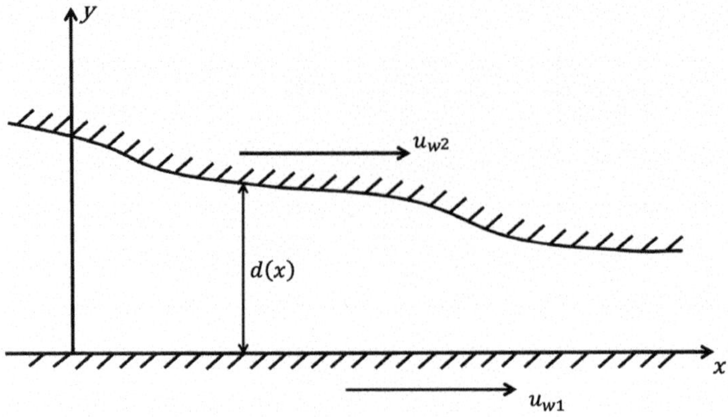

Fig. 11.1 Schematic of a two-dimensional slider geometry

$$\frac{\partial(\rho d)}{\partial t}+\frac{\partial \dot{M}_x}{\partial x}+\frac{\partial \dot{M}_z}{\partial z}=0, \tag{11.2}$$

where $\dot{M}_x=\int_0^d \rho u dy$ and $\dot{M}_z=\int_0^d \rho w dy$ are, respectively, the MFRs in the x and z-directions, along the thickness of film.

In Reynolds' original derivation, the continuum flow is considered. Ignoring the inertial effect, the momentum equations in Eq. (1.1) with Newton's viscosity law (1.2) are $\frac{\partial p}{\partial x}=\mu\frac{\partial^2 u}{\partial y^2}$ and $\frac{\partial p}{\partial z}=\mu\frac{\partial^2 w}{\partial y^2}$. With the thin-film assumption, the partial derivatives of pressure p do not vary across y. Thus, $u=\frac{1}{2\mu}\frac{\partial p}{\partial x}(y^2-yd)+\left(1-\frac{y}{d}\right)u_{w1}+\frac{y}{d}u_{w2}$ and $w=\frac{1}{2\mu}\frac{\partial p}{\partial z}(y^2-yd)$, where the non-slip boundary conditions have been applied. Therefore, Eq. (11.2) becomes

$$\frac{\partial(\rho d)}{\partial t}+\frac{\partial}{\partial x}\left(\underbrace{-\frac{d^3\rho}{12\mu}\frac{\partial p}{\partial x}}_{\text{Poiseuille}}+\underbrace{\frac{u_{w1}+u_{w2}}{2}\rho d}_{\text{Couette}}\right)+\frac{\partial}{\partial z}\left(\underbrace{-\frac{d^3\rho}{12\mu}\frac{\partial p}{\partial z}}_{\text{Poiseuille}}\right)=0. \tag{11.3}$$

If the rarefied gas effects are to be considered, MFRs in Eq. (11.2) should be calculated from gas kinetic equations at arbitrary Knudsen number, i.e., the Poiseuille and Couette flow through a cross section; also, the effect of thermal transpiration can be incorporated if the wall has temperature gradient [2–5], see the corresponding expression of MFR in Chap. 5.

11.2 Experiments and Upscaling

Typical geometry in most low-speed experiments designed to measure the MFR of Poiseuille flow and thermal transpiration [6–9] is shown in Fig. 11.2, where two

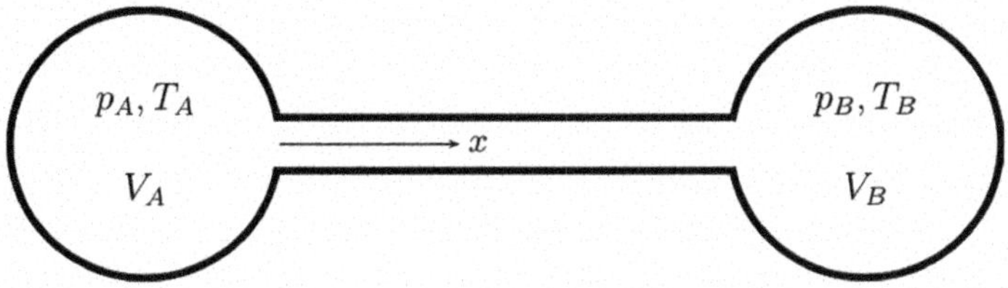

Fig. 11.2 Poiseuille flow and thermal transpiration along a long micro-channel that connects two large reservoirs

reservoirs are connected by a long micro-channel with a cross-sectional area of S. With Eq. (11.2), the time-dependent flow behaviors can be analyzed as

$$S\frac{\partial \rho}{\partial t} = -\frac{\partial \dot{M}}{\partial x}, \tag{11.4}$$

where the MFR is

$$\dot{M} = \frac{2LpS}{v_m}\left(-G_P\frac{\partial \ln p}{\partial x} + G_T\frac{\partial \ln T}{\partial x}\right), \tag{11.5}$$

with G_P and G_T the normalized MFR in Poiseuille flow and thermal transpiration, respectively, and L the characteristic flow length, see Chap. 5 for details.

Poiseuille Flow

Suppose the two reservoirs A and B are maintained at the pressure p_A and p_B, respectively, while the temperature of the whole system is T_0. When the steady-state is established, the MFR of the Poiseuille flow becomes

$$\dot{M} = -\frac{2LS}{v_m}G_P\frac{\partial p}{\partial x}, \tag{11.6}$$

or equivalently, if integrated along the long-channel,

$$\dot{M}L_t = -\frac{2LS}{v_m}\int_{p_A}^{p_B} G_P dp \equiv -2S\mu\int_{\delta_{rp}^A}^{\delta_{rp}^B} G_P(\delta_{rp})d\delta_{rp}, \tag{11.7}$$

where L_t is the length of the channel along x-direction. In analogy to Eq. (5.13), if we define the normalized MFR $G_P(\delta_{rp}^m)$ as

$$\dot{M} = -\frac{2LS}{v_m}\frac{p_B - p_A}{L_t}G_P(\delta_{rp}^m), \tag{11.8}$$

where $p_m = (p_A + p_B)/2$ is the mean pressure and δ_{rp}^m is the rarefaction parameter defined at this mean pressure and characteristic length L, we have

$$G_P(\delta_{rp}^m) = \frac{p_m}{(p_A - p_B)\delta_{rp}^m} \int_{\delta_{rp}^B}^{\delta_{rp}^A} G_P(\delta_{rp}) d\delta_{rp}. \tag{11.9}$$

In Chap. 5, we have solved the LBE to obtain the MFR $G_P(\delta_{rp})$ for various gas kinetic BCs. Then with Eq. (11.9) we shall assess the accuracy of BCs when compared to the experimental MFR.

Thermal Transpiration

Due to the thermal transpiration along the micro-channel that connects two reservoirs, the gas moves from cold region to hot region, which increase the pressure of the hot reservoir. This induces a Poiseuille flow which drives the gas from hot region to cold region. Eventually the balance is reached, and the pressure in hot reservoir is higher than that in the cold reservoir. The thermal molecular pressure ratio (TPR) can be measured, which is usually expressed by the following equation

$$\frac{p_A}{p_B} = \left(\frac{T_A}{T_B}\right)^{\gamma}, \tag{11.10}$$

where γ is the TPD exponent.

Theoretically, the TPR is calculated as follows. When the steady state is reached, we know the MFR is zero. Therefore, from Eq. (11.5) we know that the gas pressure p along the channel satisfies

$$\frac{dp}{dx} = \frac{G_T(\delta_{rp})}{G_P(\delta_{rp})} \frac{p}{T} \frac{dT}{dx}, \tag{11.11}$$

thus the pressure distribution (and hence TPR) can be obtained easily, provided that the temperature difference along the long micro-channel is known.

In case of small temperature difference between two reservoirs, the TPD exponent can be expressed as

$$\gamma = \frac{G_T}{G_P}. \tag{11.12}$$

On the other hand, when the two reservoirs are initially held at the same pressure, the MFR of thermal transpiration can also be extracted from the time evolution of reservoir pressure [9]. That is, during a small experiment duration Δt, the counter Poiseuille flow is negligible, thus

$$\dot{M} = \frac{V}{RT} \frac{dp}{\Delta t}, \tag{11.13}$$

where V and T are the volume and temperature of the reservoir, respectively. In analogy to Eq. (5.20), if we define the MFR in experiment as

$$\dot{M} = \frac{2LpS}{\sqrt{2k_B T_m/m}} G(\delta_{rp}^m) \frac{T_B - T_A}{L_t}, \tag{11.14}$$

then when the temperature difference $T_B - T_A$ is small we have [7, 9]

$$G(\delta_{rp}^m) = 2G_T(\delta_{rp}^m)\left(\sqrt{\frac{T_m}{T_A}} - \sqrt{\frac{T_m}{T_B}}\right), \tag{11.15}$$

where δ_{rp}^m is evaluated at the initial gas pressure p and $(T_A + T_B)/2$.

Therefore, in the thermal transpiration experiment, we can use not only the TPD exponent but also the MFR to assess various kinetic BCs.

11.3 Approximate Velocity Slip Coefficients

The velocity slip and temperature jump BCs of the Maxwell and Cercignani-Lampis BCs are presented in Chap. 10; those of the Epstein BC (2.76) are not included because it contains huge degrees of freedom. Here, the approximate solutions of VSC and TSC are presented for the Epstein BC:

$$\sigma_P = \frac{\frac{\Theta_0}{(1+\alpha)^3} + \Theta_1 - \frac{\Theta_1}{(1+\beta)^3}}{1 + \frac{\Theta_0}{2(1+\alpha)^3}\left(1 - \frac{1}{\sqrt{1+\alpha}}\right) - \frac{\Theta_1}{2(1+\beta)^3}\left(1 - \frac{1}{\sqrt{1+\beta}}\right)}, \tag{11.16}$$

and

$$\sigma_T = 1 - \frac{6\left[\frac{\alpha}{(1+\alpha)^4}\Theta_0 - \frac{\beta}{(1+\beta)^4}\Theta_1\right]}{\frac{\Theta_0}{(1+\alpha)^3} + \left[1 - \frac{1}{(1+\beta)^3}\right]\Theta_1}, \tag{11.17}$$

which have error of 10% or so [10]. Figure 11.3 shows examples of effective TMAC $\chi = 2/(1 + \sigma_P)$ and TSC σ_T in the Epstein BC. It is seen that by adjusting the

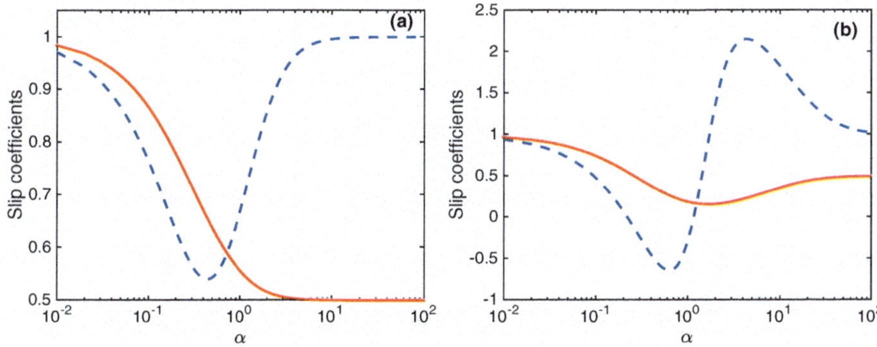

Fig. 11.3 The effective TMAC χ (solid lines) and TSC σ_T (dashed lines) in the Epstein BC [11]. Parameters in Eq. (2.77) are $\Theta_0 = 1$, $\Theta_1 = 0.5$, and **a** $\alpha = \beta$, **b** $\alpha = 20\beta$

parameter in the $\Theta(\boldsymbol{v})$ function (2.77), we can modify the VSC and TSC over a wide range. With these approximation solutions we are able to choose the right parameters to compare with experimental data.

11.4 Comparison with Experiment

The LBE (2.40), where the collision kernel is directly calculated from the Lennard-Jones potential (2.10), is solved by the FSM and GSIS. The MFRs in the Poiseuille flow and thermal transpiration are compared with experimental data to assess the performance of Maxwell, Cercignani-Lampis and Epstein BCs.

11.4.1 Poiseuille Flow Through a Rectangular Duct

In the experiment of Ewart et al., the helium flow in a silicon micro-channel with aspect ratio 52.45 is measured. The pressure ratio of the two reservoirs is 5, so that the measurement of MFR has sufficient accuracy even in the free-molecular flow regime where the gas pressure is small.

In the slip flow regime, the TMAC is 0.92, so we choose $\alpha_M = 0.92$ and $\alpha_t = 0.92$, $\alpha_n = 1$ in the Maxwell and Cercignani-Lampis BCs, respectively, while in the Epstein BC, $\chi = 0.92$, and

$$\Theta_0 = 1, \quad \Theta_1 = 0.1, \quad \alpha = \beta = 0.0293. \tag{11.18}$$

Figure 11.4 shows that, although the three BCs predict the MFR well in the slip flow regime, discrepancy emerges in the free-molecular flow regime, where the Cercignani-Lampis and Epstein BCs perform slightly better than the Maxwell BC.

11.4.2 Thermal Transpiration in a Rectangular Duct

The thermal transpiration of helium through a long rectangular channel made of polyether ether ketone, where the aspect ratio of the cross section is 27.27 [6, 7], is studied. It should be noted that, unlike the Poiseuille flow with large pressure ratio, the MFR induced by thermal transpiration is relative weak, so the experimental data might have large errors when $\delta_{rp} < 1$: the accuracy of the measurement decreases with the pressure.

When $\delta_{rp} < 3.5$, it is seen from Fig. 11.5 that the diffuse BC predicts a higher MFR. Since the data in Table 5.4 show that, in this region of rarefaction parameter, the MFR in Maxwell's diffuse-specular BC increases when the α_M decreases, in order to explain the experimental data, the effective TMAC should be larger than one, which is not possible in Maxwell's BC.

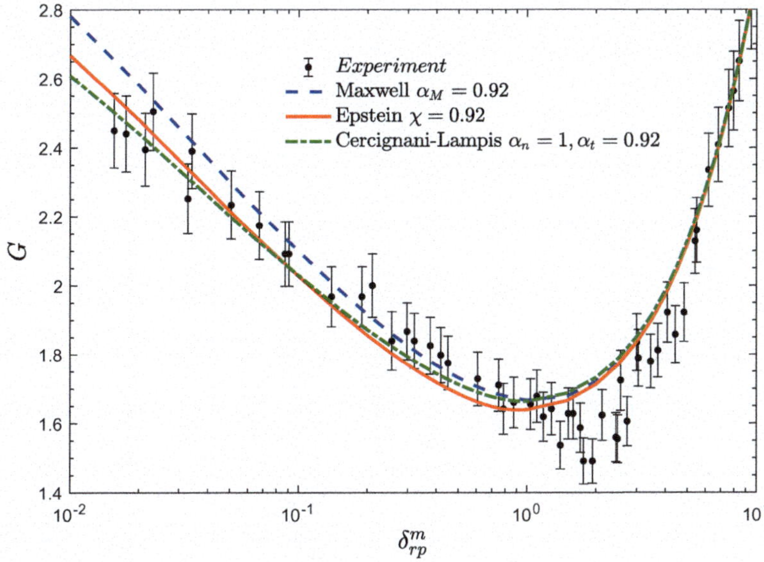

Fig. 11.4 The mass flow rate in the Poiseuille flow [11]

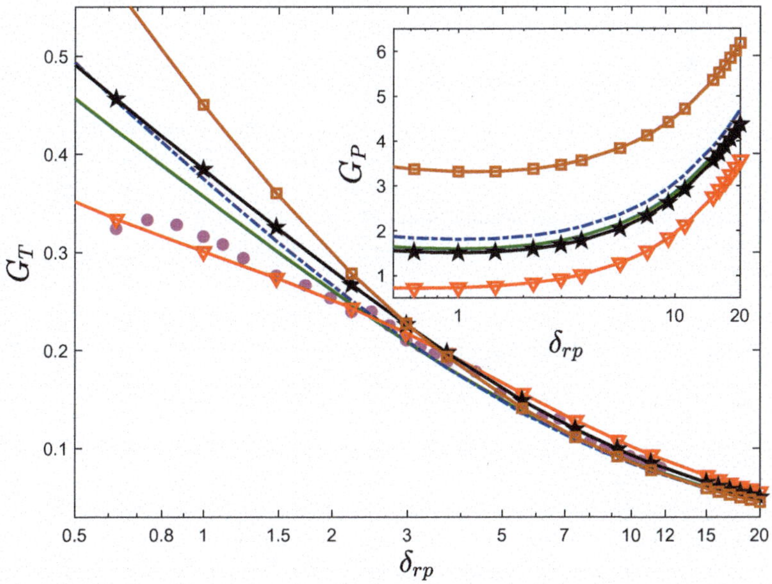

Fig. 11.5 Mass flow rates in the thermal transpiration of helium through a rectangular cross section of aspect ratio 27.27 [11]. Dots: experimental data collected from Table 1 in Ref. [7]. Pentagrams: diffuse BC. Squares: Cercignani-Lampis BC with $\alpha_n = 0.25$ and $\alpha_t = 0.5$. Triangles: Cercignani-Lampis BC with $\alpha_n = 0.25$ and $\alpha_t = 1.75$. Lines without symbols: Epstein BC (2.76) with the TSC $\sigma_T = 0.9$: dash-dotted line for $\Theta_0 = 0.9$ and $\alpha = 0.0192$, while solid line for $\Theta_0 = 1$ and $\alpha = 0.019$. Other parameters are $\Theta_1 = 0.1$ and $\beta = \alpha$, so that according to Eq. (11.16), the effective TMAC χ are 0.85 and 0.95 when $\Theta_0 = 0.9$ and 1, respectively

Table 11.1 The MFR in the thermal transpiration of helium between two infinite parallel plates [11]. Experimental data are collected from the Table 1 in Ref. [7]. Parameters in various BCs are given in Fig. 11.5. Cercignani-Lampis1: $\alpha_t = 0.5$. Cercignani-Lampis2: $\alpha_t = 1.75$. Epstein1: $\Theta_0 = 0.9$. Epstein2 $\Theta_0 = 1$

δ_{rp}	0.998	1.48	2.22	2.98	3.71	5.57	7.41	11.1
Exp.	0.316	0.276	0.239	0.210	0.188	0.140	0.107	0.0803
Diffuse	0.385	0.326	0.267	0.226	0.198	0.149	0.119	0.0851
CL1	0.450	0.360	0.278	0.227	0.193	0.141	0.111	0.0779
CL2	0.301	0.274	0.243	0.217	0.197	0.156	0.128	0.0932
Epstein1	0.355	0.300	0.245	0.208	0.181	0.137	0.110	0.0784
Epstein2	0.375	0.312	0.251	0.210	0.182	0.136	0.109	0.0771
Relative error $100 \times (G_T^{\text{numerical}}/G_T^{\text{exp}} - 1)$								
Diffuse	21.66	17.94	11.67	7.73	5.04	6.35	11.59	5.98
CL1	42.41	30.43	16.32	8.10	2.66	0.71	3.74	−2.99
CL2	−4.85	−0.87	1.62	3.50	4.55	11.34	19.63	16.07
Epstein1	12.22	8.59	2.66	−1.02	−3.49	−2.21	2.69	−2.33
Epstein2	18.60	12.86	4.91	0.02	−3.18	−2.93	1.42	−4.01

We then consider the Cercignani-Lampis BC. When δ_{rp} is large, better agreement between the experiment and simulation is achieved when small values of α_t and α_n are chosen, see the squares in Fig. 11.5 and data denoted by "Cercignani-Lampis1" in Table 11.1, where $\alpha_n = 0.25$ and $\alpha_t = 0.5$. However, when $\delta_{rp} < 2$, the agreement becomes worse than that of the diffuse BC. One can significantly reduce this difference by choosing large values of α_t (see triangles in Fig. 11.5 and data corresponding to "Cercignani-Lampis2" in Table 11.1, where $\alpha_n = 0.25$ and $\alpha_t = 1.75$), but this increases the error when δ_{rp} is large. For example, when $\delta_{rp} > 5$, numerical results overpredicts the MFR by about 20%. This large difference in the slip regime is caused by the fact that the TSC of Cercignani-Lampis BC is larger than one [12], which is much larger than the experimental measured value of about 0.9.

The Epstein BC allows more flexibility to choose the VSCs and is believed to give better agreement to experiment. To illustrate this, we choose the effective TSC σ_T to be the experimentally measured value [7]. To achieve this, we select $\alpha = \beta$ in Eq. (2.77). We first fix the values of Θ_0 and Θ_1, and obtain α and β from Eq. (11.17) by setting $\sigma_T = 0.9$. We then vary the values of Θ_0 and Θ_1, to see the possible influence of different parameters. Our numerical results show that Θ_1 has very small influence on the MFR, so in Fig. 11.5 and Table 11.1 only the results for $\Theta_1 = 0.1$ are shown. It is clear that when $\sigma_T = 0.9$, the simulation results agree well with the experimental data when $\delta_{rp} \gtrsim 2$, for a wide range of Θ_0, see $\Theta_0 = 0.9$ and 1 in Fig. 11.5 and Table 11.1. When $\delta_{rp} \lesssim 2$, the value of Θ_0 begins to have a strong influence on the MFR: the larger the value of Θ_0, the smaller the MFR in thermal transpiration flow.

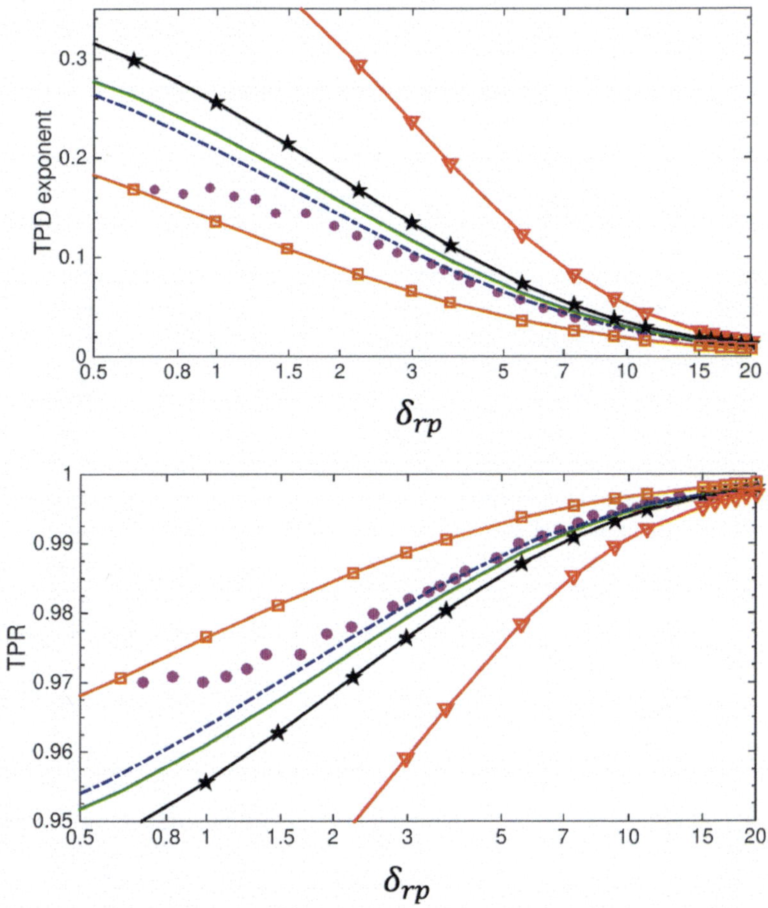

Fig. 11.6 The TPD exponent and TPR in the thermal transpiration of helium through a rectangular cross section of aspect ratio 27.27 [11]. Solid dots: experimental data collected from Fig. 8 in Ref. [6]. Pentagrams, squares, triangles, and lines without symbols: see the parameters in Fig. 11.5

11.4.2.1 A Tough Assessment of BC

It is still hard to conclude which BC is the best based solely on the comparison in Fig. 11.5 and Table 11.1. In the inset of Fig. 11.5, we find that the MFR in Poiseuille flow varies significantly among different BCs. Therefore, the TPD exponent and TPR also vary significantly among various BCs. Fortunately, these two parameters are measured experimentally using the same gas and solid surface [6], which provides an ideal and tough case to assess the accuracy of BCs.

Figure 11.6 compares the TPD exponent and TPR between the experimental and numerical results. It becomes clear that the Epstein BC with $\Theta_0 = 0.9$ (other parameters shown in Fig. 11.5) is the best, while the Cercignani-Lampis BC has large errors in the prediction of TPR and TPD exponent. Note that the discrepancies between the Epstein BC and experimental data at $\delta_{rp} < 2$ might due to the experimental error, which becomes larger and larger as the gas pressure reduces.

11.4.3 Thermal Transpiration Through a Long Tube

The MFR, TPD exponent, and TPR are measured in the thermal transpiration flow through a glass tube [9], which also provides a tough assessment of the BC.

We first consider the diffuse BC. Figure 11.7 shows that, although the TPR predicted by the numerical solutions agrees well with the experimental data, the MFR is higher. Like the thermal transpiration through the rectangular channel studied in Sect. 11.4.2, the Cercignani-Lampis BC has difficulty to predict both MFR and TPR correctly. Therefore, we solve the LBE with the Epstein BC to obtain a good agreement in G_T first. This is easily achieved when we choose $\sigma_T = 0.9$, however, the TPR from the numerical simulation is higher than the experimental ones when $\delta_{rp} > 2$. This means that, according to Eq. (11.11), we have to make the effective TMAC in the Epstein BC larger than unity, which is impossible to guarantee the positivity of VDF. Thus, a new kinetic BC has to be developed to describe this experimental data.

11.4.3.1 The Combined Boundary Condition

When the experimental results indicate VSC and TSC smaller than 1, no existing kinetic model can be used. Therefore, Wu and Struchtrup proposed to linearly combine the Cercignani-Lampis and Epstein BCs together [11]:

$$R_{Mix}(\boldsymbol{v}' \to \boldsymbol{v}) = \varpi R_{CL}(\boldsymbol{v}' \to \boldsymbol{v}) + (1 - \varpi) R_E(\boldsymbol{v}' \to \boldsymbol{v}), \tag{11.19}$$

where $\varpi \in [0, 1]$ is a constant. Since the only reason to include the Cercignani-Lampis BC is to make the effective TMAC larger than unity, the EAC α_n, which has little effect to the effective TMAC, is chosen as unity.

Approximate analytical solutions for the effective TMAC and TSC can be simply given as [10]:

$$\begin{aligned} \frac{1}{\chi_{Mix}} &= \frac{\varpi}{\chi_{CL}} + \frac{1-\varpi}{\chi_{Epstein}}, \\ \sigma_{T,Mix} &= \varpi \sigma_{T,CL} + (1-\varpi)\sigma_{T,Epstein}. \end{aligned} \tag{11.20}$$

To compare with the experimental data on helium [9], we consider an equal mixture of the Epstein and Cercignani-Lampis BCs, i.e., $\varpi = 0.5$ in Eq. (11.19). We choose $\alpha_t = 1.5$ in the Cercignani-Lampis BC, and $\chi = 0.85$ and $\sigma_T = 0.7$ in the Epstein BC, so that according to Eq. (11.20), the effective TMAC is $\chi_{Mix} \approx 1.1$

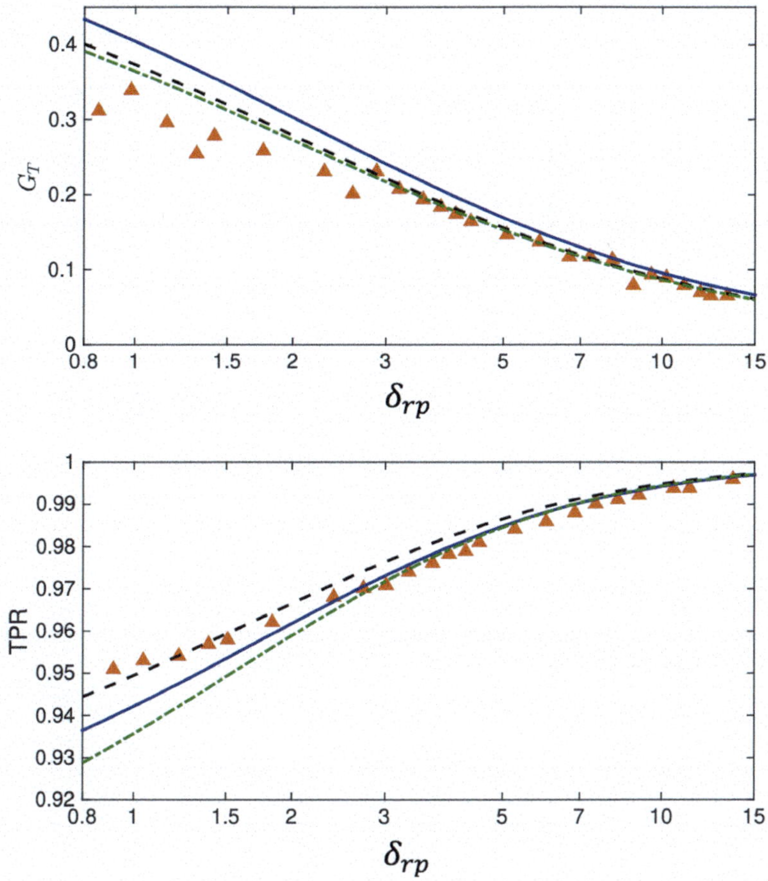

Fig. 11.7 The MFR and TPR in the thermal transpiration of helium through a circular glass tube [11]. Solid dots: the experimental data [9]. Solid lines: diffuse BC. Dashed lines: the Epstein BC with $\Theta_0 = 1$, $\Theta_1 = 0.1$, and $\alpha = \beta = 0.019$, so that the effect TSC $\sigma_T = 0.9$ and effective TMAC is $\chi = 0.95$. Dash-dotted lines: the combined BC (11.19), where $\alpha_n = 1$ and $\alpha_t = 1.5$ in the Cercignani-Lampis BC, while $\Theta_0 = 1$, $\Theta_1 = 0.1$, and $\alpha = \beta = 0.06$, so that the TSC $\sigma_T = 0.7$ and the effective TMAC is $\chi = 0.85$ in the Epstein BC

and the TSC is $\sigma_{T,Mix} \approx 0.9$. Good agreement between the numerical simulation and experiment is now achieved, see the dash-dotted lines in Fig. 11.7.

11.5 Implication in Hypersonic Flows

In the previous section, it is shown that the Epstein BC is better than the Maxwell's specular-diffusion boundary condition in low-speed flows. However, the difference

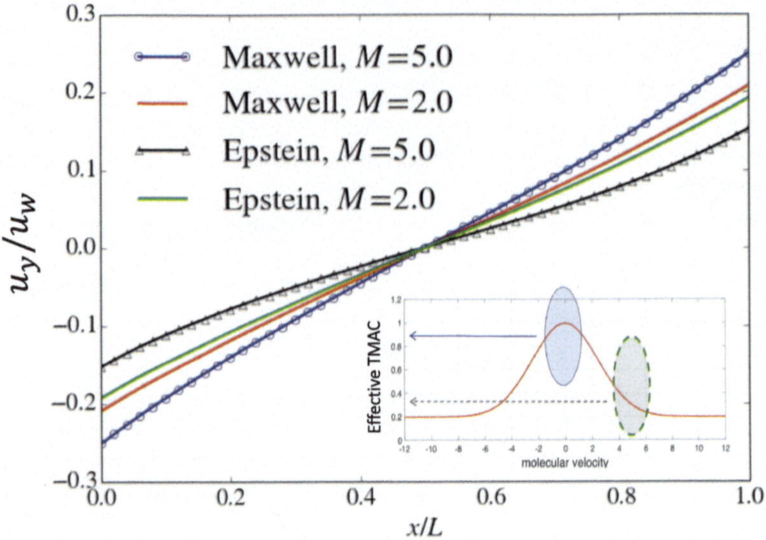

Fig. 11.8 The flow velocity in the planar Couette flow at different values of the wall speed, and different BCs, when Kn = 1. The wall speed is $\pm M$ times of the most probable speed. When the flow speed is small, the Epstein BC has an equivalent TMAC of 0.8. (Inset) Schematic of the effective TMAC as a function of the molecular velocity

between different BCs is not very significant. Here we show that in hypersonic flows, the RGD with the Epstein boundary condition could differ significantly from that of the Maxwell's one. In the Maxwell BC, we set the TMAC to be 0.8, while in the Epstein's BC, we make sure that the effective TMAC (11.16) is 0.8 when the flow velocity is much smaller than the most probable speed.

Consider the planar Couette flow for an example. Figure 11.8 shows the velocity profiles when the wall velocity is M times of the most probable speed. When the wall speed is small, the velocity profiles obtained from the two BCs are nearly identical. Also, the wall stress, normalized by the product of mean pressure and M, are 0.24 and 0.29 for the Epstein and Maxwell BCs, respectively. However, at large wall speed, the slip velocity of the Epstein BC is larger than that of the Maxwell one, and the normalized stress is 0.16, which is about twice smaller than the Maxwell's BC, 0.37.

The numerical results suggest that at large wall speed, the Epstein BC has an effective TMAC smaller than 0.8. This can be qualitatively explained as follows. When the wall speed is high, most of the gas molecules has large velocity, and hence smaller value of TMAC , see the inset in Fig. 11.8. Therefore, if in reality the gas-surface interaction is better described by the Epstein BC, in hypersonic flight, one may achieve much smaller drag force. More studies are required to test this conjecture.

References

1. Reynolds O (1886) On the theory of lubrication and its application to Mr. Beauchamp Tower's experiments, including an experimental determination of the viscosity of olive oil. Proc R Soc Lond 40(242–245):191–203
2. Fukui S, Kaneko R (1988) Analysis of ultra-thin gas film lubrication based on linearized Boltzmann equation: first report–derivation of a generalized lubrication equation including thermal creep flow. J Tribol 110:253–261
3. Sharipov F, Fahrenbach P, Zipp A (2005) Numerical modeling of holweck pump. J Vac Sci Technol A 23:1331–1339
4. Cercignani C, Lampis M, Lorenzani S (2007) On the Reynolds equation for linearized models of the Boltzmann operator. Transp Theory Stat Phys 36(4–6):257–280
5. Su W, Liu HH, Zhang YH, Wu L (2017) Rarefaction cloaking: influence of the fractal rough surface in gas slider bearings. Phys Fluids 29:102003
6. Yamaguchi H, Rojas-Cárdenas M, Perrier P, Graur I, Niimi T (2014) Thermal transpiration flow through a single rectangular channel. J Fluid Mech 744:169–182
7. Yamaguchi H, Perrier P, Ho MT, Méolans JG, Niimi T, Graur IA (2016) Mass flow rate measurement of thermal creep flow from transitional to slip flow regime. J Fluid Mech 795:690–707
8. Ewart T, Perrier P, Graur IA, Méolans JG (2007) Mass flow rate measurements in a microchannel, from hydrodynamic to near free molecular regimes. J Fluid Mech 584:337–356
9. Rojas-Cárdenas M, Graur IA, Perrier P, Méolans JG (2013) Time-dependent experimental analysis of a thermal transpiration rarefied gas flow. Phys Fluids 25:072002
10. Struchtrup H (2013) Maxwell boundary condition and velocity dependent accommodation coefficient. Phys Fluids 25:112001
11. Wu L, Struchtrup H (2017) Assessment and development of the gas kinetic boundary condition for the Boltzmann equation. J Fluid Mech 823:511–537
12. Sharipov F (2003) Application of the Cercignani–Lampis scattering kernel to calculations of rarefied gas flows. II. Slip and jump coefficients. Eur J Mech B-Fluids 22:133–143

Chapter 12
Porous Media Flow

Unconventional gas extraction has attracted significant attentions due to the shale gas revolution, but it is a grand challenge to reliably predict the gas production. The permeability of porous medium is a key parameter used in the upscaling methods to predict the gas production. Therefore, it is necessary to investigate the factors affecting the permeability. This chapter is dedicated to the interpretation of Klinkenberg's experimental data on the relation of gas permeability and reciprocal gas pressure.

12.1 Apparent Gas Permeability

For flows in the porous media, the Darcy law is widely used, which states that the discharge of the fluid in a permeable porous media is proportional to the cross-section area, pressure gradient, and inverse viscosity of the fluid. The proportionality constant is called the intrinsic permeability k_∞, which is determined by the porous media.

For gas flows in porous media, however, the apparent gas permeability (k_a, AGP) increases with the reciprocal mean gas pressure as [1]:

$$k_a = k_\infty \left(1 + \frac{b}{\bar{p}}\right), \tag{12.1}$$

where b is the correction factor. Constant values of b are widely used, although in the original experiment, Klinkenberg found that it is "evident the value of b increases with increasing pressure", that is, b decreases when Kn increases, see Fig. 12.1.

L. Wu, *Rarefied Gas Dynamics*,
https://doi.org/10.1007/978-981-19-2872-7_12

$\bar{p}=\frac{p_1+p_2}{2}$ (Atmospheres)	p_1-p_2 (Atmospheres)	$\frac{1}{\bar{p}}$ (Atmospheres)$^{-1}$	K_a (Millidarcys)	$\frac{K_a}{K}$	b
0.01665	0.02717	60.1	900	5.29	0.071
0.01779	0.02652	56.2	863	5.07	0.072
0.02130	0.02528	46.96	758	4.44	0.073
0.04890	0.01821	20.46	455	2.67	0.082
0.1738	0.00843	5.75	268	1.58	0.082
1.701	0.02017	0.587	184.9	1.087	0.15
2.659	0.01625	0.3761	180.0	1.058	0.16
3.641	0.01511	0.2748	179.8	1.057	0.21
4.619	0.01412	0.2165	179.1	1.052	0.24
7.04	0.00821	0.1420	176.7	1.038	0.27
12.89	0.00582	0.0776	173.9	1.022	0.29

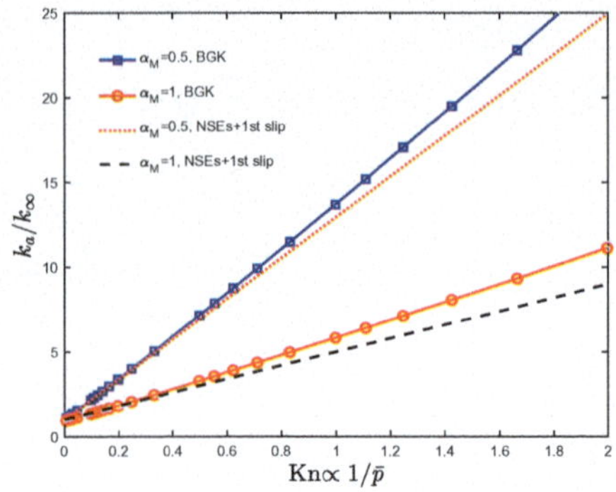

Fig. 12.1 (Left) Klinkenberg's experimental data show the correction factor b decreases when the Knudsen number increases [1]. (Right) The AGP in a straight tube [2], obtained from the numerical simulation of the BGK equation with Maxwell's BC. NSEs+1st slip: analytical solutions of the Navier-Stokes equations with first-order velocity slip BC

The increase of AGP with the reciprocal gas pressure is caused by the rarefaction effects, e.g., the velocity slip at solid walls, the high-order constitutive relations beyond that of the NSF equations. By solving the gas kinetic equation on the exact porous media used by Klinkenberg, the nonlinear relation between k_a/k_∞ and the reciprocal gas pressure (or equivalently, Kn) should be recovered. However, since the numerical simulations, based on the gas kinetic theory, on complicated porous media is time-consuming, the porous medium is often simplified by a single straight tube, or by the pore-network in which each neighboring nodes are connected by straight tubes [3]. Then, using the Maxwell's diffuse-specular BC, the AGP of a straight tube can be obtained; and that of pore-network can also be found using the Kirchhoff's law. Results in Fig. 12.1 show that, however, the correction factor b increases with the Knudsen number.

12.2 Kinetic Formulation

The discrepancy between the experiment and gas kinetic theory may be caused by the oversimplified geometry. Therefore, kinetic theory-based numerical simulation on complex porous media should be carried out. To this end, we study the gas flow in 2D periodic porous media shown in the inset of Fig. 12.2. Since the flow speed in porous media is usually small, the rarefied gas flow can be investigated in the unit rectangular cell ABCD.

The linearized BGK equation is used to describe the evolution of the perturbation VDF:

$$\frac{\partial h}{\partial t}+v_1\frac{\partial h}{\partial x_1}+v_2\frac{\partial h}{\partial x_2}=\frac{\sqrt{\pi}}{2\text{Kn}}\left(\varrho+2u_1v_1+2u_2v_2-h\right), \tag{12.2}$$

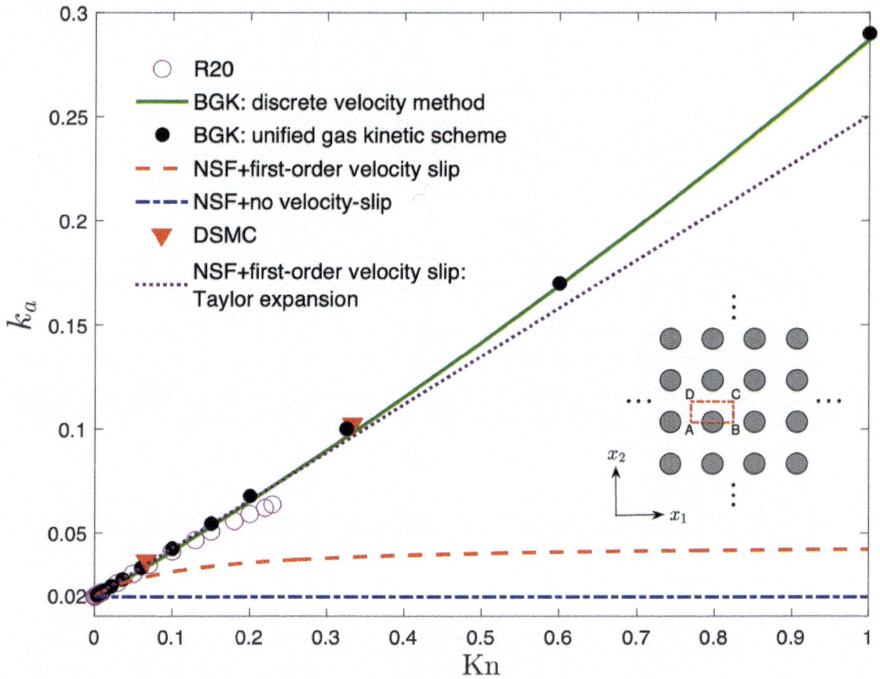

Fig. 12.2 The AGP as a function of the Knudsen number for a gas passing through porous medium of porosity 0.8 and consists of square cylinders [2]

where the definition and normalization are the same as those in Chap. 5, when the characteristic length L is the length of side AB.

A pressure gradient is applied in the horizontal direction, so that the kinetic BC at the inlet and outlet is

$$h(\mp 0.5, x_2, v_1, v_2, v_3) = \pm 1 + h(\pm 0.5, x_2, v_1, v_2, v_3), \text{ when } v_1 \gtrless 0, \tag{12.3}$$

while at the lines AB and CD, the specular reflection BC is used as a result of periodic structure. At the solid surface, the diffuse-specular BC is used.

The AGP, which is normalized by L^2, is calculated by

$$k_a = 2\sqrt{\frac{1}{\pi}}\text{Kn}G_P, \tag{12.4}$$

where $G_p = 2\int_0^{1/2} u_1(x_2)dx_2$ is the dimensionless mass flow rate.

12.3 Accuracy of Navier-Stokes Equations

Applying the Navier-Stokes equations and the first-order velocity slip BC to the geometry in Fig. 12.2, the AGP is found analytically [4]:

$$k_a = \frac{-\ln\phi - \frac{3}{2} + 2\phi - \frac{\phi^2}{2} + 4\sigma_P \mathrm{Kn}\sqrt{\frac{1}{\phi}}\left(-\ln\phi - \frac{1}{2} + \frac{\phi^2}{2}\right)}{8\pi\left(1 + 4\sigma_P \mathrm{Kn}\sqrt{\frac{1}{\phi}}\right)}, \tag{12.5}$$

where $\epsilon = 1 - \pi D^2/4L^2$ is the porosity with D being the disk diameter, $\phi = 1 - \epsilon$ is the solid fraction, and σ_P is the VSC in Chap. 10.

We solve the kinetic equation using the discrete velocity method. Different numerical methods, including the unified gas kinetic scheme for the BGK equation, the DSMC that is equivalent to the Boltzmann equation, and the regularized 20 moment equations that is more accurate than the Navier-Stokes equations, are also used. Figure 12.2 shows that they agree with each other, which demonstrates the accuracy of our numerical scheme [2, 5–7].

Thus, our numerical solution is used to assess the accuracy of the analytical solution (12.5). Figure 12.2 shows that the analytical permeability increases linearly with the Knudsen number when Kn $\lesssim 0.02$, and then quickly saturates when Kn $\gtrsim 0.2$. Such a behavior is used to explain Klinkenberg's experimental result [8], but it is wrong. Since the Navier-Stokes equations with first-order velocity slip BC are only accurate to the first-order of Knudsen number, only the linear relation between the k_a and Kn should be valid. This example shows that the Navier-Stokes equations and the first-order velocity slip BC should be carefully used when the curved boundary is concerned.

The Slip-Corrected AGP

Since the Navier-Stokes equation is only accurate to the first-order of Knudsen number, the analytical solution (12.5) is expanded into the Taylor's series of Kn, and only the zeroth- and first-order terms are kept. The resulting slip-corrected AGP reads

$$k_{a,\mathrm{slip}} = \frac{1}{8\pi}\left[-\ln\phi - \frac{3}{2} + 2\phi - \frac{\phi^2}{2} + 4\sigma_P \mathrm{Kn}\sqrt{\frac{1}{\phi}}\left(1 - 2\phi + \phi^2\right)\right]. \tag{12.6}$$

It is surprising to note in Fig. 12.2 that this slip-corrected AGP agrees well with the BGK solution when Kn $\lesssim 0.35$; it is only 15% smaller than that from the BGK solution even at Kn $= 1$.

We further test on the more complicated geometry shown in Fig. 12.3, which is fitted into the unit cell in Fig. 12.2. Figure 12.3 shows that the AGP predicted by the Navier-Stokes equations is only accurate when Kn $\lesssim 0.002$. As the Knudsen number increases, the AGP quickly saturates to a value that is about 3 times larger than the intrinsic permeability.

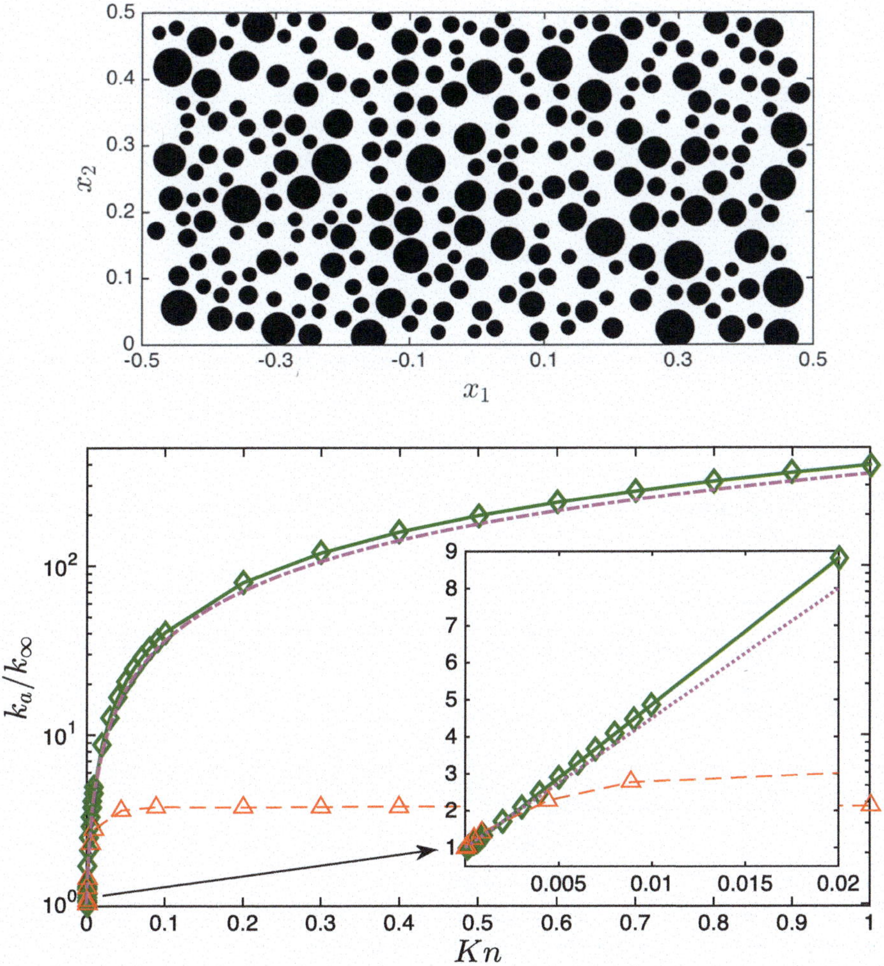

Fig. 12.3 The AGP in the porous media made of random disks with porosity 0.6 [2]. Solid (dashed) lines: BGK (Navier-Stokes equations with velocity slip BC). Dotted lines: the slip-corrected permeability

Interestingly, like Eq. (12.6), when the numerical solution of the Navier-Stokes equations with first-order velocity slip BC is Taylor expanded and only the linear term is kept, the obtained slip-corrected AGP is accurate up to Kn ≈ 1, where k_a/k_∞ is about 400. If we introduce the effective Knudsen number as [8]:

$$L^* = L\sqrt{\frac{12k_\infty}{\epsilon}}, \quad \text{Kn}^* = \frac{\lambda}{L^*} = \sqrt{\frac{\epsilon}{12k_\infty}}\text{Kn}, \tag{12.7}$$

then we find $\text{Kn}^* = 73\text{Kn}$, which means that the slip-corrected AGP works up to an effective Knudsen number of 73.

12.4 Interpretation of Experiment

Numerical simulations in the previous section still cannot explain Klinkenberg's experimental result, even when the porous media are already very complex. This drives us to investigate the role of gas-surface kinetic boundary conditions. To this end, the disks in Fig. 12.2 are replaced by the squares, so that the specular reflection can be accurately implemented in the numerical simulation.

Since L^* is representative of the characteristic flow length, we rewrite Klinkenberg's equation (12.1) in the dimensionless form

$$\frac{k_a}{k_\infty} = 1 + \frac{b}{\bar{p}} \equiv 1 + b'\mathrm{Kn}^*, \tag{12.8}$$

where $b = b'\mu v_m\sqrt{\pi\epsilon/48k_\infty}$. Therefore, the relation between b and Kn can be studied from the relation between b' and Kn^*.

The BGK solution is shown in Fig. 12.4. When Kn is fixed, the AGP increases when TMAC decreases. However, at different value of Kn, the amount of increase is different. When $\alpha_M = 1$, the correction factor b' increases with Kn, which contradicts Klinkenberg's experimental results. However, when $\alpha_M = 0.5$ and 0.1, at small Knudsen number, b' decreases when Kn increases. This behavior is in qualitatively

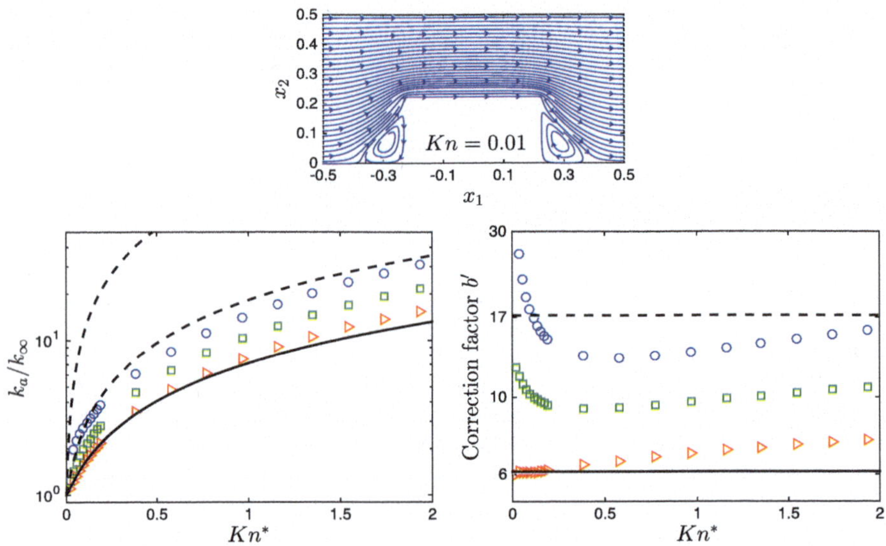

Fig. 12.4 Streamlines, AGP, and correction factor for the rarefied gas flow in period array of square cylinders, obtained from numerical results of the BGK equation [2]. Triangles, squares, and circles are for the TMAC $\alpha_M = 1$, 0.5, and 0.1, respectively. The porosity is $\epsilon = 0.8$, the intrinsic permeability is $k_\infty = 0.018$, and $\mathrm{Kn}^* = 1.94\mathrm{Kn}$. Lines in the bottom left figure show the slip-corrected AGP

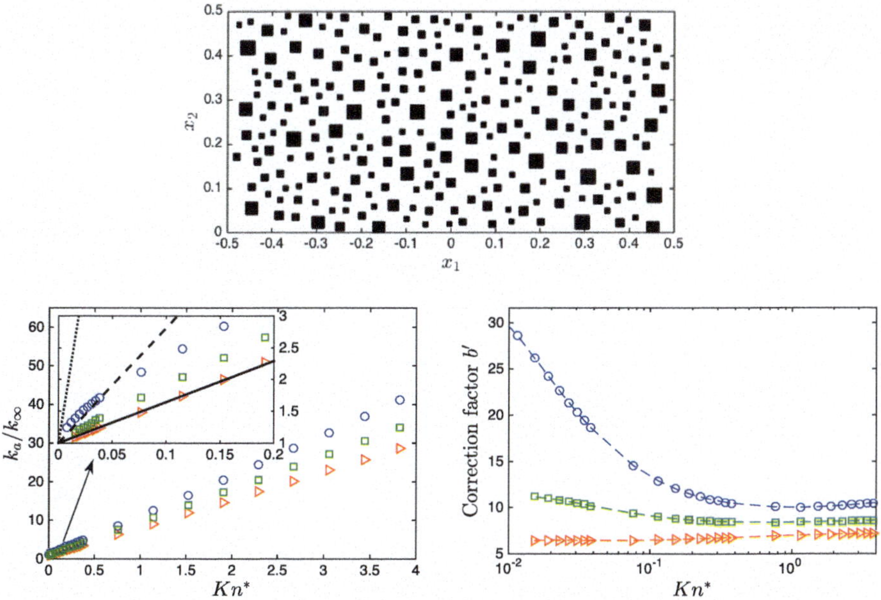

Fig. 12.5 The AGP and the correction factor in rarefied gas flows through the porous medium made of random squares, obtained from numerical results of the BGK equation [2]. Triangles, squares, and circles are for the TMAC $\alpha_M = 1$, 0.5, and 0.1, respectively. In the inset, the solid, dashed, dotted lines are the slip-corrected permeabilities. Note that the intrinsic permeability is $k_\infty = 4.77 \times 10^{-5}$, the porosity is 0.84, and $\mathrm{Kn}^* = 38.3\mathrm{Kn}$. Lines in the bottom left figure show the slip-corrected AGP

agreements[1] with experimental results summarized in Fig. 12.1. This discovery is also observed in the gas flow through a porous medium consisting of squares of random size and position in Fig. 12.5.

So two key factors that could explain Klinkenberg's experimental observation have been pinpointed. First, the TMAC for the gas-surface interaction should be less than unity. Second, the flow streamlines must be tortuous. Both factors should be combined together to explain Klinkenberg's observations, as the variation of TMAC in the straight tube (Fig. 12.1) and the variation of porous media but with unity TMAC (Figs. 12.2 and 12.3) yield the result contradictory to experimental observations.

12.5 Asymptotic Behavior at Large Kn

In the free-molecular flow regime, the normalized MFR G_p is a constant for Poiseuille flow through straight tube. Similar trends of G_p are observed in all the cases investi-

[1] Note that a quantitative comparison between the experiment and our simulation is currently impossible because we do not have the detailed structure data of the porous media.

gated here. Therefore, according to Eq. (12.4), at large Kn, the AGP is proportional to Kn. Therefore, the equivalent correction factor b'

$$b' = \frac{k_a/k_\infty - 1}{\text{Kn}^*} = \frac{2\text{Kn}G_p/\sqrt{\pi}k_\infty}{\text{Kn}^*} - \frac{1}{\text{Kn}^*} \equiv \text{Const} - \frac{1}{\text{Kn}^*} \tag{12.9}$$

increases with Kn, no matter what the TMAC is.

This is not reported in Klinkenberg's original paper. This is probably because the Knudsen number is not large enough in his experiments, as from Tables 5–7 in his paper, the maximum value of k_a/κ_∞ is within 30. In this region of k_a/κ_∞, from Figs. 12.4 and 12.5 we see the correction factor decreases when the Knudsen number increases.

Finally, we would like to comment on the influence of TMAC on the MFR G_P (or equivalently, the correction factor) in the free-molecular flow regime. In the straight tube, G_P is roughly proportional to $(2 - \alpha_M)/\alpha_M$. However, where there are obstacles along the flow passage, G_P obtained from the kinetic equation with small value of TMAC α_M will be much reduced, to a value that is very close to that of $\alpha_M = 1$. For example, for the flow passing through random array of squares, Fig. 12.5 shows that G_P with $\alpha_M = 0.1$ is only 1.3 times larger than that with $\alpha_M = 1$. This may be used to explain why the enhancement (i.e., k_a/k_∞) inside the graphene channel decreases when the channel height increases in the free-molecular flow [9]: when the channel height is large, polymethyl methacrylate is absorbed in the graphene surface, which acts as the barrier in the flow passage.

References

1. Klinkenberg LJ (1941) The permeability of porous media to liquids and gases. American Petroleum Institute, pp API–41–200
2. Wu L, Ho MH, Germanou L, Gu XJ, Liu C, Xu K, Zhang YH (2017) On the apparent permeability of porous media in rarefied gas flows. J Fluid Mech 822:398–417
3. Ma J, Sanchez JP, Wu K, Couples GD, Jiang Z (2014) A pore network model for simulating non-ideal gas flow in micro- and nano-porous materials. Fuel 116:498–508
4. Chai Z, Lu J, Shi B, Guo Z (2011) Gas slippage effect on the permeability of circular cylinders in a square array. Int J Heat Mass Transf 54:3009–3014
5. Ho MT, Zhu LH, Wu L, Wang P, Guo ZL, Li ZH, Zhang YH (2019) A multi-level parallel solver for rarefied gas flows in porous media. Comput Phys Commun 234:14–25
6. Ho MT, Zhu LH, Wu L, Wang P, Guo Z, Ma J, Zhang YH (2019) Pore-scale simulations of rarefied gas flows in ultra-tight porous media. Fuel 249:341–351
7. Su W, Ho MT, Zhang YH, Wu L (2020) GSIS: an efficient and accurate numerical method to obtain the apparent gas permeability of porous media. Comput Fluids 206:104576
8. Lasseux D, Valdes Parada FJ, Porter ML (2016) An improved macroscale model for gas slip flows in porous media. J Fluid Mech 805:118–146
9. Keerthi A, Geim AK, Janardanan A, Rooney AP, Esfandiar A, Hu S, Dar SA, Grigorieva IV, Haigh SJ, Wang FC, Radha B (2018) Ballistic molecular transport through two-dimensional channels. Nature 558:420–424

Chapter 13
Gas Mixture

We extend the fast spectral method to mixture of monatomic gases, focusing on the calculation of cross-collision operators between molecules with different masses. We also briefly introduce the general synthetic iterative scheme to find the solution of Poiseuille flow for gases interacting through the Lennard-Jones potentials, and assess the accuracy of McCormack model.

13.1 Boltzmann Equation for Gas Mixture

Consider the binary mixture of monatomic gases, where the component $\iota(= A$ or $B)$ has a molecular mass m^ι and VDF $f^\iota(t, \boldsymbol{x}, \boldsymbol{v})$. The Boltzmann equation takes the form of

$$\frac{\partial f^\iota}{\partial t} + \boldsymbol{v} \cdot \frac{\partial f^\iota}{\partial \boldsymbol{x}} = \sum_{\jmath=1,2} Q^{\iota\jmath}(f^\iota, f_*^\jmath), \tag{13.1}$$

where the BCO is

$$Q^{\iota\jmath}(f^\iota, f_*^\jmath) = \iint B^{\iota\jmath}(\cos\theta, v_r)\left[f^\jmath(\boldsymbol{v}_*')f^\iota(\boldsymbol{v}') - f^\jmath(\boldsymbol{v}_*)f^\iota(\boldsymbol{v})\right]d\Omega d\boldsymbol{v}_*, \tag{13.2}$$

with the post-collision velocities determined from the conservation of momentum and kinetic energy as:

$$\begin{aligned} \boldsymbol{v}' &= \boldsymbol{v} + \frac{m^\jmath}{m^\iota + m^\jmath}(v_r\boldsymbol{\Omega} - \boldsymbol{v}_r), \\ \boldsymbol{v}_*' &= \boldsymbol{v}_* - \frac{m^\iota}{m^\iota + m^\jmath}(v_r\boldsymbol{\Omega} - \boldsymbol{v}_r). \end{aligned} \tag{13.3}$$

L. Wu, *Rarefied Gas Dynamics*,
https://doi.org/10.1007/978-981-19-2872-7_13

When $\iota \neq \jmath$, $Q^{\iota\jmath}$ is the cross-collision operator, which describes the change of VDF f^ι due to the collision with the component $\jmath$. It conserves the mass ($\int Q^{\iota\jmath} d\boldsymbol{v} = 0$), total momentum ($\int m^\iota \boldsymbol{v} Q^{\iota\jmath} d\boldsymbol{v} + \int m^\jmath \boldsymbol{v} Q^{\jmath\iota} d\boldsymbol{v} = 0$), and total energy ($\int m^\iota |\boldsymbol{v}|^2 Q^{\iota\jmath} d\boldsymbol{v} + \int m^\jmath |\boldsymbol{v}|^2 Q^{\jmath\iota} d\boldsymbol{v} = 0$), instead of the momentum and energy of individual species.

The number density n^ι, bulk velocity $\boldsymbol{u}^\iota$, pressure tensor p^ι_{ij}, and heat flux $\boldsymbol{q}^\iota$ of each component are calculated as:

$$n^\iota = \int f^\iota d\boldsymbol{v}, \quad \boldsymbol{u}^\iota = \frac{\int \boldsymbol{v} f^\iota d\boldsymbol{v}}{n^\iota}, \quad p^\iota_{ij} = m^\iota \int (v_i - u^\iota_i)(v_j - u^\iota_j) f^\iota d\boldsymbol{v},$$
$$q^\iota_i = \frac{m^\iota}{2} \int (v_i - u^\iota_i)|\boldsymbol{v} - \boldsymbol{u}^\iota|^2 f^\iota d\boldsymbol{v}, \tag{13.4}$$

while these of the mixture are calculated as

$$n = \sum_\iota n^\iota, \quad \boldsymbol{u} = \frac{\sum_\iota m^\iota n^\iota \boldsymbol{u}^\iota}{\sum_\iota m^\iota n^\iota}, \quad p_{ij} = \sum_\iota m^\iota \int \bar{c}_i \bar{c}_j f^\iota d\boldsymbol{v},$$
$$T = \frac{p_{kk}}{3nk_B}, \quad \boldsymbol{q} = \sum_\iota \frac{m^\iota}{2} \int \bar{\boldsymbol{c}} \bar{c}^2 f^\iota d\boldsymbol{v}, \tag{13.5}$$

where $\bar{\boldsymbol{c}} = \boldsymbol{v} - \boldsymbol{u}$ is the peculiar velocity relative to the bulk velocity of the mixture.

13.2 Fast Spectral Method

Under the Carleman representation, the gain part of the BCO (13.2) is expressed as

$$Q^+_{\iota\jmath} = 4 \iint \Theta \delta(\boldsymbol{y} \cdot \boldsymbol{z}) f^\jmath(\boldsymbol{v} + \boldsymbol{z} + b^{\iota\jmath} \boldsymbol{y}) f^\iota(\boldsymbol{v} + a^{\iota\jmath} \boldsymbol{y}) d\boldsymbol{y} d\boldsymbol{z}, \tag{13.6}$$

where $a^{\iota\jmath} = 2m^\jmath/(m^\iota + m^\jmath)$, $b^{\iota\jmath} = (m^\jmath - m^\iota)/(m^\iota + m^\jmath)$, and $\Theta = C^{\iota\jmath}/v_r$ is the differential cross section. Applying the Fourier transform, the j-th mode of the BCO is related to the spectrum of the VDFs $\hat{f}^\iota$ and $\hat{f}^\jmath$ as

$$\widehat{Q}^{\iota\jmath}_j = \sum_{\substack{\boldsymbol{l}+\boldsymbol{m}=j \\ \boldsymbol{l},\boldsymbol{m}=-(N_1,N_2,N_3)/2}}^{(N_1,N_2,N_3)/2-1} \hat{f}^\iota_{\boldsymbol{l}} \hat{f}^\jmath_{\boldsymbol{m}} \beta^{\iota\jmath}_1(\boldsymbol{l}, \boldsymbol{m}) - \hat{f}^\iota_{\boldsymbol{l}} \hat{f}^\jmath_{\boldsymbol{m}} \beta_2(\boldsymbol{m}), \tag{13.7}$$

where $\boldsymbol{l} = (l_1, l_2, l_3)$, $\boldsymbol{m} = (m_1, m_2, m_3)$, the kernel mode $\beta^{\iota\jmath}_1(\boldsymbol{l}, \boldsymbol{m})$ is

$$\beta^{\iota\jmath}_1(\boldsymbol{l}, \boldsymbol{m}) = 4 \int_{\mathcal{B}_R} \int_{\mathcal{B}_R} d\boldsymbol{y} d\boldsymbol{z} \delta(\boldsymbol{y} \cdot \boldsymbol{z}) \Theta'(\boldsymbol{y}, \boldsymbol{z})$$
$$\times \exp[i a^{\iota\jmath} \boldsymbol{\xi}_{\boldsymbol{l}} \cdot \boldsymbol{y} + i \boldsymbol{\xi}_{\boldsymbol{m}} \cdot (\boldsymbol{z} + b^{\iota\jmath} \boldsymbol{y})], \tag{13.8}$$

and $\beta_2(\boldsymbol{m})$ is equal to $\beta_1(\boldsymbol{m}, \boldsymbol{m})$ with $a^{\iota J} = 1$ and $b^{\iota J} = 0$.

The right-hand side of Eq. (13.8) is expressed in spherical coordinates as $4 \int\int \delta(\boldsymbol{e}\cdot\boldsymbol{e}') \int_0^R \int_0^R \rho\rho'\Theta'(\rho, \rho') \cos[\rho(a^{\iota J}\boldsymbol{\xi}_l + b^{\iota J}\boldsymbol{\xi}_m)\cdot\boldsymbol{e}] \cos(\rho'\boldsymbol{\xi}_m \cdot \boldsymbol{e}')d\rho' d\rho d\boldsymbol{e}' d\boldsymbol{e}$, where $\boldsymbol{e}, \boldsymbol{e}'$ are vectors in the unit sphere. The Gauss-Legendre quadrature is adopted to approximate the integration with respect to ρ: suppose ρ_r and ω_r ($r = 1, 2, \ldots, M_2$) are the abscissas and weights of the quadrature for $\rho \in [0, R]$, we have

$$\beta_1^{\iota J}(\boldsymbol{l}, \boldsymbol{m}) = \int I^{\iota J}(\boldsymbol{\xi}_l, \boldsymbol{\xi}_m, \boldsymbol{e}) \left[\int \delta(\boldsymbol{e}\cdot\boldsymbol{e}')\phi(\boldsymbol{\xi}_m, \boldsymbol{e}')d\boldsymbol{e}'\right] d\boldsymbol{e}, \tag{13.9}$$

where $\phi(\boldsymbol{\xi}_m, e') = 2\int_0^R \rho'\Theta'(\rho_r, \rho')\cos(\rho'\boldsymbol{\xi}_m \cdot e')d\rho'$ and

$$\begin{aligned} I^{\iota J}(\boldsymbol{\xi}_l, \boldsymbol{\xi}_m, e) &= 2\sum_{r=1}^{M_2} \omega_r \rho_r \cos[\rho_r(a^{\iota J}\boldsymbol{\xi}_l + b^{\iota J}\boldsymbol{\xi}_m)\cdot \boldsymbol{e}] \\ &= 2\sum_{r=1}^{M_2} \omega_r \rho_r \Big[\cos(\rho_r a^{\iota J}\boldsymbol{\xi}_l \cdot \boldsymbol{e}_{\theta_p,\varphi_q}) \cos(\rho_r b^{\iota J}\boldsymbol{\xi}_m \cdot \boldsymbol{e}_{\theta_p,\varphi_q}) \\ &\quad - \sin(\rho_r a^{\iota J}\boldsymbol{\xi}_l \cdot \boldsymbol{e}_{\theta_p,\varphi_q}) \sin(\rho_r b^{\iota J}\boldsymbol{\xi}_m \cdot \boldsymbol{e}_{\theta_p,\varphi_q}) \Big]. \end{aligned} \tag{13.10}$$

Following a similar procedure in Sect. 4.3.2, we have

$$\beta_1^{\iota J}(\boldsymbol{l}, \boldsymbol{m}) = \frac{4\pi^2}{M^2} \sum_{p,q=1}^{M-1,M} I^{\iota J}(\boldsymbol{\xi}_l, \boldsymbol{\xi}_m, \boldsymbol{e}_{\theta_p,\varphi_q})\psi\left(\xi_l^{\perp}\right) \sin\theta_p, \tag{13.11}$$

with $\psi(s) = 2\pi \int_0^R \rho'\Theta'(\rho_r, \rho')J_0(\rho' s)d\rho'$.

The computational cost for the mixture BCO is $O(M^2 M_2 N^3 \log N)$ if the FFT-based convolution is applied, which is the same as that for the BCO of single-species monatomic gas when the collision kernel is directly calculated from the Lennard-Jones potential, see Sect. 5.5.1. The cosine function in Eq. (13.10) oscillates at most $O(N)$ times, so M_2 should be roughly of the order N, see the discussion around Eq. (4.38). The performance of this FSM (accuracy, conservation property) is the same with that for single-species monatomic gas.

13.2.1 Accuracy Analysis

We assess the FSM in the homogeneous relaxation of Maxwellian molecules with the following collision kernel

$$C^{\iota J} = C_0^{\iota J} \sin^{-1}\left(\frac{\theta}{2}\right). \tag{13.12}$$

The exact Bobylev-Krook-Wu solutions are (for simplicity, we take $k_B = 1$) [1]:

$$f^\iota(v,t) = n^\iota \left(\frac{m^\iota}{2\pi K}\right)^{3/2} \exp\left(-\frac{m^\iota v^2}{2K}\right)\left(1 - 3rp^\iota + \frac{rp^\iota}{K} m^\iota v^2\right), \quad (13.13)$$

where

$$\begin{aligned}
p^A &= \frac{4n^B}{5}[2C_0^{BB} - m_0 C_0^{AB}(5 - 3m_0)],\\
p^B &= \frac{4n^A}{5}[2C_0^{AA} - m_0 C_0^{AB}(5 - 3m_0)],\\
\widetilde{A} &= \frac{4n^A C_0^{AA} + 2n^B m_0 C_0^{AB}(5 - 3m_0 \frac{p^B}{p^A})}{15},\\
\widetilde{B} &= \frac{8n^A C_0^{AA} p_1 + 4n^B m_0 C_0^{AB}(5 - 3m_0) p^B}{15},
\end{aligned} \quad (13.14)$$

and

$$\begin{aligned}
m_0 &= \frac{4m^A m^B}{(m^A + m^B)^2},\\
r &= \frac{\widetilde{A}}{\widetilde{A}\exp[4\pi\widetilde{A}(t + t_0)] - \widetilde{B}},\\
K &= \frac{n^A + n^B}{n^A + n^B + 2(n^A p^A + n^B p^B) r}.
\end{aligned} \quad (13.15)$$

In addition to the above relations, type I solution exists when the following equation holds:

$$\frac{n^B}{n^A} = \frac{2C_0^{AA} - m_0(5 - 3m_0)C_0^{AB}}{2C_0^{BB} - m_0(5 - 3m_0)C_0^{AB}},$$

so that $p^A = p^B$ and there is no energy exchange between the two components, while the type II solution exists for any values of n^A and n^B, as long as the following relation is satisfied:

$$\frac{1}{2C_0^{AA} - m_0 C_0^{AB}(5 - 3m_0)} + \frac{1}{2C_0^{BB} - m_0 C_0^{AB}(5 - 3m_0)} = \frac{1}{3m_0^2 C_0^{AB}}.$$

The FSM is validated in Fig. 13.1, where the Maxwellian molecules with a mass ratio $m^A/m^B = 4$ are used. In Chap. 4, we have shown that the $32 \times 32 \times 32$ velocity grid is adequate, therefore the number of velocity grids is doubled in each direction because the velocity domain $[-L_v, L_v]^3$ with $L_v = 16$ is enlarged to cover the VDF of the B-component. The solid angle is still discretized by $M(M-1) = 20$ points, and $M_2 = 7$ in Eq. (13.10).

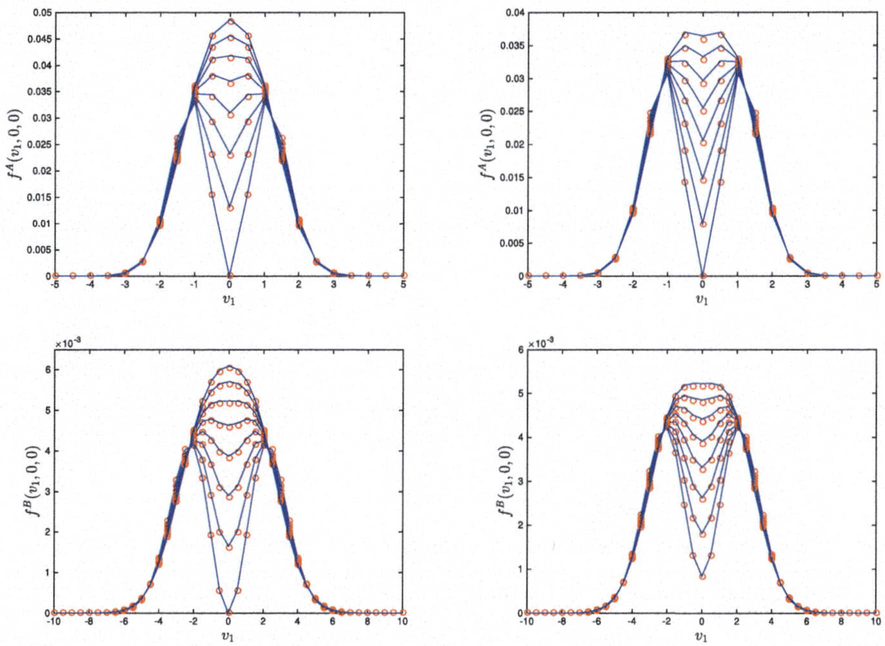

Fig. 13.1 Homogeneous relaxation in a mixture of Maxwell gases [2]. Left column: type I solution. parameters are $C_0^{BB} = C_0^{AA} = 4C_0^{AB} = 1/4\pi$, and $n^A = n^B = 1$. Right column: type II solution. parameters are $C_0^{AB} = 1/16\pi, C_0^{BB} = C_0^{AA} = C_0^{AB}(3m_0^2 + 5m_0)/2, n^A = 0.95$, and $n^B = 1$. Solid lines: analytical solutions. Circles: FSM. In each figure, from bottom to top (near $v_1 = 0$), the time for each line is $(0, 1, 2, \ldots, 7) \times 0.2$. The initial time is $t_0 = \max\{\log[(B + 3Ap_1)/A], \log[(B + 3Ap_2)/A]\}/4A\pi$

13.2.2 Efficient Algorithm for Large Mass Ratio

Note that the same velocity discretization is used for both VDFs of the two components. This is acceptable when the two components have similar molecular masses, but is unaffordable for molecules with disparate mass, as the memory usage and computational time are proportional to $m_r^{3/2}$ and m_r^2, respectively, where $m_r > 1$ is the mass ratio.

To reduce the memory cost, the scaled velocity is used: suppose the truncated velocity of the A-component $[-L_v, L_v]^3$ is uniformly discretized by $N_1 \times N_2 \times N_3$ points and the corresponding frequency space is uniformly discretized by the same number of points:

$$\boldsymbol{v}^A = \frac{2L_v}{N_v}\boldsymbol{j}, \quad \boldsymbol{\xi}^A = \frac{\pi}{L_v}\boldsymbol{j}, \tag{13.16}$$

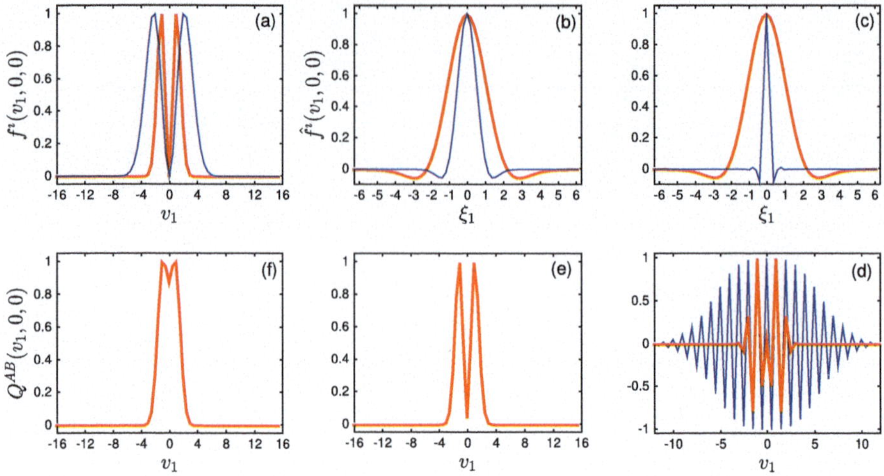

Fig. 13.2 Calculation of Q^{AB} when both VDFs are discretized using the same velocity grids [2]. **a** the VDFs; **b** the spectra of VDFs; **c** weighted spectra, when $\theta_p, \varphi_q = \pi/5$ and $\rho_r = 0.8582$; **d** spectra of the weighted spectra; **e** the product of the two lines in **d**; **f** Q^{AB}. Thick red (thin blue) lines: component A (B). Each curve is normalized by its maximum value for clarity

that of the B-component are discretized as

$$\boldsymbol{v}^B = M_{AB}\boldsymbol{v}^A, \quad \boldsymbol{\xi}^B = \frac{\boldsymbol{\xi}^A}{M_{AB}}, \tag{13.17}$$

where M_{AB} is chosen to be the closest integer to $\sqrt{m_r}$.

Efficient Calculation of Q^{AB}

In order to develop efficient methods to reduce the computational time, one has to know how the FSM works, when the same velocity discretization is used for both components. Figure 13.2 shows how Q^{AB} is calculated at one of the discretized solid angles. It is seen that the calculation of the spectrum of the weighted spectra in (c,d) is the most time-consuming part, because it is repeated M_2M^2 times. The weighted spectra for B-component is confined in $|\xi_1| < 3$, which means that it is sufficient to use the central 32 frequency components. The weighted spectrum of A-component in (c) is smooth, so the computational cost can be reduced by using fewer frequency components but with a larger frequency step. Further, for the VDF of B-component in (d), only the central velocity region $-8 < v_1 < 8$ contributes to Q^{AB}.

Based on these observations, an efficient algorithm for Q^{AB} is proposed:

- apply the FFT to calculate the spectra f^A and f^B in their own discretized frequency spaces $\boldsymbol{\xi}^A$ and $\boldsymbol{\xi}^B$, respectively.
- for given values of θ_p, φ_q, and r:

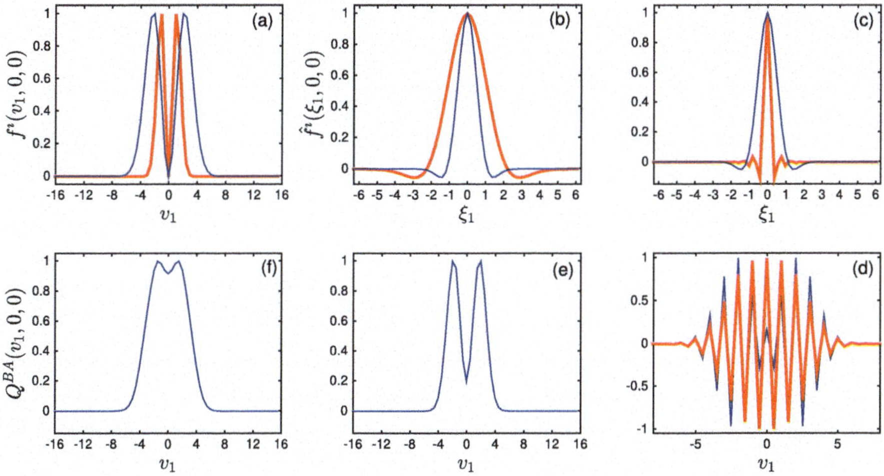

Fig. 13.3 Calculation of Q^{BA} when both VDFs are discretized using the same velocity grids [2]. **a** the VDFs; **b** the spectra of VDFs; **c** weighted spectra, when $\theta_p, \varphi_q = \pi/5$ and $\rho_r = 0.8582$; **d** spectra of the weighted spectra; **e** the product of the two lines in **d**; **f** Q^{BA}. Thick red (thin blue) lines: component A (B). Each curve is normalized by its maximum value

1. calculate $t_1 = \mathcal{SF}[\hat{f}_l^A \beta_1^{AB}(l)]$, where $\mathcal{S}$ and $\mathcal{F}$ stand for the FFT shift and FFT in Matlab, at a cost of $O(N^3 \log N)$. Note that $\beta_1^{AB}(\boldsymbol{l})$ is the kernel mode $\beta_1^{AB}(\boldsymbol{l}, \boldsymbol{m})$, which involves only l.
2. let $t_2 = \hat{f}_l^B \beta_1^{AB}(\boldsymbol{m})$.
3. zero padding t_2 to the size of $M_{AB}N_1 \times N_2 \times N_3$, calculate its frequency spectrum in ξ_1 direction through FFT, and assign the central $N_1 \times N_2 \times N_3$ frequency components to t_2. The computational cost is $O(M_{AB}N^3 \log N)$; do the same in the other two frequency directions.
4. let $t_3 = t_1 t_2$.

- sum up all t_3 for each set of θ_p, φ_q, r, and the final result is the gain part of the collision operator Q^{AB}, that is, Q^{AB+}; the loss part of Q^{AB} is calculated as that of Q^{BB}.

Efficient Calculation of Q^{BA}

Similarly, based on the observation in Fig. 13.3, Q^{BA} is calculated as

- refine $\hat{f}^A$ by FFT interpolation, such that the distance of discrete frequency components is the same as that of component B. The cost is $O(M_{AB}N^3 \log N)$.
- select central frequency components of $\hat{f}_m^A$: the total number of chosen frequency components is $(M_{ex,1}N_1)(M_{ex,2}N_2)(M_{ex,3}N_3)$, where the integer $M_{ex,k}$ satisfies

$$1 \le M_{ex,k} \le M_{AB}, \tag{13.18}$$

with $k = 1, 2, 3$.

- for given values of θ_p, φ_q, and r:

  1. calculate $t_2 = \mathcal{FS}[\hat{f}_l^B \beta_1^{BA}(\boldsymbol{l})]$.
  2. apply fftshift to $\hat{f}_{\boldsymbol{m}}^A$ and $\beta_1^{BA}(\boldsymbol{m})$, and let $t_1 = \hat{f}_{\boldsymbol{m}}^A \beta_1^{BA}(\boldsymbol{m})$.
  3. in ξ_1 direction, calculate $t_1 = \mathcal{SF}(t_1)$; select N_1 points from the first frequency node with a step $M_{ex,1}$, apply the fftshift to the chosen data, and assign the final result to t_1; do the same in the other two frequency directions.
  4. let $t_3 = t_1 t_2$.

- sum up all t_3 for each set of θ_p, φ_q, r, and the final result is the gain part of the collision operator Q^{BA}, that is, Q^{BA+}; the loss part of Q^{BA} is calculated as that of Q^{AA}.

Compared to the FSM using same velocity discretization for both VDFs, the computational efficiency is improved significantly in this new algorithm. First of all,

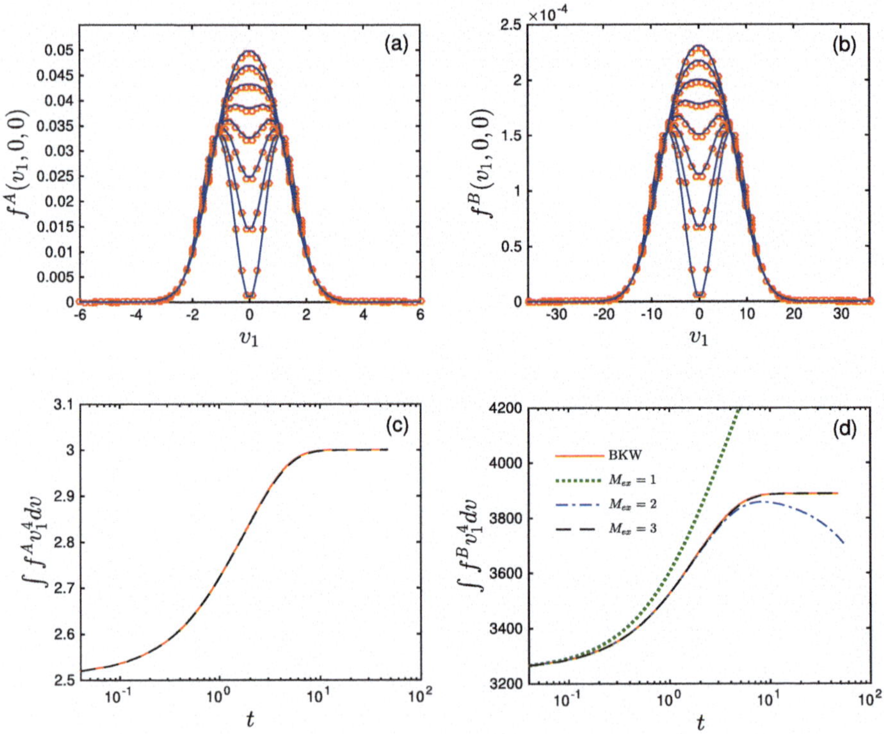

Fig. 13.4 Homogeneous relaxation in the mixture of Maxwell gas with the mass ratio $m^A/m^B = 36$ [2]. Solid lines in **a**, **b**: type I solution. Circles: FSM with $M_{ex} = 2$. In each figure, from bottom to top, the time is $(0, 1, 2, \ldots, 7) \times 0.25$. Same parameters as those used in the left column of Fig. 13.1. The velocity space is discretized by $64 \times 32 \times 32$ grids, while the frequency space is discretized by $32 \times 32 \times 32$ grids. We take $M = 5$, $M_2 = 14$ in the calculation of Q^{AB}, while $M_2 = 32$ in the calculation of Q^{BA}. **c**, **d** The time evolution of fourth-order moments

now the memory usage does not depend on the mass ratio. Second, the computational cost for Q^{AB} is $O(\sqrt{m_r}M_2M^2N^3\log N)$. Since $\max(\xi_l^A)=N\pi/2L_v$, $\max(\xi_m^B)=N\pi/2L_vM_{AB}$, and $\max(\rho_r)=L_vM_{AB}$, $\cos[\rho_r(a^{AB}\xi_l^A+b^{AB}\xi_m^B)\cdot e]$ in Eq. (13.10) oscillates $O(N)$ times, so M_2 should be $O(N)$, and the overall computational cost for Q^{AB} is $O(\sqrt{m_r}M^2N^4\log N)$. In the calculation of Q^{BA}, since $\cos[\rho_r(a^{BA}\xi_l^B+b^{BA}\xi_m^A)\cdot e]$ oscillates $O(NM_{ex})$ times, M_2 should be of the order of NM_{ex}. For m_r up to 100, we find that the case of

$$M_{ex}^3\sim m_r$$

provides high accuracy, and the computational cost of Q^{BA} is slightly higher than Q^{AB}.

Figure 13.4 shows the accuracy of the improved algorithm in the calculation of cross-collision operators. The VDFs are well captured when $M_{ex}=2$ (i.e., only 1/27 of the frequency components of $\hat{f}^A$ is used in the calculation of Q^{BA}). However, to capture the long-time behavior of the fourth- and sixth-order moments of component B, M_{ex} has to be increased to 3.

13.3 Accuracy in Inhomogeneous Problems

The spatial coordinate is normalized by L, the molecular velocity is normalized by $v_m^A=\sqrt{2k_BT_0/m^A}$ at the temperature T_0, the number density is normalized by n_0, and the VDF is normalized by $n_0/(v_m^A)^3$. Then the BCO for the power-law potential is given by

$$\begin{aligned}Q^{IJ}&=\frac{1}{\mathrm{Kn}_1^{IJ}}\iint\sin^{\alpha^{IJ}+\gamma^{IJ}-1}\left(\frac{\theta}{2}\right)\cos^{-\gamma^{IJ}}\left(\frac{\theta}{2}\right)\\&\quad\times v_r^{\alpha^{IJ}}f^J({v'}_*^{IJ})f^I({v'}^{IJ})d\Omega d\boldsymbol{v}_*-\nu^{IJ}f^I(v),\\\nu^{IJ}&=\frac{1}{\mathrm{Kn}_1^{IJ}}\iint\sin^{\alpha^{IJ}+\gamma^{IJ}-1}\left(\frac{\theta}{2}\right)\cos^{-\gamma^{IJ}}\left(\frac{\theta}{2}\right)v_r^{\alpha^{IJ}}f^J(v_*)d\Omega d\boldsymbol{v}_*,\end{aligned}\tag{13.19}$$

with

$$\begin{aligned}\mathrm{Kn}_1^{IJ}&=\frac{64(2)^{\alpha^{IJ}/2}}{5}\Gamma\left(\frac{\alpha^{IJ}+\gamma^{IJ}+3}{2}\right)\Gamma\left(2-\frac{\gamma^{IJ}}{2}\right)\mathrm{Kn}_{un}^{IJ},\\\mathrm{Kn}_{un}^{IJ}&=\frac{\mu^{IJ}(T=T_0)}{n_0L}\sqrt{\frac{\pi}{2m^Ak_BT_0}}.\end{aligned}\tag{13.20}$$

Note that μ^{AB} should be understood as the shear viscosity when the collision kernel is given by Eq. (4.5) and $m^A=m^B$.

The normalized molecular number density, flow velocity, and temperature T' are defined as:

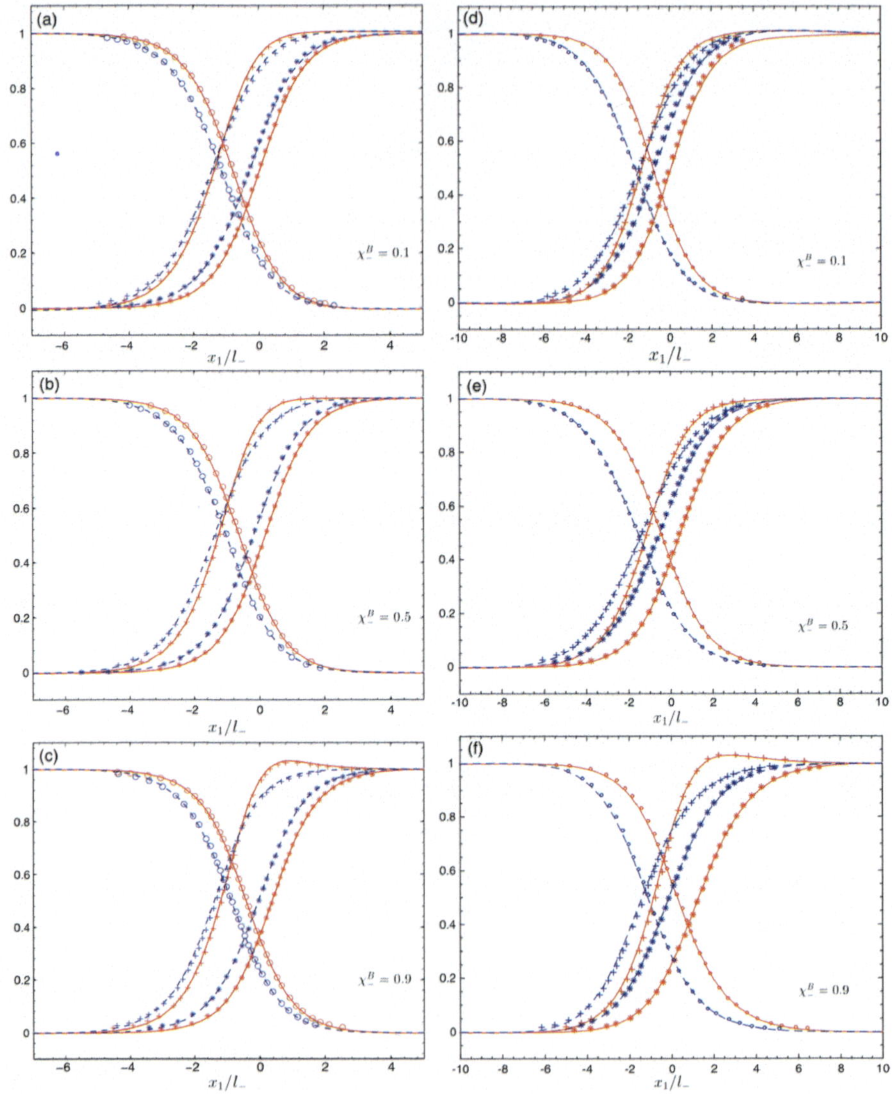

Fig. 13.5 Structures of the normal shock wave in monatomic gas mixture [2]. (Left) Ma = 3, $m^A/m^B = 2$, and $d_m^B/d_m^A = 1$. (Right) Ma = 2, $m^A/m^B = 4$, and $d_m^B/d_m^A = 1$. Note that χ_-^B is the concentration of the component B at the upstream infinity. Stars, circles, and crosses stand for the scaled number density, flow velocity, and temperature. Lines: FSM (solid: component A, dashed: component B). Symbols: numerical kernel method [3]

$$n^\iota = \int f^\iota d\boldsymbol{v}, \quad \boldsymbol{u}^\iota = \frac{1}{n^\iota}\int \boldsymbol{v} f^\iota d\boldsymbol{v}, \quad T^\iota = \frac{2m^\iota}{3m^A n^\iota}\int |\boldsymbol{v}-\boldsymbol{u}^\iota|^2 f^\iota d\boldsymbol{v}. \quad (13.21)$$

The CIS is adopted to find the steady-state solution of the Boltzmann equation:

$$[\nu^{AA}(f_k^A)+\nu^{AB}(f_k^B)]f_{k+1}^A + \boldsymbol{v}\cdot\frac{\partial f_{k+1}^A}{\partial \boldsymbol{x}} = Q^{AA+}(f_k^A, f_k^A)+Q^{AB+}(f_k^A, f_k^B),$$
$$[\nu^{BB}(f_k^B)+\nu^{BA}(f_k^A)]f_{k+1}^B + \boldsymbol{v}\cdot\frac{\partial f_{k+1}^B}{\partial \boldsymbol{x}} = Q^{BB+}(f_k^B, f_k^B)+Q^{BA+}(f_k^B, f_k^A),$$

where the subscript k is the iteration step.

Normal Shock Waves

HS molecules with diameters d_m^ι are considered. The reference number density n_0 is chosen to be the number density n_- at the upstream infinity, while L equals the MFP $l_- = [\sqrt{2}\pi(d_m^A)^2 n_-]^{-1}$ of the component A. Therefore, in Eq. (13.19), $\text{Kn}_1^{AA} = 4\sqrt{2}\pi$, $\text{Kn}_1^{BB} = \text{Kn}_1^{AA}/(d_m^B/d_m^A)^2$, and $\text{Kn}_1^{AB} = \text{Kn}_1^{BA} =$

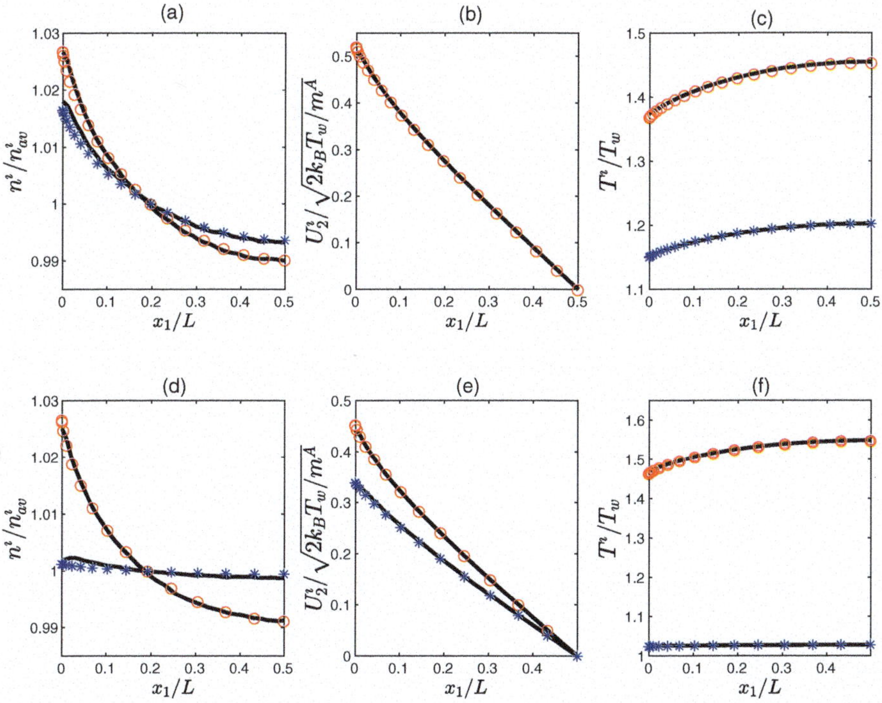

Fig. 13.6 Molecular number density, velocity, and temperature profiles in planar Couette flow with Kn = 1 and $n_{av}^B = n_{av}^A$ [2]. **a–c** $m^A/m^B = 4$, $d_m^A = d_m^B$. **d–f** $m^A/m^B = 36$, $d_m^A = 2d_m^B$. Lines: DSMC. Circles and stars: FSM results (when $M_{ex} = 2$) for the components A and B, respectively. Note that in **b** the velocity of the two components nearly overlap

$4\text{Kn}_1^{AA}/(1+d_m^B/d_m^A)^2$. Figure 13.5 shows that the FSM captures the shock structure well, when compared to the numerical kernel method [3].

Planar Couette Flow
HS gases between two parallel plates moving in opposite directions with the speed $\sqrt{2k_B T_w/m^A}$ are considered. The velocity of component A $[-6,6]^3$ is respectively discretized by 64 non-uniform grids in the direction perpendicular to the plate, and 32×32 grids in the other two directions; that of component B is given by Eq. (13.17). $32\times 32\times 32$ uniformly-distributed frequency grid points are used in FSM. The comparison with DSMC in Fig. 13.6 demonstrates the accuracy of FSM for a mass ratio of $m_r=36$.

13.4 Linearization and GSIS

We introduce the LBE for gas mixtures as well as the general synthetic iterative scheme. For simplicity, only the Poiseuille flow is considered [4], but the method of construction of GSIS is similar to that of monatomic gas in Chap. 8.

Let the molar fractions of components A and B be χ^A and $\chi^B=1-\chi^A$, respectively. When the normalization around Eq. (13.19) is used, the equilibrium VDF is

$$f_{eq}^{\iota}(\boldsymbol{v})=\chi^{\iota}\left(\frac{m^{\iota}}{\pi m^A}\right)^{3/2}\exp\left(-\frac{m^{\iota}v^2}{2m^A}\right). \tag{13.22}$$

Expressing the VDF as $f^{\iota}=f^{\iota,eq}+h^{\iota}$, the LBE for the perturbation VDFs h^{ι} is

$$v_1\frac{\partial h^{\iota}}{\partial x_1}+v_2\frac{\partial h^{\iota}}{\partial x_2}=L^{\iota}-v_3 f_{eq}^{\iota}, \tag{13.23}$$

with the linearized collision operators $L^{\iota}=\sum_{\beta=1,2}Q^{\iota\beta}(f_{eq}^{\iota},h_{\beta})+Q^{\iota\beta}(h^{\iota},f_{eq}^{\beta})$.

The normalized flow velocity is calculated as $u^{\iota}=\int h^{\iota}v_3 d\boldsymbol{v}/\chi_{\iota}$, the shear stress normalized by the total gas pressure p_0 is $p_{ij}^{\iota}=2m^{\iota}\int h^{\iota}v_{<i}v_{j>}d\boldsymbol{v}/m^A$, and the heat flux normalized by $p_0 v_m^A$ is $q^{\iota}=\int\left(m^{\iota}v^2/m^A-5/2\right)v_3 h d\boldsymbol{v}$. The dimensionless MFR and HFR are calculated as

$$\begin{aligned} G^{\iota}&=-\frac{1}{A}\sqrt{\frac{m}{m^A}}\iint u^{\iota}dx_1dx_2,\\ Q^{\iota}&=\frac{1}{A}\sqrt{\frac{m}{m^A}}\iint q^{\iota}dx_1dx_2, \end{aligned} \tag{13.24}$$

where $m=\chi^A m^A+(1-\chi^A)m^B$.

Following the basic steps in developing the macroscopic synthetic equation, the following two equations can be derived for the flow velocity of each component:

$$\chi^\iota\left(\frac{\partial^2}{\partial x_1^2}+\frac{\partial^2}{\partial x_2^2}\right)u^\iota = 2\frac{\partial}{\partial x_1}\int v_1 v_3 L^\iota d\boldsymbol{v} + 2\frac{\partial}{\partial x_2}\int v_2 v_3 L^\iota d\boldsymbol{v} - \frac{1}{4}\left(\frac{\partial^2 F^\iota_{2,0,1}}{\partial x_1^2} + 2\frac{\partial^2 F^\iota_{1,1,1}}{\partial x_1 \partial x_2} + \frac{\partial^2 F^\iota_{0,2,1}}{\partial x_2^2}\right), \quad (13.25)$$

where $F^\iota_{m,n,l} = \int h^\iota H_m(v_1)H_n(v_2)H_l(v_3)d\boldsymbol{v}$, with $H_n(v)$ the n-th order physicists' Hermite polynomial.

To obtain the diffusion-type equations which facilitate fast convergence, we rewrite the linearized BCO as: $L^\iota = (L^\iota + \delta_{rp}h^\iota) - \delta_{rp}h^\iota$, and let $2\int v_{<i}v_{j>}L^\iota d\boldsymbol{v} = \int v_{<i}v_{j>}(L^\iota + \delta_{rp}h^\iota)d\boldsymbol{v} - \delta_{rp}m^A p^\iota_{ij}/m^\iota$. Then, for component B, Eq. (13.25) is transformed to

$$\begin{aligned}\chi^B\left(\frac{\partial^2}{\partial x_1^2}+\frac{\partial^2}{\partial x_2^2}\right)u^B =& -\delta_{rp}\frac{m^A}{m^B}\left(\frac{\partial p^B_{13}}{\partial x_1}+\frac{\partial p^B_{23}}{\partial x_2}\right)\\ &-\frac{1}{4}\left(\frac{\partial^2 F^B_{2,0,1}}{\partial x_1^2} + 2\frac{\partial^2 F^B_{1,1,1}}{\partial x_1 \partial x_2} + \frac{\partial^2 F^B_{0,2,1}}{\partial x_2^2}\right)\\ &+2\frac{\partial}{\partial x_1}\int v_1 v_3(L^B - \delta_{rp}h^B)d\boldsymbol{v}\\ &+2\frac{\partial}{\partial x_2}\int v_2 v_3(L^B - \delta_{rp}h^B)d\boldsymbol{v},\end{aligned} \quad (13.26)$$

while for component A, by using the relation $\partial p_{13}/\partial x_1 + \partial p_{23}/\partial x_2 = -1$ with $p_{13} = p^A_{13} + p^B_{13}$ and $p_{23} = p^A_{23} + p^B_{23}$, Eq. (13.25) is transformed to

$$\begin{aligned}\chi^A\left(\frac{\partial^2}{\partial x_1^2}+\frac{\partial^2}{\partial x_2^2}\right)u^A =& \delta_{rp}\left(\frac{\partial p^B_{13}}{\partial x_1}+\frac{\partial p^B_{23}}{\partial x_2}+1\right)\\ &-\frac{1}{4}\left(\frac{\partial^2 F^A_{2,0,1}}{\partial x_1^2} + 2\frac{\partial^2 F^A_{1,1,1}}{\partial x_1 \partial x_2} + \frac{\partial^2 F^A_{0,2,1}}{\partial x_2^2}\right)\\ &+2\frac{\partial}{\partial x_1}\int v_1 v_3(L^A - \delta_{rp}h^A)d\boldsymbol{v}\\ &+2\frac{\partial}{\partial x_2}\int v_2 v_3(L^A - \delta_{rp}h^A)d\boldsymbol{v}.\end{aligned} \quad (13.27)$$

Table 13.1 Mass and heat flow rates in Poiseuille flow of Ne-Ar gas mixtures between two parallel plates. The LBE with the HS model is used

δ_{rp}	$\chi_{Ne} = 0.1$				$\chi_{Ne} = 0.5$				$\chi_{Ne} = 0.9$			
	G_{Ne}	G_{Ar}	$\frac{Q_{Ne}}{\chi_{Ne}}$	$\frac{Q_{Ar}}{\chi_{Ar}}$	G_{Ne}	G_{Ar}	$\frac{Q_{Ne}}{\chi_{Ne}}$	$\frac{Q_{Ar}}{\chi_{Ar}}$	G_{Ne}	G_{Ar}	$\frac{Q_{Ne}}{\chi_{Ne}}$	$\frac{Q_{Ar}}{\chi_{Ar}}$
0.01	2.131	1.416	0.999	0.639	1.858	1.227	0.861	0.544	1.547	1.016	0.705	0.440
0.05	1.593	1.071	0.720	0.459	1.393	0.936	0.619	0.391	1.167	0.787	0.506	0.316
0.1	1.386	0.951	0.607	0.391	1.218	0.840	0.522	0.334	1.030	0.717	0.428	0.272
0.5	1.005	0.766	0.372	0.251	0.912	0.709	0.323	0.218	0.808	0.647	0.269	0.180
1	0.909	0.744	0.284	0.196	0.846	0.707	0.248	0.170	0.773	0.666	0.208	0.141
2	0.879	0.776	0.204	0.142	0.841	0.756	0.179	0.123	0.794	0.730	0.151	0.102
5	1.016	0.968	0.115	0.080	1.002	0.964	0.101	0.068	0.977	0.949	0.085	0.056
10	1.371	1.346	0.067	0.045	1.367	1.348	0.058	0.039	1.352	1.337	0.049	0.031
20	2.153	2.141	0.036	0.024	2.155	2.145	0.031	0.021	2.145	2.137	0.026	0.017
30	2.955	2.946	0.024	0.017	2.958	2.951	0.021	0.014	2.950	2.943	0.018	0.012
50	4.570	4.565	0.015	0.010	4.574	4.569	0.013	0.009	4.568	4.562	0.011	0.007
100	8.624	8.620	0.008	0.006	8.625	8.622	0.007	0.005	8.623	8.614	0.007	0.004

Consider the Poiseuille flow of a Ne-Ar mixture between two parallel plates. First, the HS molecular potential is considered. At an equilibrium temperature $T_0 = 300$ K, the molecular mass (diameter) ratio is $m_2/m_1 = 1.979$ ($d_2/d_1 = 1.406$). The mass and heat flow rates are shown in Table 13.1. While in the continuum flow regime the MFRs of each component are the same, in the transition and free-molecular flow regimes the MFR of Ne is always larger than that of Ar.

We then simulate the Ne-Ar mixture flow when the collision kernels are directly calculated from the Lennard-Jones potentials, to investigate the influence of intermolecular potentials on the velocity and heat flux. Figure 13.7 shows that the HS model overestimates the velocity and heat flux in the free-molecular regime.

Finally, the Poiseuille flows of an equimolar Ne-Ar mixture passing through a rectangular cross section of aspect ratio 1 and 2 are simulated based on the HS model. For each rarefaction parameter, numerical results are obtained within 60 iterations with the error less than 10^{-7}, demonstrating again the efficiency of GSIS. Table 13.2 shows that, the normalized mass flow rates through the cross section of aspect ratio 2 are always larger than those of aspect ratio 1. However, although the heat flow rates through the cross section of aspect ratio 2 are larger than those of aspect ratio 1 when $\delta_{rp} < 20$, they are roughly same in the continuum flow regime.

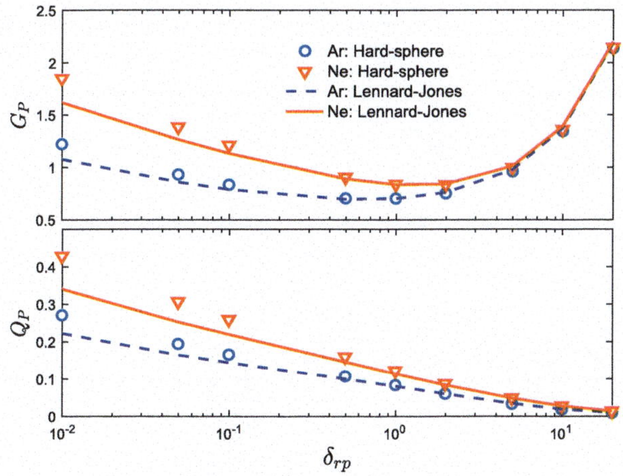

(a) Mass and heat flow rates

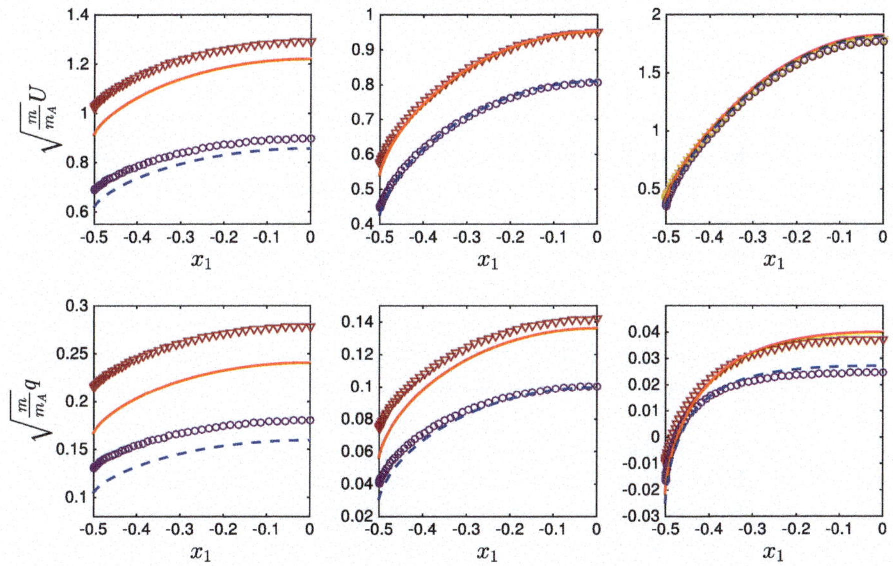

(b) Velocity and heat flux. From left to right, $\delta_{rp} = 0.1$, 1, and 10

Fig. 13.7 Poiseuille flow of an equimolar Ne-Ar mixture between two parallel plates [4], where the linearized Boltzmann equations are solved by GSIS for the HS and Lennard-Jones potentials

Table 13.2 Mass and heat flow rates in Poiseuille flow of an equimolar Ne-Ar gas mixture along channels of rectangular cross section, of aspect ratios one and two [4]. The LBE with the HS model is used

δ_{rp}	Aspect ratio 1				Aspect ratio 2			
	G_{Ne}	G_{Ar}	$2Q_{Ne}$	$2Q_{Ar}$	G_{Ne}	G_{Ar}	$2Q_{Ne}$	$2Q_{Ar}$
0.01	0.506	0.358	0.251	0.177	0.692	0.490	0.343	0.241
0.05	0.493	0.349	0.241	0.167	0.669	0.474	0.324	0.224
0.1	0.483	0.344	0.232	0.160	0.653	0.466	0.309	0.212
0.5	0.452	0.338	0.195	0.134	0.604	0.459	0.249	0.170
1	0.441	0.347	0.170	0.117	0.589	0.475	0.210	0.144
2	0.442	0.373	0.139	0.096	0.597	0.518	0.164	0.112
5	0.497	0.459	0.091	0.061	0.699	0.659	0.101	0.067
10	0.630	0.610	0.057	0.038	0.926	0.904	0.060	0.040
20	0.928	0.917	0.032	0.021	1.417	1.406	0.033	0.022
30	1.235	1.228	0.022	0.015	1.920	1.913	0.023	0.015
50	1.856	1.852	0.014	0.009	2.932	2.929	0.014	0.009
100	3.416	3.416	0.007	0.005	5.474	5.474	0.007	0.005

13.5 McCormack Model

Several kinetic models are proposed to replace the BCO for mixtures of monatomic gas, including the Andries-Aoki-Perthame model [6], the Kosuge model [7], and the McCormack model [8]. Note that the linearized collision operator in the McCormack model was designed by requiring that its first three velocity moments be the same as these of the linearized BCO. It reduces to the linearized Shakhov model when mixtures of same properties are considered.

We use the FSM to assess the accuracy of the McCormack model in the Fourier flow [5]. The equal-mole mixtures Ne-Ar and He-Xe are selected in order to investigate the influence of molecular mass ratio. For the He-Xe mixture at $T_0 = 300$ K, the molecular mass ratio and diameter ratio are $m_2/m_1 = 32.8$ and $d_2/d_1 = 2.226$, respectively. Figure 13.8 plots the number density. For the Ne-Ar mixture, the Ne number density from the McCormack model is larger than that from the LBE by 31%, when $\delta_{rp} = 0.1$. This difference is reduced when δ_{rp} increases, e.g., to only 5% when $\delta_{rp} = 10$. This kind of difference increases with the molecular mass ratio, e.g., for the He-Xe mixture with $m_2/m_1 = 32.8$, the Ne number density from the McCormack model is 1.8 times of that from the LBE.

This example shows that there is plenty of work need to be done to improve the accuracy of kinetic models for gas mixtures, especially for mixture with disparate mass ratio, mixture of molecular gas with complicated internal energy and heat flux relaxations. These works shall have important applications in the hypersonic flows (ionization) and inertial confinement fusion, where the dynamics of electron (whose mass is much smaller that other species) plays a significant role.

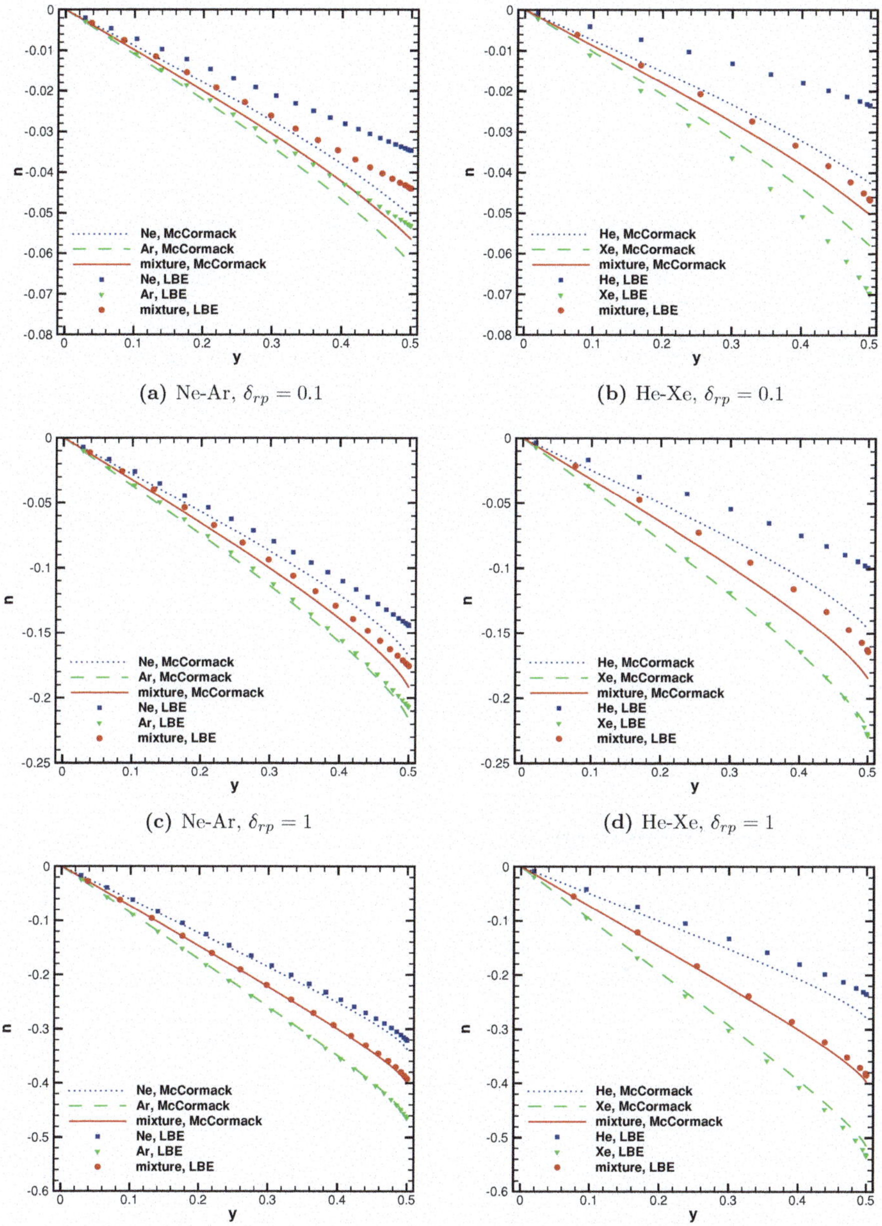

(a) Ne-Ar, $\delta_{rp} = 0.1$ (b) He-Xe, $\delta_{rp} = 0.1$

(c) Ne-Ar, $\delta_{rp} = 1$ (d) He-Xe, $\delta_{rp} = 1$

(e) Ne-Ar, $\delta_{rp} = 10$ (f) He-Xe, $\delta_{rp} = 10$

Fig. 13.8 The number density in the Fourier flow of gas mixtures [5]

References

1. Krook M, Wu TT (1997) Exact solution of Boltzmann equation for multicomponent systems. Phys Rev Lett 38:991–993
2. Wu L, Zhang J, Reese JM, Zhang YH (2015) A fast spectral method for the Boltzmann equation for monatomic gas mixtures. J Comput Phys 298:602–621
3. Kosuge S, Aoki K, Takata S (2001) Shock-wave structure for a binary gas mixture: finite-difference analysis of the Boltzmann equation for hard-sphere molecules. Eur J Mech B/Fluids 20(1):87–126
4. Wu L, Zhang J, Liu HH, Zhang YH, Reese JM (2017) A fast iterative scheme for the linearized Boltzmann equation. J Comput Phys 338:431–451
5. Ho MT, Wu L, Graur I, Zhang YH, Reese JM (2016) Comparative study of the Boltzmann and McCormack equations for Couette and Fourier flows of binary gaseous mixtures. Int J Heat Mass Transfer 96:29–41
6. Pierre A, Kazuo A, Benoit P (2002) A consistent BGK-type model for gas mixtures. J Stat Phys 106:516
7. Kosuge S (2009) Model Boltzmann equation for gas mixtures: construction and numerical comparison. Eur J Mech B/Fluids 28:170–184
8. McCormack FJ (1973) Construction of linearized kinetic models for gaseous mixture and molecular gases. Phys Fluids 16:2095–2105

Chapter 14
Dense Gas Flow

The fast spectral method is extended to solve the generalized Enskog equation, where the numerical method is validated by analytical solutions, event-driven molecular dynamics simulations, and direct simulation Monte Carlo method. The dense (granular) flow in a long channel is investigated both at the macroscopic and mesoscopic levels.

14.1 Fast Spectral Method

To develop an FSM for the inelastic and nonlocal collision operator (2.69), the weak form of the collision operator is expressed in the Carleman representation. By the relation $2^{d-1}(\boldsymbol{v}_r \cdot \boldsymbol{k})d\boldsymbol{k} = v_r \sin^{3-d}(\theta/2)\, d\boldsymbol{\Omega}$, Eq. (2.71) is rewritten as

$$\int Q\Psi(\boldsymbol{v})d\boldsymbol{v} = \left(\frac{\sigma}{2}\right)^{d-1} \iiint d\Omega d\boldsymbol{v}_* d\boldsymbol{v} v_r \sin^{3-d}\left(\frac{\theta}{2}\right) \times g(-\boldsymbol{k}, \boldsymbol{v}_*) f(\boldsymbol{v}) \left[\Psi(\boldsymbol{v}') - \Psi(\boldsymbol{v})\right]. \tag{14.1}$$

With $\int_{\mathbb{S}^{d-1}} f\left(\frac{v_r\Omega - \boldsymbol{v}_r}{2}\right) d\boldsymbol{\Omega} = \frac{2^{d-1}}{v_r^{d-2}} \int_{\mathbb{R}^d} \delta(\boldsymbol{y}\cdot\boldsymbol{v}_r + y^2) f(\boldsymbol{y}) d\boldsymbol{y}$, Eq. (14.1) is rewritten as

$$\int Q\Psi(\boldsymbol{v})d\boldsymbol{v} = \sigma^{d-1} \iiint \delta(\boldsymbol{y}\cdot\boldsymbol{z})|\boldsymbol{y}|^{3-d} g\left(\frac{\boldsymbol{y}}{|\boldsymbol{y}|}, \boldsymbol{v}+\boldsymbol{y}+\boldsymbol{z}\right) f(\boldsymbol{v}) \times \left[\Psi\left(\boldsymbol{v} + \frac{1+\alpha}{2}\boldsymbol{y}\right) - \Psi(\boldsymbol{v})\right] d\boldsymbol{y} d\boldsymbol{z} d\boldsymbol{v}. \tag{14.2}$$

Let $\Psi(\boldsymbol{v}) = \exp(-i\xi_j \boldsymbol{v})/(2L_v)^d$, the spectrum of the Enskog collision operator (14.2) is expressed as [1]:

L. Wu, *Rarefied Gas Dynamics*,
https://doi.org/10.1007/978-981-19-2872-7_14

$$\widehat{Q}_j(\boldsymbol{x}) = \sigma^{d-1} \int \sum_{l+m=j} \hat{f}_l(\boldsymbol{x})\hat{g}_m(\boldsymbol{x}, \boldsymbol{e})[\beta(\boldsymbol{l}, \boldsymbol{m}) - \beta(-\boldsymbol{m}, \boldsymbol{m})]d\boldsymbol{e}, \tag{14.3}$$

where $\boldsymbol{l} = (l_1, l_2, \ldots, l_d)$, $\boldsymbol{m} = (m_1, m_2, \ldots, m_d)$,

$$\hat{g}_m(\boldsymbol{x}, \boldsymbol{e}) = \chi\left(\boldsymbol{x} + \frac{\sigma}{2}\boldsymbol{e}\right) \hat{f}_m(\boldsymbol{x} + \sigma\boldsymbol{e}) \tag{14.4}$$

and

$$\begin{aligned}\beta(\boldsymbol{l}, \boldsymbol{m}) = & \int \int_0^R \int_0^R d\rho d\rho' d\boldsymbol{e}' \delta(\boldsymbol{e} \cdot \boldsymbol{e}')\rho(\rho')^{d-2} \\ & \times \exp\left\{i\rho\left[-\frac{1+\alpha(\rho)}{2}\boldsymbol{\xi}_l + \frac{1-\alpha(\rho)}{2}\boldsymbol{\xi}_m\right] \cdot \boldsymbol{e} + i\rho'\boldsymbol{\xi}_m \cdot \boldsymbol{e}'\right\},\end{aligned} \tag{14.5}$$

with $\boldsymbol{e}, \boldsymbol{e}'$ the unit vectors in the sphere ($d = 3$) or circle ($d = 2$). Note that α is a function of ρ because in general the restitution coefficient is a function of the normal relative velocity $|\boldsymbol{v}_r \cdot \boldsymbol{k}|$. Also note that the integration with respect to $\boldsymbol{e}$ is not put into $\beta(\boldsymbol{l}, \boldsymbol{m})$ as that in Chap. 4, because due to the nonlocal collision, $\boldsymbol{e}$ appears in $\hat{g}_m$.

The kernel mode $\beta(\boldsymbol{l}, \boldsymbol{m})$ should be decomposed as $\sum_p A_p(\boldsymbol{l})B_p(\boldsymbol{m})$, so that Eq. (14.3) can be efficiently computed by FFT-based convolution. To this end, the integral with respect to ρ in Eq. (14.5) is approximated by Gauss-Legendre quadrature, resulting in

$$\beta(\boldsymbol{l}, \boldsymbol{m}) = \frac{1}{2} \sum_r \phi(\boldsymbol{\xi}_l \cdot \boldsymbol{e}) \int \delta(\boldsymbol{e} \cdot \boldsymbol{e}')\psi(\boldsymbol{\xi}_m \cdot \boldsymbol{e}')d\boldsymbol{e}', \tag{14.6}$$

where ρ_r and ω_r ($r = 1, 2, \ldots, M_r$) are, respectively, the abscissas and weights of the Gauss-Legendre quadrature for $\rho \in [0, R]$, and

$$\phi(s) = \omega_r \rho_r \exp\left[-i\rho_r \frac{1+\alpha(\rho_r)}{2} s\right], \tag{14.7}$$

$$\psi(s) = 2\exp\left[i\rho_r \frac{1-\alpha(\rho_r)}{2}\boldsymbol{\xi}_m \cdot \boldsymbol{e}\right] \int_0^R (\rho')^{d-2} \exp(i\rho' s)d\rho'. \tag{14.8}$$

Similar to the development of FSM for monatomic gas in Sect. 4.3.2, the kernel mode can be written as $\beta(\boldsymbol{l}, \boldsymbol{m}) = \phi(\boldsymbol{\xi}_l \cdot \boldsymbol{e})\psi\left(\boldsymbol{\xi}_l^{\perp}\right)$. When $d = 3$, we have

$$\psi(s) = 2\pi \left[i\rho_r \frac{1-\alpha(\rho_r)}{2}\boldsymbol{\xi}_m \cdot \boldsymbol{e}\right] R\frac{J_1(Rs)}{s}, \tag{14.9}$$

and when $d = 2$, we have

$$\psi(s) = 2\exp\left[i\rho_r \frac{1-\alpha(\rho_r)}{2}\boldsymbol{\xi}_m \cdot \boldsymbol{e}\right]\frac{\sin(Rs)}{s}. \tag{14.10}$$

We use the trapezoidal rule to approximate the integration in Eq. (14.3). When $d = 3$, the polar (azimuthal) angle of the unit vector $\boldsymbol{e}$ is divided uniformly into M_1 (M_2) sections, i.e., $\theta_p = \pi/M_1$ and $\varphi_q = 2\pi/M_2$ with $p = 1, 2, \ldots, M_1$ and $q = 1, 2, \ldots, M_2$. Then Eq. (14.3) becomes

$$\begin{aligned}\widehat{Q}_j(\boldsymbol{x}) = &\frac{2\pi^2\sigma^2}{M_1 M_2}\sum_{p,q,r}\sum_{l+m=j}\hat{f}_l(\boldsymbol{x})\phi(\boldsymbol{\xi}_l\cdot\boldsymbol{e}_{pq})\hat{g}_m(\boldsymbol{x},\boldsymbol{e}_{pq})\psi\left(\boldsymbol{\xi}_l^{\perp}\right)\sin\theta_p \\ &-\frac{2\pi^2\sigma^2}{M_1 M_2}\sum_{l+m=j}\sum_{p,q,r}\hat{f}_l(\boldsymbol{x})\hat{g}_m(\boldsymbol{x},\boldsymbol{e}_{pq})\phi(-\boldsymbol{\xi}_m\cdot\boldsymbol{e}_{pq})\psi\left(\boldsymbol{\xi}_l^{\perp}\right)\sin\theta_p,\end{aligned} \tag{14.11}$$

where $\boldsymbol{e}_{pq} = (\sin\theta_p\cos\varphi_q, \sin\theta_p\sin\varphi_q, \cos\theta_p)$.

When $d = 2$, the polar angle of the unit vector $\boldsymbol{e}$ is divided uniformly as $\theta_p = 2\pi/M$, with $p = 1, 2, \ldots, M$. Then Eq. (14.3) is approximated by

$$\begin{aligned}\widehat{Q}_j(\boldsymbol{x}) = &\frac{2\pi\sigma}{M}\sum_{p,r}\sum_{l+m=j}\hat{f}_l(\boldsymbol{x})\phi(\boldsymbol{\xi}_l\cdot\boldsymbol{e}_p)\hat{g}_m(\boldsymbol{x},\boldsymbol{e}_p)\psi\left(\boldsymbol{\xi}_l^{\perp}\right) \\ &-\frac{2\pi\sigma}{M}\sum_{l+m=j}\sum_{p,r}\hat{f}_l(\boldsymbol{x})\hat{g}_m(\boldsymbol{x},\boldsymbol{e}_p)\phi(-\boldsymbol{\xi}_m\cdot\boldsymbol{e}_p)\psi\left(\boldsymbol{\xi}_l^{\perp}\right),\end{aligned} \tag{14.12}$$

where $\boldsymbol{e}_p = (\cos\theta_p, \sin\theta_p)$.

If the FFT-based convolution is used, the computational cost for Enskog collision operator is $O(M_1 M_2 M_r N^3 \log N)$ when $d = 3$ and $O(M M_r N^2 \log N)$ when $d = 2$. For elastic collisions, we have $\alpha = 1$, and the integration with respect to ρ in Eq. (14.5) can be calculated analytically. Therefore, the computational cost will be reduced by M_r times [1]. Finally, it should be noted that, although the HS gas is considered here, the method can be extended to other intermolecular force straightforwardly.

14.2 Heated Granular Gas

The spatial-homogeneous heat of a granular gas is used to assess the accuracy of FSM, where the VDF is governed by

$$\frac{\partial f}{\partial t} - \epsilon\Delta_{\boldsymbol{v}} f = Q,$$

$$\int Q\Psi(\boldsymbol{v})d\boldsymbol{v} = \chi C_0 \iiint v_r^{\gamma} f(\boldsymbol{v}_*)f(\boldsymbol{v})\left[\Psi(\boldsymbol{v}') - \Psi(\boldsymbol{v})\right]d\Omega d\boldsymbol{v}_* d\boldsymbol{v}, \tag{14.13}$$

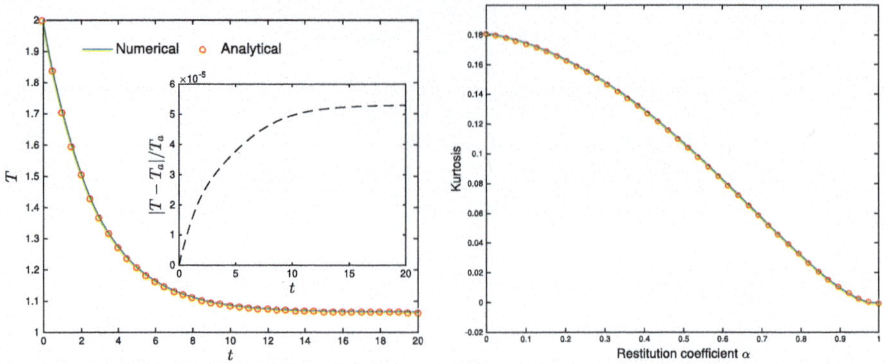

Fig. 14.1 Spatial-homogeneous heating in 2D Maxwell granular gas [1]. (Left) Temperature evolution. (Right) Kurtosis. The restitution coefficient is $\alpha = 0.5$, the energy gain coefficient is $\epsilon = 0.1$, and $\chi C_0 = 1/\pi$

where $\epsilon > 0$ is the energy gain coefficient, χ is a constant, and the term $C_0 v_r^\gamma$ represents the VHS collision kernel. For the hard-disk (HS) and Maxwell gases, $\gamma = 1$ and 0, respectively.

For Maxwell gas, when the restitution coefficient α is a constant, the evolution of temperature $T = (2/d) \int c^2 f d\boldsymbol{v}$ is governed by

$$\frac{dT}{dt} = 4\epsilon - \chi C_0 \pi 2^d \frac{1-\alpha^2}{8} T, \tag{14.14}$$

under the initial condition $n = \int f d\boldsymbol{v} = 1$ and $nu = \int \boldsymbol{v} f d\boldsymbol{v} = 0$. Also, the exact kurtosis in a 2D Maxwell gas is [2]

$$a_2 = \frac{\int f v^4 d\boldsymbol{v}}{2T^2} - 1 = \frac{6(1-\alpha)^2(1+\alpha)}{33 - 25\alpha + 3\alpha^2(1-\alpha)}. \tag{14.15}$$

Figure 14.1 shows that the FSM with $M, M_r = 12$ and $N = 32$ has sufficient accuracy. Figure 14.2 further shows that the FSM can capture the asymptotic high-energy tails [3–5]

$$f(v, t = \infty) \sim \exp\left(-A|v|^{1+\frac{\gamma}{2}}\right), \quad |v| \to \infty, \tag{14.16}$$

when the steady state is reached, even for the soft-potential gases with negative values of γ (i.e., molecules with larger relative velocity have smaller collision frequency, see Fig. 2.2). It is interesting to note that these high-energy tails in the distribution function are ubiquitous in energy-losing collisions, for example, in turbulence where the kinetic energy is gradually dissipated.

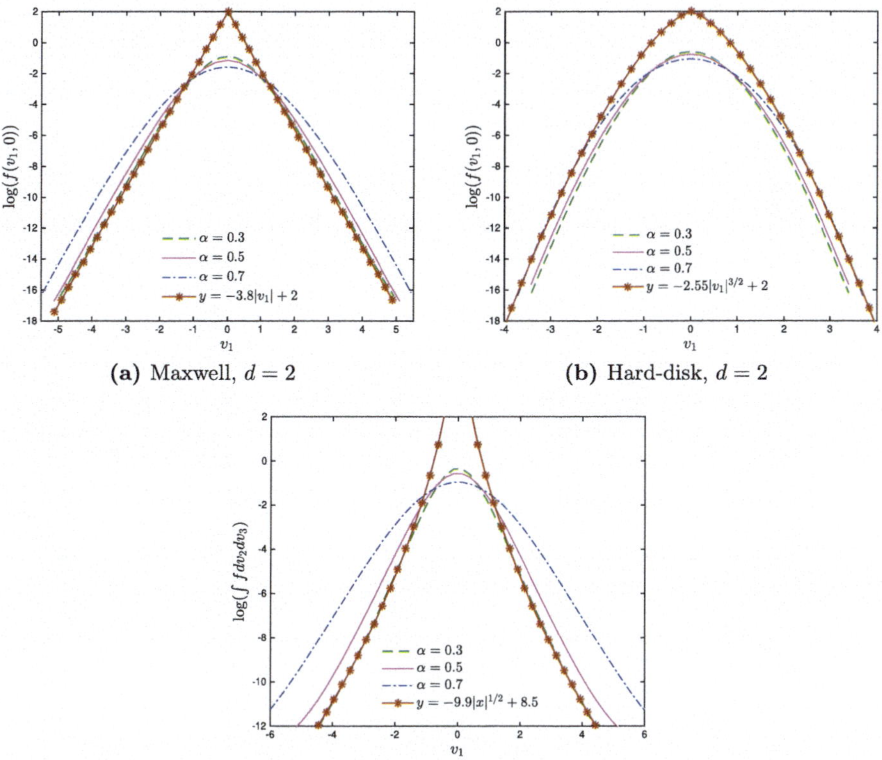

(a) Maxwell, $d = 2$ **(b)** Hard-disk, $d = 2$

(c) Soft-potential ($\gamma = -1$), $d = 3$

Fig. 14.2 High energy tails in steady-state VDFs of heated granular gases for various values of restitution coefficient [1], at $\epsilon = 0.1$ and various values of restitution coefficient α. Stars are the numerical fit according to Eq. (14.16)

14.3 Force-Driven Poiseuille Flow

Consider the flow of hard disks between two parallel plates located at $x_2 = -\sigma/2$ and $x_2 = L + \sigma/2$, respectively. Each disk is subject to an external acceleration a in the x_1-direction. The hard disks are diffusely reflected at the wall with temperature T_w.

14.3.1 *Mass Flow Rate of Dense Gas*

We first consider the Enskog equation with elastic collision, in the 2D molecular velocity space. The steady-state solution can be found when the acceleration is small and hence the flow is laminar. The velocity and temperature profiles of dense gas

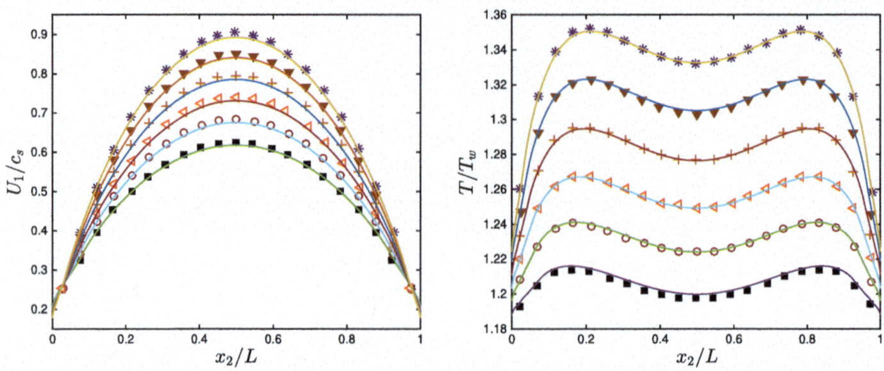

Fig. 14.3 Velocity and temperature profiles in the force-driven Poiseuille flow of hard disks [1], when $Fr = 0.45$ and $\eta_2(n_0) = 0.0873$. From bottom to top, the Knudsen number for each line is 0.185, 0.154, 0.132, 0.116, 0.103, and 0.092. Lines: FSM; Symbols: event-driven MD [7]

flow are determined by the reduced density $\eta_2(n) = \pi n \sigma^2/4$, the Knudsen number $\text{Kn} = \lambda/L$, and the dimensionless acceleration,

$$Fr = \frac{maL}{2k_B T_w}, \tag{14.17}$$

where $\lambda = 1/2\sqrt{2} n_0 \sigma \chi(n_0)$ is the equilibrium MFP of a hard disk, and $\chi(n) = (1 - \eta_2/2)/(1 - \eta_2)^2$ is the 2D pair correlation function [6].

The spatial coordinate x_2, molecular velocity $\boldsymbol{v}$, the polar angle for the vector $\boldsymbol{e}$ is uniformly discretized by 201, 32×64, and 12 grids, respectively. In calculating the Enskog collision operator, as $\boldsymbol{e}$ rotates, $\chi(\boldsymbol{x} + \sigma \boldsymbol{e}/2)$ and $\hat{f}(\boldsymbol{x} + \sigma \boldsymbol{e})$ in Eq. (14.4) are calculated by linear interpolation, and they are assigned to be zero if $\boldsymbol{x} + \sigma \boldsymbol{e}$ falls out of the computational domain. The CIS (2.84) is used to find the steady-state solution. The iterations are terminated when the maximum relative difference in velocity and temperature between two consecutive steps is less than 10^{-5}. The flow velocity and temperature are calculated as

$$u_1 = \frac{1}{n} \int f v_1 d\boldsymbol{v}, \quad T = \frac{m}{2k_B n} \int f[(v_1 - u_1)^2 + v_2^2] d\boldsymbol{v}. \tag{14.18}$$

The same problem is simulated by the event-driven MD simulations [7]. Since the hard disk only occupies regions half a diameter away from the walls, to make a fair comparison with the numerical setting here, η_2 and Fr therein are multiplied by $W/(W-1)$ and $(W-1)/W$, respectively, where W is the wall distance normalized by the molecular diameter. The Knudsen number is also adjusted accordingly. Figure 14.3 shows that the velocity and temperature obtained from the FSM compare well with those MD from event-drievn simulations. The FSM is very efficient, since

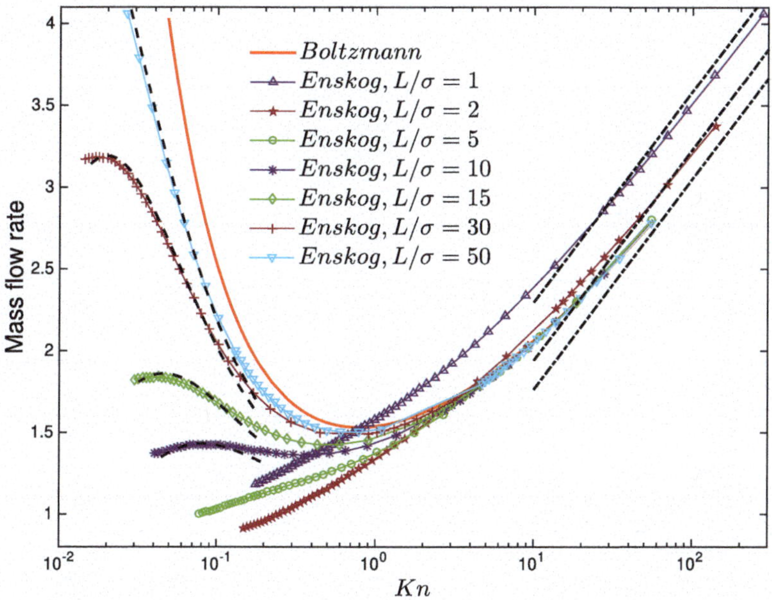

Fig. 14.4 The MFR as a function of the Knudsen number, for various values of L/σ, when the collision between the hard disks is elastic [8]

it finds the steady-state solutions within 5 min when our MATLAB program runs on a PC with a single Intel Xeon 3.3 GHz CPU.

We investigate how the dense effect affect the MFR [8]:

$$M = \int_{-1/2}^{1/2} n(y)u_x(y)dy/Fr, \tag{14.19}$$

when the normalized acceleration is $Fr = 0.0001$ so that the viscous heating is negligible. Figure 14.4 shows that the MFR is significantly influenced by L/σ, the ratio between the channel width and disk diameter. When $L/\sigma \geq 50$, the MFR for Kn < 1 is smaller than that from the Boltzmann equation, and the difference increases as Kn decreases. As L/σ decreases, the MFR becomes smaller. When $L/\sigma \leq 5$, the Knudsen minimum even disappears.

14.3.2 Influence of Restitution Coefficient

We then consider the Enskog equation with inelastic collision. Numerical results in Fig. 14.5 show that, as the restitution coefficient α decreases, the gas temperature decreases monotonically from the wall to channel center, and the gas density near

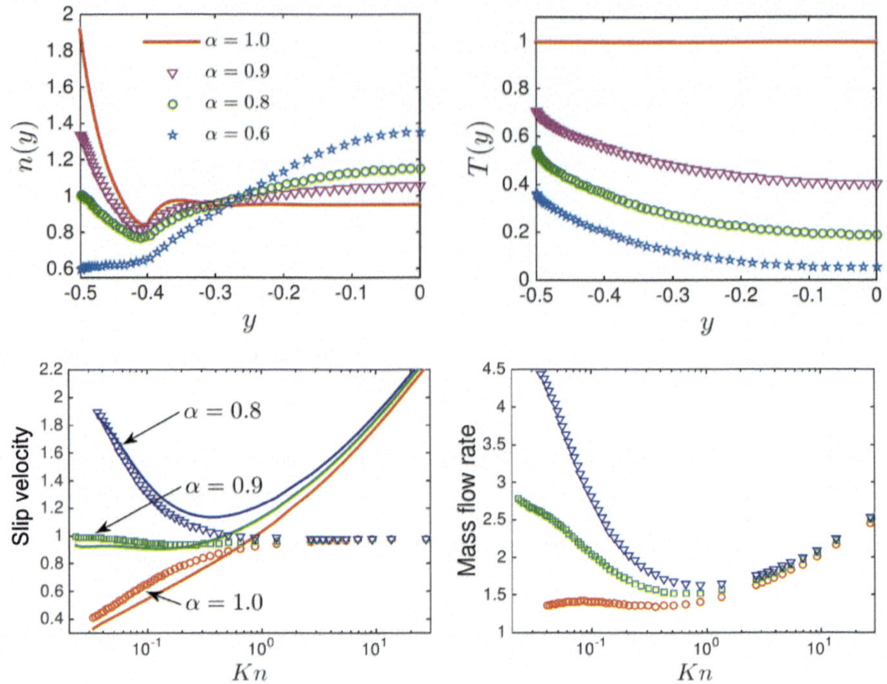

Fig. 14.5 Poiseuille flow of dense granular gas under tight confinement $L/\sigma = 10$ [8]. (Top) Normalized density and temperature. (Bottom) The slip velocity and MFR. Left column: lines: slip velocity obtained from the numerical solution of the generalized Enskog equation; symbols: calculated from the numerical results according to Eq. (14.20). The normalized acceleration is $Fr = 0.0001$

the wall (at the channel center) decreases (increases). At small α, gas concentrates in the channel center and its density could even be larger than that near the wall.

For elastic collision, the slip velocity increases with the Knudsen number. However, for inelastic collision, the anomalous velocity slip occurs, where the slip velocity decreases when Kn increases. We find that the slip velocity is described by [8].

$$\Delta u \approx \frac{1}{n(y = \pm 1/2)} \sqrt{\frac{T(y = \pm 1/2)}{T(y = 0)}}. \tag{14.20}$$

The variation of the MFR with restitution coefficient is also shown. In the free-molecular flow regime, the MFR is insensitive to the restitution coefficient due to the absence of collisions between hard disks. Therefore, the MFR is a logarithmic function of the Knudsen number. In the transition and slip flow regimes, when Kn is fixed, the MFR increases as α decreases, because (i) the velocity slip increases when α decreases and (ii) at small α the hard disks concentrate in the channel center making the effective Knudsen number smaller and hence velocity profile steeper.

14.4 Heat Transfer

The planar heat transfer is investigated based on the Enskog equation in 3D molecular velocity space. The configuration is the same as that in the force-driven Poiseuille flow, except the plate at $x_2 = L + \sigma/2$ has a temperature of $2T_w$. We set $L = 20\lambda$, the reduced density $\eta_3(n_0) = 0.1$ and 0.2, where the MFP $\lambda = [\sqrt{2}\pi n_0 \sigma^2 \chi(n_0)]^{-1}$ is equal to and one-third of the molecular diameter, respectively.

The velocity space and spatial coordinate are uniformly discretized by $32 \times 48 \times 32$ and 200 grids, respectively, while in the discretization of solid angle, we choose $M_1 = 8$ and $M_2 = 12$. Macroscopic flow quantities including the molecular number density n, the temperature T, the kinetic pressure $p_{22}^{kin} = m \int v_2^2 f dv$, and the kinetic heat flux $Q_2^{kin} = (m/2) \int v_2 v^2 f dv$ are considered. Note that since the momentum and energy are not conserved in the Enskog collision operator, there are potential parts of the pressure and heat flux; in this problem they are calculated as

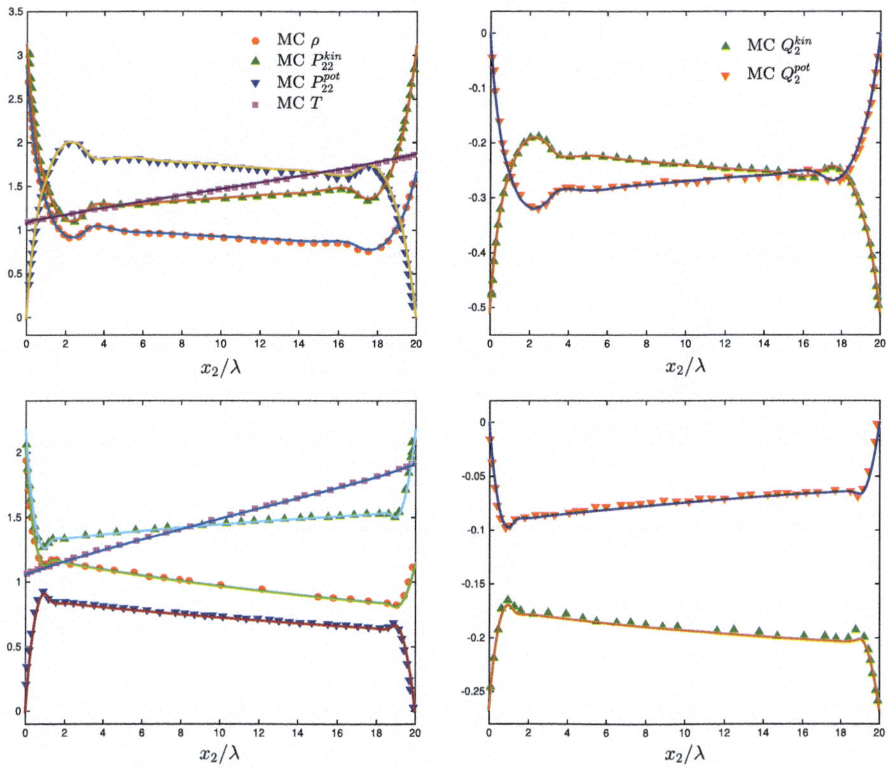

Fig. 14.6 Planar heat conduction in 3D dense HS gas [1]. Lines: FSM; Symbols: DSMC. First row: $\eta_3(n_0) = 0.2$. Second row: $\eta_3(n_0) = 0.1$. The number density, temperature, pressure, and heat flux are normalized by n_0, T_w, $n_0 k_B T_w$, and $n_0 k_B T_w \sqrt{k_B T_w / m}$, respectively

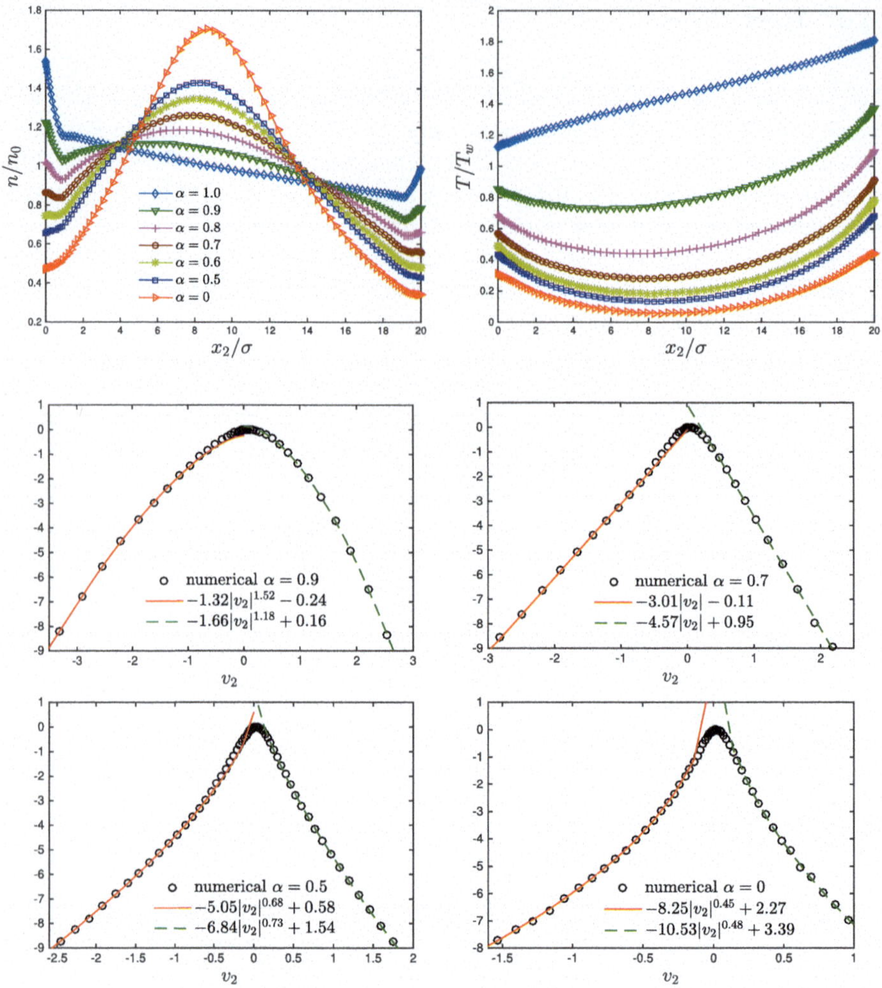

Fig. 14.7 Planar heat conduction in 2D granular gas of hard disks [1]. The marginal VDF is calculated as $\log[\int f dv_1 / \max(\int f dv_1)]$. Circles: FSM results. Lines: power-law fit of the high-energy tail. The velocity has been normalized by $\sqrt{2k_B T_w/m}$

$$
\begin{aligned}
p_{22}^{pot}(x_2) &= p_{22}^{kin}(0) - p_{22}^{kin}(x_2), \\
Q_2^{pot}(x_2) &= Q_2^{kin}(0) - Q_2^{kin}(x_2).
\end{aligned}
\tag{14.21}
$$

Figure 14.6 shows the good agreement between the FSM and DSMC [9].

Influence of Restitution Coefficient

The 2D Enskog equation is considered. The channel width is 20 times the disk diameter and the reduced density is $\eta_2(n_0) = 0.1$. The Knudsen number is 0.118 and the disk diameter is 0.422 times the MFP.

Numerical results are summarized in Fig. 14.7. For elastic collisions, the density (temperature) in the bulk region (i.e., about one disk diameter away from the two plates) decreases (increases) monotonically from left to right. However, when $\alpha < 0.9$, the bulk density (temperature) becomes a concave (convex) function of x_2. As α decreases, the temperature becomes lower, while the hard disks increasingly concentrate around the central region of the channel. In the limiting case of $\alpha = 0$, the peak hard-disk density is about 5 times larger than the density near the warmer plate, while the minimum temperature is about 17 times smaller than that of the cold plate.

In the homogeneous relaxation of a heated granular gas, Eq. (14.16) shows that the high-energy tail is determined by the collision kernel, irrespective of the restitution coefficient. For hard disk, $\log(f) \sim -|v|^{3/2}$ as $|v| \to \infty$. Figure 14.7 show that, however, the high-energy tail is strongly affected by the restitution coefficient. When the binary collision is inelastic, although the high-energy tail of the VDF in the channel center can be well fitted by power-law functions, the exponent decreases from 2 as α decreases. In the limiting case $\alpha = 0$, the high-energy behavior of the HS gas is like that of the soft-potential gas in Fig. 14.2.

14.5 Kinetic Model for Dense Gas

We briefly introduce the kinetic model to simplify the complicated collision operator in the Enskog equation when the collision is elastic. In the derivation of transport coefficients [10], the Enskog collision operator (2.59) is first expanded into Taylor's series of the molecular diameter:

$$Q \approx Q^0 + Q_a^1 + Q_b^1 \equiv Q^0 + \Omega_E, \tag{14.22}$$

where

$$\begin{aligned} Q^0 &= \chi(\boldsymbol{x}) \int \left[f(\boldsymbol{v}') f(\boldsymbol{v}'_*) - f(\boldsymbol{v}) f(\boldsymbol{v}_*) \right] \sigma^2 g \cdot \boldsymbol{k} d\boldsymbol{k} d\boldsymbol{v}_*, \\ Q_a^1 &= \frac{\sigma}{2} \int \boldsymbol{k} \cdot \nabla \chi(\boldsymbol{x}) \left[f(\boldsymbol{v}') f(\boldsymbol{v}'_*) + f(\boldsymbol{v}) f(\boldsymbol{v}_*) \right] \sigma^2 g \cdot \boldsymbol{k} d\boldsymbol{k} d\boldsymbol{v}_*, \\ Q_b^1 &= \sigma \chi(\boldsymbol{x}) \int \boldsymbol{k} \cdot \left[f(\boldsymbol{v}') \nabla f(\boldsymbol{v}'_*) + f(\boldsymbol{v}) \nabla f(\boldsymbol{v}_*) \right] \sigma^2 g \cdot \boldsymbol{k} d\boldsymbol{k} d\boldsymbol{v}_*. \end{aligned} \tag{14.23}$$

Then, f in Q_a^1 and Q_b^1 is replaced by F_{eq}, resulting in

$$\Omega_E \approx -bn\chi F_{eq}\left\{ \boldsymbol{c}\cdot\left[\nabla \ln\left(n^2\chi T\right) + \frac{3}{5}\left(\frac{mc^2}{2k_BT} - \frac{5}{2}\right)\nabla \ln T\right] + \frac{2}{5}\frac{\boldsymbol{cc}}{k_BT} : \nabla \boldsymbol{u} + \frac{2}{5}\left(\frac{mc^2}{2k_bT} - \frac{5}{2}\right)\nabla\cdot\boldsymbol{u}\right\}, \tag{14.24}$$

where $b = 2\pi\sigma^3/3$. Finally, using the Chapman-Enskog expansion, it is found that the shear viscosity and thermal conductivity of the dense gas described by the Enskog equation are

$$\begin{aligned} \mu &= \frac{5\sqrt{mk_BT}}{16\sqrt{\pi}\sigma^2}\left[\frac{1}{\chi} + 0.8bn + 0.7614(bn)^2\chi\right], \\ \kappa &= \frac{75k_B\sqrt{mk_BT}}{64m\sqrt{\pi}\sigma^2}\left[\frac{1}{\chi} + 1.2bn + 0.7574(bn)^2\chi\right], \end{aligned} \tag{14.25}$$

where terms in the brackets are due to the dense gas effect.

The BCO Q^0 can be replaced by simple kinetic models in Chap. 6. We adopt the Shakhov model (6.22) based on its relative good performance [11]; as a result, the Shakhov-Enskog kinetic model for dense gas reads

$$Q = \chi\nu\left\{F_{eq}\left[1 + (1 - \mathrm{Pr})\frac{2m\boldsymbol{q}\cdot\boldsymbol{c}}{5n(k_BT)^2}\left(\frac{mc^2}{2k_BT} - \frac{5}{2}\right)\right] - f\right\} + \Omega_E. \tag{14.26}$$

The velocity-independent collision frequency and the Prandtl number are determined by equating the shear viscosity and thermal conductivity of this kinetic model to these of the Enskog equation Eq. (14.25):

$$\begin{aligned} \nu &= \frac{n_0k_BT}{\mu}\left(\frac{1}{\chi} + \frac{2}{5}bn\right), \\ \mathrm{Pr} &= \frac{5k_B}{2m}\frac{n_0k_BT}{\kappa}\left(\frac{1}{\chi} + \frac{3}{5}bn\right). \end{aligned} \tag{14.27}$$

Alternatively, the ν-model (6.44) may be used to better approximate the BCO. Also, the results in Chap. 7 can be borrowed to build kinetic models for the dense molecular gas flows; in this case, one may assume the translational collision is non-local, while the internal-translational energy relaxation is local.

References

1. Wu L, Zhang Y, Reese JM (2015) Fast spectral solution of the generalized Enskog equation for dense gases. J Comput Phys 303:66–79
2. Santos A (2003) Transport coefficients of d-dimensional inelastic Maxwell models. Phys A 321:442–466

3. van Noije TPC, Ernst MH (1998) Velocity distributions in homogeneous granular fluids: the free and the heated case. Granular Matter 1:57–64
4. Ernst MH, Brito R (2002) High-energy tails for inelastic Maxwell models. Europhys Lett 58:182–187
5. Ernst MH, Trizac E, Barrat A (2006) The Boltzmann equation for driven systems of inelastic soft spheres. J Stat Phys 124:549–586
6. Sanchez IC (1994) Virial coefficients and close-packing of hard spheres and discs. J Chem Phys 101:7003–7006
7. Shen G, Ge W (2010) Simulation of hard-disk flow in microchannels. Phys Rev E 81:011201
8. Wu L, Liu HH, Reese JM, Zhang YH (2016) Non-equilibrium dynamics of dense gas under tight confinement. J Fluid Mech 794:252–266
9. Frezzotti A (1999) Monte Carlo simulation of the heat flow in a dense hard sphere gas. Eur J Mech B/Fluids 18:103–119
10. Chapman S, Cowling TG (1970) The mathematical theory of non-uniform gases. Cambridge University Press
11. Wang P, Wu L, Ho MT, Li J, Li ZH, Zhang YH (2020) The kinetic Shakhov-Enskog model for non-equilibrium flow of dense gases. J Fluid Mech 883:A48

Chapter 15
Fluctuation and Light Scattering

The Boltzmann equation describes the gas dynamics in an average sense, where the fluctuations to the VDF around its average value are ignored, but they have an effect on, e.g., the spectrum of the Rayleigh-Brillouin scattering of laser light. These fluctuations can be considered formally in the linearized Boltzmann equation. In this chapter, the theory, numerical method, as well as applications of Rayleigh-Brillouin scattering are introduced, especially in the extraction of bulk viscosity and translational thermal conductivity of molecular gas.

15.1 Rayleigh-Brillouin Scattering

Light propagating in the gas is scattered by the motion of gas molecules. In most general circumstances, the spectrum of scattered light has several peaks corresponding to different scattering mechanisms. The RBS spectrum consists of the Rayleigh part due to the Doppler shift by the thermal motion of individual gas molecules and the Brillouin part related to the acoustic effect of gas molecules. Nowadays, RBS has become an invaluable non-destructive optical diagnostic technique for measuring the gas properties, such as the sound speed, flow velocity, temperature, and bulk viscosity. For instances, the satellite ADM-Aeolus launched in 2018 by the European Space Agency measures global wind profile from the surface of the planet to the stratosphere 30 km high and hence provides much-needed information to improve weather forecasting.

L. Wu, *Rarefied Gas Dynamics*,
https://doi.org/10.1007/978-981-19-2872-7_15

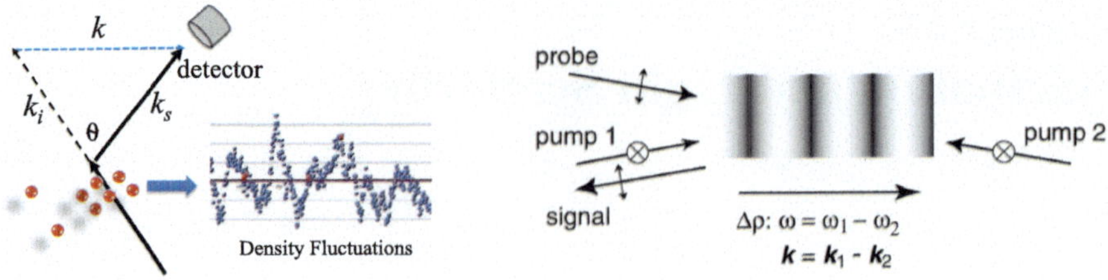

Fig. 15.1 Schematic of the (left) spontaneous and (right) coherent RBS [1]

15.1.1 Spontaneous RBS

The spontaneous fluctuation of gas density scatters the light, see the left figure in Fig. 15.1. The intensity of scattered light is proportional to the spectral density function $S(f_s)$ [2], i.e., the spatial and temporal Fourier transform of the density perturbation correlation function $G(\boldsymbol{r}, t)$. Suppose the scattering wave is propagating along the x_2-direction, the correlation function is [2, 3]:

$$G(x_2, t) = \int_{-\infty}^{\infty} \varrho(l + x_2, t)\varrho(l, 0)dl, \tag{15.1}$$

and hence the spectrum is

$$S(f_s) = \int_{-\infty}^{\infty}\int_{-\infty}^{\infty} G(x_2, t)e^{i(kx_2 - 2\pi f_s t)}dx_2 dt, \tag{15.2}$$

where $\varrho(x_2, t)$ is the perturbation density of the gas, f_s is the scattering frequency, and k is the scattering wavenumber

$$k = |\boldsymbol{k}_i - \boldsymbol{k}_s| = 2|\boldsymbol{k}_i| \sin\left(\frac{\theta}{2}\right), \tag{15.3}$$

with $\boldsymbol{k}_i$ being the wave vector of an incident light and θ being the angle of scattering.

Due to the short wavelength and high scattering frequency, the kinetic theory should be applied to describe the evolution of density perturbation. The perturbation density is related to the perturbation VDF $h(t, x_2, \boldsymbol{v})$ as per Eq. (5.6), and the evolution of h for monatomic gas is governed by the LBE (5.3). The initial density impulse is represented by [3, 4]

$$h(t = 0, x_2, \boldsymbol{v}) \propto \delta(x_2) f_{eq}(\boldsymbol{v}). \tag{15.4}$$

Suppose $\hat{h}(\boldsymbol{v})$ is the Fourier and Laplace transform of h in the spatial and temporal directions, respectively, the LBE for $\hat{h}(\boldsymbol{v})$ becomes

$$2\pi i(f_s - v_2)\hat{h} = \mathcal{L}^+(\hat{h}) - \nu_{eq}\hat{h} + f_{eq}, \tag{15.5}$$

and the spectrum of spontaneous RBS is calculated as

$$S(f_s) = \Re\left(\int \hat{h} d\boldsymbol{v}\right). \tag{15.6}$$

Note that the scattering wavelength $L = 2\pi/k$ is chosen as the characteristic flow length, the velocity is normalized by the most probable speed v_m, and the scattering frequency f_s is normalized by v_m/L. Also, the VDF can be solved by the CIS as

$$\hat{h}^{k+1}(\boldsymbol{v}) = \frac{\mathcal{L}^+(\hat{h}^k) + f_{eq}(\boldsymbol{v})}{2\pi i(f_s - v_2) + \nu_{eq}(\boldsymbol{v})}. \tag{15.7}$$

Figure 15.2a depicts the spectra of spontaneous RBS. When Kn = 0.00125, the spectrum consists of the Rayleigh peak at $f_s = 0$ and Brillouin peak at

$$f_s = \frac{\sqrt{5k_B T_0/3m}}{v_m} = \sqrt{\frac{5}{6}}, \tag{15.8}$$

which is due to the acoustic effect of gas molecules. When Kn = 0.0125, both the Brillouin and Rayleigh parts broaden, due to the increased dissipation in sound propagation. When Kn further increases, the Rayleigh and Brillouin parts of the spectrum mix, and eventually the whole spectrum is nearly Gaussian $S = \exp(-f_s^2)$. This Gaussian line shape is predicted theoretically and observed experimentally [5, 6], in the free-molecular regime. Here, it is observed at Kn = 0.627, because the gas dynamics is in strong non-equilibrium: not only the spatial Knudsen number Kn is large but also the temporal Knudsen number (1.14)

$$\text{Kn}_t = \frac{2\pi f_s v_m/L}{\bar{\nu}} = 4\sqrt{2}\text{Kn} f_s$$

is large.

Note that the RBS spectrum is obtained when the flow velocity is zero. If the gas has a macroscopic velocity u, however, the shape of the RBS spectrum will be shifted horizontally by a factor of $\Delta f_s = u/v_m$. Therefore, by measuring $\int_{-\infty}^{0} S(f_s) df_s$ and $\int_0^\infty S(f_s) df_s$ separately, the gas velocity can be determined. This is the basic principle used in the satellite ADM-Aeolus to map the global wind profile.

15.1.2 Coherent RBS

The right figure in Fig. 15.1 shows that the gas density in coherent RBS is perturbed by the moving optical lattice induced by two pump laser beams [8]. The coherent RBS has higher signal-to-noise ratio than the spontaneous RBS because, by the phase matching method, the scattered light propagates in a certain direction, rather than

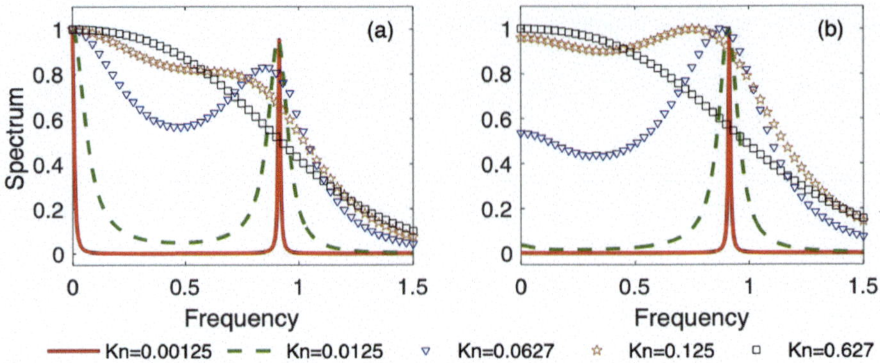

Fig. 15.2 Spectra of the **a** spontaneous and **b** coherent RBS from the LBE for Maxwellian gases [7]. Note that in this figure and following figures, each spectrum has been normalized by the maximum value

scattered everywhere. The spectrum of the coherent RBS is related to the density perturbation as $S = |\delta\bar{\rho}|^2$, with

$$\delta\bar{\rho} = \iint_{-\infty}^{\infty} e^{i(kx_2 - 2\pi f_s t)} \varrho(x_2, t) dx_2 dt. \tag{15.9}$$

Suppose the acceleration $a_2 \propto \cos[2\pi(x_2 - f_s t)]$ generated by the two pump beams propagates along the x_2-direction. The pump beam intensities are so low that the gas dynamics is described by the following LBE [1]:

$$\frac{\partial h}{\partial t} + v_2 \frac{\partial h}{\partial x_2} - \cos[2\pi(x_2 - f_s t)] v_2 f_{eq}(\boldsymbol{v}) = \mathcal{L}^+(h) - \nu_{eq}(\boldsymbol{v}) h. \tag{15.10}$$

Taking the Fourier transform in the temporal-spatial domain, the spectrum of coherent RBS is calculated as

$$S(f_s) = \left| \int \hat{h} d\boldsymbol{v} \right|^2, \tag{15.11}$$

where $\hat{h}$, the spatial-temporal Fourier transform of h, can be solved by the CIS [4]:

$$\hat{h}^{k+1}(\boldsymbol{v}) = \frac{v_2 f_{eq}(\boldsymbol{v}) + \mathcal{L}^+(\hat{h}^k)}{2\pi i (f_s - v_2) + \nu_{eq}(\boldsymbol{v})}. \tag{15.12}$$

Typical spectra are shown in Fig. 15.2(b). When Kn $= 0.00125$, the spectrum consists of the Brillouin peak only, where the sound waves are resonantly pumped by the optical force. As Kn increases, the relative intensity of the Rayleigh peak to the Brillouin peak increases, and eventually the whole spectrum is Gaussian: $S \approx \exp\left(-f_s^2/1.21\right)$.

15.2 Numerical Methods

The CIS in Eqs. (15.7) and (15.12) converges extremely slow when the rarefaction parameter is large. In this section, the GSIS is applied to expedite the convergence.

15.2.1 Monatomic Gas

Due to the Laplace and Fourier transforms, the time and spatial derivatives in the LBE are replaced by $2i\pi f_s$ and $-2i\pi$, respectively. In the macroscopic synthetic equations, both derivatives should be replaced too. Also, it should be noted that, in spontaneous RBS, the source term f_{eq} in Eq. (15.5) appears from the Laplace transform of initial density impulse (15.4); hence, the first synthetic equation in Eq. (8.16) should be modified to

$$\frac{\partial \varrho}{\partial t}+\frac{\partial u_i}{\partial x_i}=1. \tag{15.13}$$

In GSIS-I, the synthetic equations (8.16), (8.20) and (8.21), after the Fourier transform, is rewritten as

$$\begin{bmatrix} 2i\pi f_s & -2i\pi & 0 & 0 & 0 \\ -2i\pi & 4i\pi f_s & -2i\pi & -2i\pi & 0 \\ 0 & -2i\pi & 3i\pi f_s & 0 & -2i\pi \\ 0 & -\frac{8}{3}i\pi & 0 & 2i\pi f_s+\delta_{rp} & 0 \\ 0 & 0 & -3i\pi C_q & 0 & 2i\pi f_s+\frac{2}{3}\delta_{rp} \end{bmatrix} \begin{bmatrix} \hat{n}^{k+1} \\ \hat{u}_2^{k+1} \\ \hat{T}^{k+1} \\ \hat{\sigma}_{22}^{k+1} \\ \hat{q}_2^{k+1} \end{bmatrix} = \begin{bmatrix} 1 \\ 0 \\ 0 \\ \mathrm{R}_4 \\ \mathrm{R}_5 \end{bmatrix}, \tag{15.14}$$

where

$$\begin{aligned} \mathrm{R}_4 &= 2i\pi \mathrm{HoT}_{\sigma_{22}}^{k+1/2}+2\int(\mathcal{L}^k-\mathcal{L}_s^k)v_2^2 d\boldsymbol{v}, \\ \mathrm{R}_5 &= 2i\pi \mathrm{HoT}_{q_2}^{k+1/2}+\int(\mathcal{L}^k-\mathcal{L}_s^k)v_2 v^2 d\boldsymbol{v}. \end{aligned} \tag{15.15}$$

In the correction of VDF (8.32), we choose $\delta_{rp}^{\text{th}}=8$ in Eq. (8.33). Spectra of coherent RBS can be found if the right-hand side term in Eq. (15.14) is replaced by $[0, 1, 0, \mathrm{R}_4, \mathrm{R}_5]^\top$.

With FSM and GSIS, each RBS spectrum can be obtained in just a few seconds in a laptop. Note that if the DSMC is used, several hours of computational time are needed to obtain one line shape [9].

15.2.2 Molecular Gas

In molecular gas, the vibrational modes will not be reflected in RBS due to their large collision numbers (or relaxation time). Thus, the Wu model can be used to find the RBS spectra when only the translational and rotational degrees of freedom are considered. Introducing the linearization $G = f_{eq} + h_0$ and $R = (d_r/2) f_{eq} + h_1$, and the VDF $h_2 = h_1 - (d/2)h_0$, the Wu model in Sect. 7.4 is linearized to

$$
\begin{aligned}
\frac{\partial h_0}{\partial t} + v_2 \frac{\partial h_0}{\partial x_1} &= C_0 \equiv \mathcal{L}^+(h_0) - \nu_{eq}(\boldsymbol{v}) h_0 \\
&\quad + \delta_{rp} \frac{f_{eq}}{Z} \left[(T - T_t) \left(v^2 - \frac{3}{2} \right) + \frac{4(\omega_0 - 1)}{15} q_t v_2 \left(v^2 - \frac{5}{2} \right) \right], \\
\frac{\partial h_2}{\partial t} + v_2 \frac{\partial h_2}{\partial x_1} &= C_2 \equiv \delta_{rp} \left[\frac{d_r}{2} T_r f_{eq} - h_2 \right] + \frac{d_r \delta_{rp}}{2Z} (T - T_r) f_{eq} \\
&\quad + \delta_{rp} \frac{2(Z + \omega_1 - 1)(1 - \mathrm{Sc})}{Z} q_r v_2 f_{eq},
\end{aligned} \tag{15.16}
$$

where the off-diagonal elements in the matrix A are chosen to be zero. The rotational temperature and heat flux are $T_r = \frac{2}{d_r} \int h_2 d\boldsymbol{v}$ and $q_r = \int v_2 h_2 d\boldsymbol{v}$. The bulk viscosity is determined by Z in Eq. (7.8), the translational and internal Eucken factors are controlled by ω_0 and ω_1 in Eq. (7.32).

The spontaneous RBS spectrum $S = \int \Re(\hat{h}_0) d\boldsymbol{v}$ of a molecular gas can be obtained by the CIS [4]:

$$
\begin{aligned}
\hat{h}_0^{k+1}(\boldsymbol{v}) &= \frac{f_{eq} + \hat{C}_0^k + \nu \hat{h}_0^k}{2\pi i (f_s - v_2) + \bar{\nu}}, \\
\hat{h}_2^{k+1}(\boldsymbol{v}) &= \frac{\hat{C}_2^k + \nu \hat{h}_2^k}{2\pi i (f_s - v_2) + \bar{\nu}},
\end{aligned} \tag{15.17}
$$

while the coherent RBS spectrum $S = |\int \hat{h}_0 d\boldsymbol{v}|^2$ can be obtained by solving

$$
\begin{aligned}
\hat{h}_0^{k+1}(\boldsymbol{v}) &= \frac{f_{eq} v_2 + \hat{C}_0^k + \nu \hat{h}_0^k}{2\pi i (f_s - v_2) + \bar{\nu}}, \\
\hat{h}_2^{k+1}(\boldsymbol{v}) &= \frac{(d_r/2) f_{eq} v_2 + \hat{C}_2^k + \nu \hat{h}_2^k}{2\pi i (f_s - v_2) + \bar{\nu}},
\end{aligned} \tag{15.18}
$$

where $\bar{\nu} = 1.5\delta_{rp}(1 + 1/Z)$ is chosen to ensure the stability of iteration.

The following macroscopic synthetic equations are designed to boost convergence (note that all the variables are evaluated at the $(k+1)$-th iteration step) [10]:

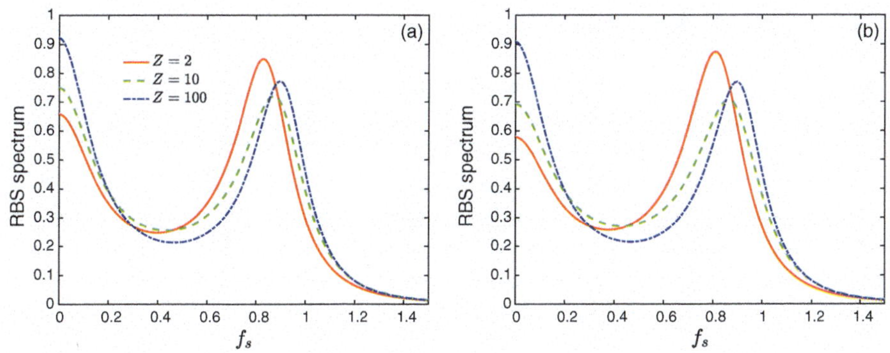

Fig. 15.3 Influence of the rotational collision number in **a** diatomic and **b** nonlinear polyatomic gases, when $f_{tr} = f_{int} = 2$, and $\delta_{rp} = 8\pi$ [11]. In this and following figures, if not further specified, the RBS spectrum is normalized by the area $\int_{-\infty}^{\infty} S(f_s)df_s$ of each line shape

$$
\begin{aligned}
\frac{\partial \varrho}{\partial t} + \frac{\partial u_j}{\partial x_j} &= 0, \\
2\frac{\partial u_i}{\partial t} + \frac{\partial \varrho}{\partial x_i} + \frac{\partial T_t}{\partial x_i} + \frac{\partial \sigma_{ij}}{\partial x_j} &= 0, \\
\frac{3}{2}\frac{\partial T_t}{\partial t} + \frac{\partial u_j}{\partial x_j} + \frac{\partial q_{t,j}}{\partial x_j} &= \frac{3\delta_{rp}}{2Z}(T - T_t), \\
\frac{d_r}{2}\frac{\partial T_r}{\partial t} + \frac{\partial q_{r,j}}{\partial x_j} &= \frac{d_r\delta_{rp}}{2Z}(T - T_r),
\end{aligned}
\tag{15.19}
$$

where, similar to the GSIS-II for monatomic gas, the constitutive relations are designed as follows:

$$
\begin{aligned}
\sigma_{ij}^{k+1} &= -\frac{2}{\delta_{rp}}\frac{\partial u_{<i}^{k+1}}{\partial x_{j>}} + \sigma_{ij}^{k+1/2} + \frac{2}{\delta_{rp}}\frac{\partial u_{<i}^{k+1/2}}{\partial x_{j>}}, \\
q_{t,i}^{k+1} &= -\frac{\kappa_t}{\delta_{rp}}\frac{\partial T_t^{k+1}}{\partial x_i} + q_{t,i}^{k+1/2} + \frac{\kappa_t}{\delta_{rp}}\frac{\partial T_t^{k+1/2}}{\partial x_i}, \\
q_{r,i}^{k+1} &= -\frac{\kappa_r}{\delta_{rp}}\frac{\partial T_r^{k+1}}{\partial x_i} + q_{r,i}^{k+1/2} + \frac{\kappa_r}{\delta_{rp}}\frac{\partial T_r^{k+1/2}}{\partial x_i}.
\end{aligned}
\tag{15.20}
$$

This GSIS for molecular gas is easy to understand, as compared to that for monatomic gas, the additional synthetic equation for the evolution of rotational temperature appears.

15.2.2.1 Role of Rotational Collision Number

Figure 15.3 shows the RBS spectra when the rarefaction parameter is $\delta_{rp} = 8\pi$, where the influence of the rotational collision number Z is clearly seen. For diatomic (nonlinear polyatomic) gases with the rotational degree of freedom $d_r = 2$ ($d_r = 3$), the position of Brillouin peak moves from $f_s = 0.83$ (0.81) to 0.90 (0.90) when Z is increased from 2 to 100. Since this position is approximately determined by the sound speed normalized by v_m, it can be concluded that the rotational degree of freedom is gradually frozen when Z increases, such that the effective sound speed of the molecular gas gradually increases from

$$\sqrt{\frac{5+d_r}{2(3+d_r)}}$$

to that of the monatomic gas $\sqrt{5/6}$.

From Eqs. (7.8) and (7.9), we know the bulk viscosity is proportional to Z in the continuum flow regime. However, when

$$Z\frac{\mu}{p} \times \frac{v_m}{L} f_s = \frac{Zf_s}{\delta_{rp}}$$

is large, the bulk viscosity approaches zero rather than infinity because the rotational-translational energy exchange is so slow that the gas system cannot feel, i.e., the rotational relaxation is frozen. In reality, this does not occur since the rotational collision number Z is usually small. However, the vibrational collision number can be very large, e.g., on the order of 10^4 for CO_2 and SF_6, and even larger for other molecules [12]. Therefore, results in Fig. 15.3 clearly show that, in the calculation of RBS spectrum, the vibrational modes should not be considered [13, 14, 22]. Similarly, the contribution of the vibrational thermal conductivity should be removed.

15.2.2.2 Role of Translational Eucken Factor

In the hydrodynamic regime, the NSF equations can be used, and the RBS spectrum is determined by the total thermal conductivity, irrespective of the allocation into its translational and internal thermal conductivities. On the other hand, the RBS spectrum is only determined by the gas temperature in the free-molecular regime. Therefore, if the translational thermal conductivity affects the RBS spectrum, it can only emerge in a certain range of the Knudsen number.

Figure 15.4 shows that the variation of f_{tr} leads to a strong change of RBS line shape, when the total thermal conductivity (or equivalently f_u) is fixed, especially when the rarefaction parameter is $\delta_{rp} \sim 2\pi$. Thus, the RBS line shape in this regime can be used to determine the translational thermal conductivity [11].

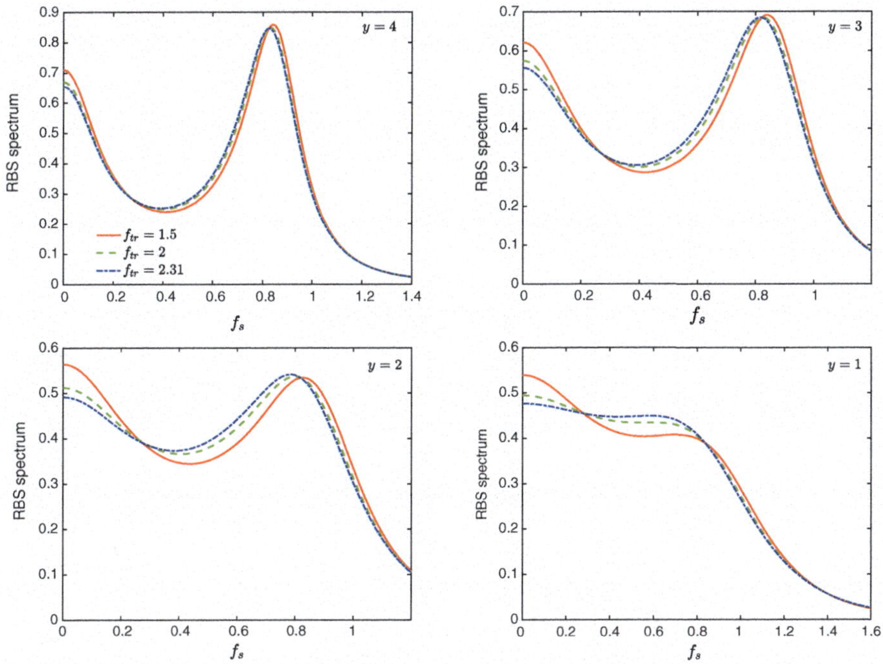

Fig. 15.4 Computed RBS spectra displaying the influence of translational Eucken factor on RBS spectra of molecular gases with $d_r = 2$, when the total Eucken factor is $f_u = 2$ and the rotational collision number is $Z = 2$ [11]. The rarefaction parameter is $\delta_{rp} = 2\pi y$

15.3 Accuracy of the Tenti Model

Before the efficient numerical method for the LBE is proposed [4, 15, 16], Tenti's s6 and s7 models are prevailing in the calculation of RBS line shape [17]. The model is called s6 model, because for molecular gas the six moments $\varrho, u_x, T_t, T_r, q_{tx}, q_{rx}$ are used, while in the s7 model the additional stress component σ_{xx} also appears.

For monatomic gas, Tenti's model can be fitted into the Gross-Jackson framework. The s6 model is just the linearized Shakhov model (6.27), while the s7 model is the original Gross-Jackson model (6.32). Although the s7 model contains additional stress in the collision operator, experimental results have revealed that the s6 model is more accurate [8].

Figure 15.5 shows the spectra of spontaneous RBS of Maxwell gas when $\text{Kn} = 0.08$. The Shakhov model yields a good agreement with the LBE, while the ESBGK and Gross-Jackson models have large deviations. Although the three kinetic models have the same value of shear viscosity and thermal conductivity, their collision frequencies are different:

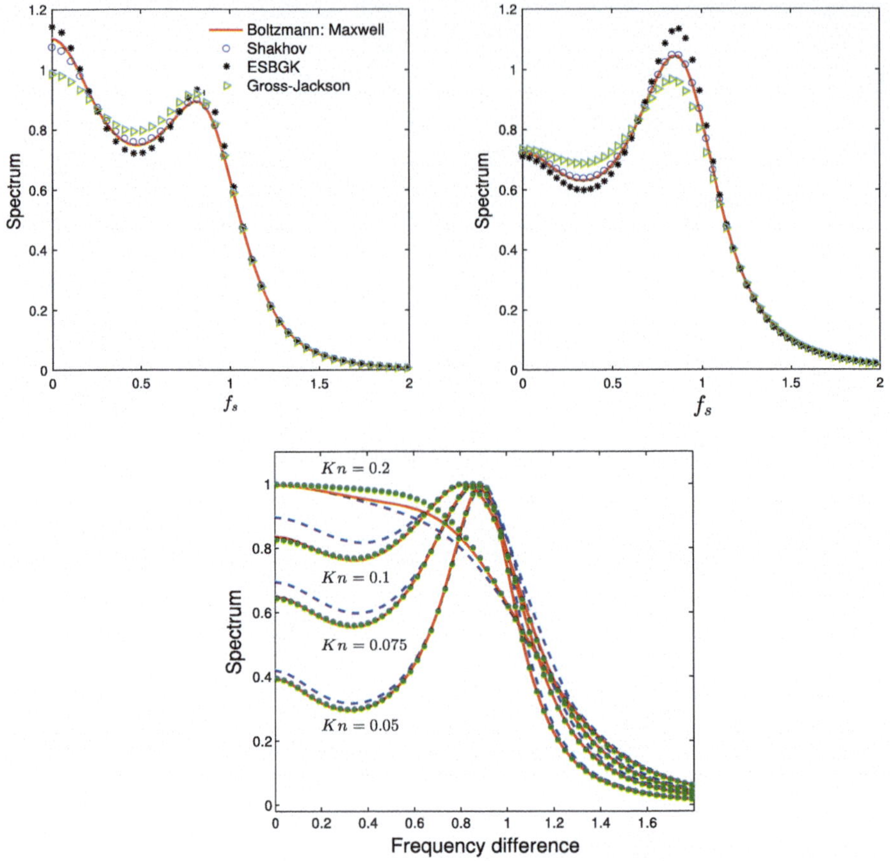

Fig. 15.5 Comparison in the spectra of (left) spontaneous and (right) coherent Rayleigh-Brillouin scattering in a gas of Maxwellian molecules with Kn = 0.08. (Bottom) The spectra of coherent RBS in monatomic gas [15], obtained from the LBE of HS gas (dashed lines), Maxwell gas (solid lines), and Tenti's s6 kinetic model (dots)

$$\nu_{\text{Shakhov}} = \frac{p}{\mu}, \quad \nu_{\text{ESBGK}} = \frac{2p}{3\mu}, \quad \nu_{\text{Gross-Jackson}} = \frac{3p}{2\mu}, \tag{15.21}$$

which means that the relaxation rates of high-order (higher than the heat flux) moments are different; and they play important roles in the accuracy of kinetic model.

The influence of power-law intermolecular potential, reflected in the viscosity index ω, is studied in Maxwell and HS gases. While the intermolecular potential has negligible influence on the spectra of spontaneous RBS, Fig. 15.5 shows that it affects the spectra of coherent RBS significantly [15]. Therefore, if the s6 model is used to extract the gas information, error will be generated.

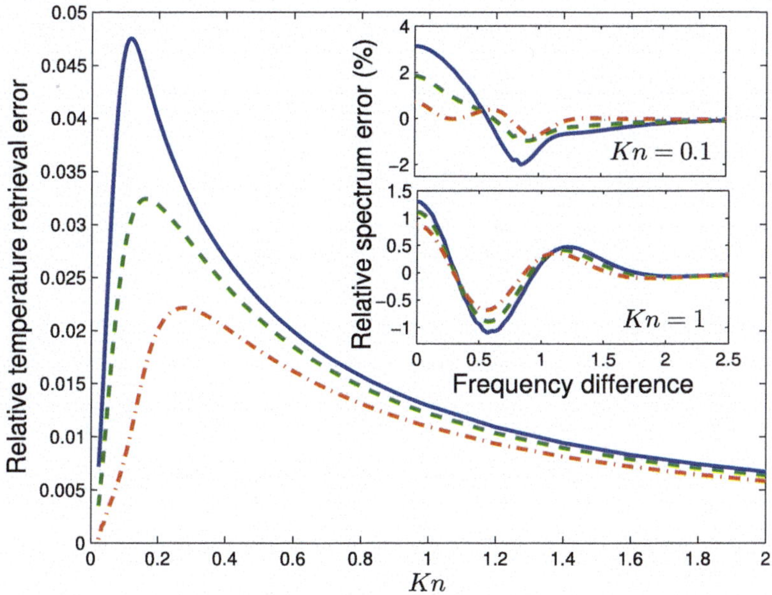

Fig. 15.6 The relative temperature retrieval error of Tenti's s6 model for coherent RBS in monatomic gases [16]. Insets: the relative error in the spectrum between the LBE and Tenti's s6 model at the same temperature, i.e., $(S_{\text{LBE}} - S_{\text{Tenti}})/\max(S_{\text{Tenti}})$. Solid lines: HS gas; dashed lines: $\omega = 0.74$; dash-dotted lines: Maxwell gas

15.3.1 *Temperature Retrieval Error*

To analyze this temperature retrieval error (TRE), we design the following "experiment". We first calculate the RBS spectrum $S_{\text{LBE}}(\text{Kn}, \omega, f_s, T_0)$ from the LBE at a temperature T_0, then calculate $S_{\text{Tenti}}(\text{Kn}, \omega, f_s, T_t)$ from the s6 model using different values of temperature T_t. Finally, the temperature is retrieved when $\int [S_{\text{LBE}}(\text{Kn}, \omega, f_s, T_0) - S_{\text{Tenti}}(\text{Kn}, \omega, f_s, T_r)]^2 df_s$ reaches a minimum value. This mimics the extraction of gas temperature from the experimental RBS spectra by using the Tenti's s6 model.

Figure 15.6 shows the relative TRE $(T_r - T_0)/T_0$ in coherent RBS. The TRE is zero in the continuum and free-molecular flow regimes. This is because, for the former, both LBE and s6 kinetic models yield the same NSF equations, while for the latter, both collision operators are negligible, so the spectra are the same. The maximum relative TRE is about 5% around Kn = 0.2 in the HS gas. When $T_0 =$ 300 K, the maximum TRE for an HS gas, the TRE would be, respectively, 6.7 K, 9.7 K, and 14.3 K in the Maxwell gas, the gas with the viscosity index $\omega = 0.74$, and HS gas. The TRE in molecular gas has a similar error when Tenti's s6 model is used.

Table 15.1 Experimental conditions in spontaneous RBS of N_2 [18] and extracted values of the bulk viscosity and translational/internal Eucken factor based on the Wu model [11]. The shear viscosity μ_s and total thermal conductivity κ are shown in SI units

Case	T (K)	$10^5\mu_s$	100κ	P (bar)	$\delta_{rp}/2\pi$	ω	μ_b/μ_s	f_{tr}	f_{int}
(a)	254.7	1.57	2.28	2.563	1.73	0.80	0.48	2.16	1.65
(b)	275.2	1.67	2.44	2.784	1.70	0.78	0.61	2.25	1.53
(c)	296.7	1.77	2.60	3.000	1.66	0.76	0.69	2.31	1.47
(d)	336.6	1.95	2.88	3.400	1.61	0.74	0.94	2.43	1.33

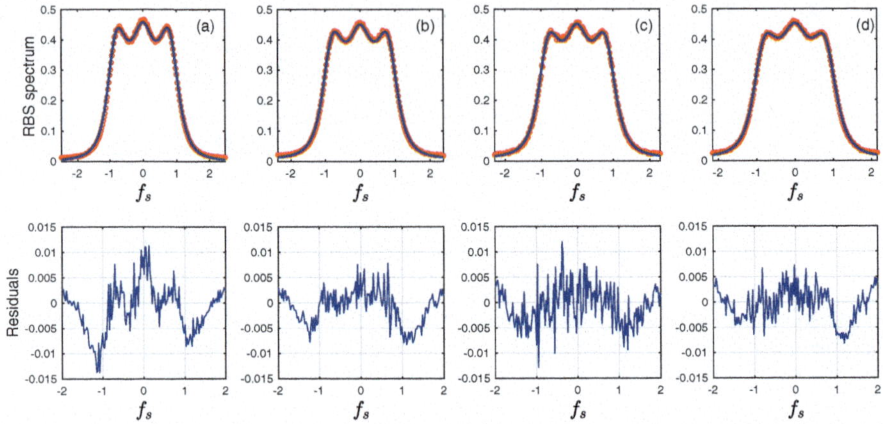

Fig. 15.7 Extraction of the bulk viscosity and phystranslational Eucken factor of nitrogen [11], from the experimental spontaneous RBS spectra (circles) [18]. Lines in the first row show the RBS spectra obtained from the Wu model. The second row shows the residuals between the experimental and theoretical spectra

15.4 Extraction of Gas Property

Although the total thermal conductivity can be measured straightforwardly, its translational and internal components are not. Since in thermal transpiration the flow rate is related to the translational Eucken factor, it can be used to extract the translational thermal conductivity, but the accuracy will be hampered by the uncertainty in gas-surface interaction. The RBS provides an ideal system to extract both translational thermal conductivity and bulk viscosity, without the influence of solid wall [11].

In analogy to the analyzing of TRE, the bulk viscosity (due to the relaxation of rotational and translational energies) and translational Eucken factor of nitrogen are extracted from the experiment [18], where the laser with a wavelength of 366.8 nm is scattered by 90°. Therefore, the effective wavelength is $L = 259.4$ nm. The rotational degrees of freedom of nitrogen is $d_r = 2$, while the vibrational modes are not excited in the experimental condition. The transport coefficients are calculated from the Sutherland formula [19], and the effective viscosity index ω is shown in Table 15.1.

Keeping the total Eucken factor fixed, by varying the rotational collision number Z and the translational Eucken factor f_{tr}, a minimum difference between the numerical and experimental RBS line shape can be found, and the corresponding values of Z and f_{tr} are chosen. Figure 15.7 shows the best fit of the experimental spectrum, where the residuals between the numerical and experimental line shapes are generally less than 1%.

According to the acoustic experiment [20] the rotational relaxation time of nitrogen at standard temperature and pressure is $\tau_r = 7.4 \times 10^{-10}$ s, therefore according to Eq. (7.8) the bulk viscosity is

$$\mu_b = 2p\tau_r \frac{d_r}{(3+d_r)^2} = 1.18 \times 10^{-5} \text{kg} \cdot \text{m}^{-1} \cdot \text{s}^{-1},$$

which agrees well with the extracted bulk viscosity 1.22×10^{-5} kg · m^{-1} · s^{-1} from the spontaneous RBS at $T = 296.7$ K. The extracted translational and internal Eucken factors are also in good agreement with the analytical solutions (7.16) from the Wang-Chang & Uhlenbeck equation [21].

References

1. Pan X, Shneider MN, Miles RB (2002) Coherent Rayleigh-Brillouin Scattering. Phys Rev Lett 89(18):183001
2. Reichl LE (2016) *A Modern Course in Statistical Physics*. John Wiley & Sons
3. Sugawara A, Yip S, Sirovich L (1968) Spectrum of density fluctuations in gases. Phys Fluids 11:925–932
4. Wu L, White C, Scanlon TJ, Reese JM, Zhang YH (2015) A kinetic model of the Boltzmann equation for non-vibrating polyatomic gases. J Fluid Mech 763:24–50
5. Muller I, Ruggeri T (1998) *Rational Extended Thermodynamics*. Springer
6. Grinstead JH, Barker PF (2000) Coherent Rayleigh scattering. Phys Rev Lett 85:1222
7. Wu L, Gu XJ (2020) On the accuracy of macroscopic equations for linearized rarefied gas flows. Adv in Aerodynamics 2:2
8. Pan X, Shneider MN, Miles RB (2004) Coherent Rayleigh-Brillouin scattering in molecular gases. Phys Rev A 69:033814
9. Bruno D, Frezzotti A, Ghiroldi GP (2017) Rayleigh-Brillouin scattering in molecular Oxygen by CT-DSMC simulations. Eur J Mech B/Fluids 64:8–16
10. Su W, Zhang YH, Wu L (2021) Multiscale simulation of molecular gas flows by the general synthetic iterative scheme. Comput Methods Appl Mech Engrg 373:113548
11. Wu L, Li Q, Liu H, Ubachs W (2020) Extraction of the translational Eucken factor from light scattering by molecular gas. J Fluid Mech 901:A23
12. Meijer AS, de Wijn AS, Peters MFE, Dam NJ, van de Water W (2010) Coherent Rayleigh-Brillouin scattering measurements of bulk viscosity of polar and nonpolar gases, and kinetic theory. J Chem Phys 133:164315
13. Wang Y, Yu Y, Liang K, Marques W Jr, van de Water W, Ubachs W (2017) Rayleigh-Brillouin scattering in SF6 in the kinetic regime. Chem Phys Lett 669:137–142
14. Wang Y, Liang K, van de Water W, Marques W Jr, Ubachs W (2018) Rayleigh-Brillouin light scattering spectroscopy of nitrous oxide (N_2O). J Quant Spectrosc Radiat Transf 206:63–69
15. Wu L, White C, Scanlon TJ, Reese JM, Zhang Y (2014) Coherent Rayleigh-Brillouin scattering: Influence of the intermolecular potential. AIP Conf Proc 1628:648

16. Wu L, Reese JM, Zhang YH (2016) Temperature retrieval error in Rayleigh-Brillouin scattering using Tenti's S6 kinetic model. AIP Conf Proc 1786:040010
17. Tenti G, Boley C, Desai R (1974) On the kinetic model description of Rayleigh-Brillouin scattering from molecular gases. Can J Phys 52:285
18. Gu Z, Ubachs W (2013) Temperature-dependent bulk viscosity of nitrogen gas determined from spontaneous Rayleigh-Brillouin scattering. Opt Lett 38(7):1110
19. White FM (1998) *Viscous Fluid Flow*. McGraw-Hill Higher Education
20. Lambert L (1977) *Vibrational and Rotational Relaxation in Gases*. Clarendon
21. Mason EA, Monchick L (1962) Heat conductivity of polyatomic and polar gases. J Chem Phys 36:1622
22. Wang Y, Ubachs W, van de Water W (2019) Bulk viscosity of CO_2 from Rayleigh-Brillouin light scattering spectroscopy at 532 nm. J Chem Phys 150:154502

Appendix A
Special Functions

Sonine polynomials are defined as

$$S^{(r)}_{l+\frac{1}{2}}(\xi^2) = \sum_{p=0}^{r} \frac{(-\xi^2)^p \Gamma(l+\frac{3}{2}+r)}{p!(r-p)!\Gamma(l+\frac{3}{2}+p)}, \tag{A.1}$$

with Γ being the gamma function, which obey the orthonormality condition

$$\int_0^\infty \exp(-\xi^2) S^{(r)}_{l+\frac{1}{2}}(\xi^2) S^{(s)}_{l+\frac{1}{2}}(\xi^2) \xi^{2l+2} d\xi = \frac{\Gamma(l+\frac{1}{2}+r)}{2r!}\delta_{rs}.$$

The first few Sonine polynomials are

$$\begin{aligned}
S^{(0)}_{l+\frac{1}{2}}(\xi^2) &= 1, \\
S^{(1)}_{l+\frac{1}{2}}(\xi^2) &= l + \frac{3}{2} - \xi^2, \\
S^{(2)}_{l+\frac{1}{2}}(\xi^2) &= \frac{1}{2}\left(l+\frac{3}{2}\right)\left(l+\frac{5}{2}\right) - \left(l+\frac{5}{2}\right)\xi^2 + \frac{1}{2}\xi^4.
\end{aligned}$$

Legendre polynomials are defined as

$$P_l(x) = \frac{1}{2^l l!} \frac{d^l}{dx^l}(x^2-1)^l, \tag{A.2}$$

which satisfy the following orthonormality condition:

$$\int_{-1}^{1} P_m(x) P_n(x) dx = \frac{2}{2n+1}\delta_{mn}. \tag{A.3}$$

The first few Legendre polynomials are

L. Wu, *Rarefied Gas Dynamics*,
https://doi.org/10.1007/978-981-19-2872-7

$$P_0(x) = 1, \quad P_1(x) = x, \quad P_2(x) = \frac{3x^2 - 1}{2}, \quad P_3(x) = \frac{5x^3 - 3x}{2}.$$

Spherical harmonics are defined as

$$Y_l^m(\theta, \varphi) = \sqrt{\frac{2l+1}{4\pi}\frac{(l-m)!}{(l+m)!}} P_l^m(\cos\theta)\exp(im\varphi), \tag{A.4}$$

which obey the following orthonormality condition:

$$\int_0^\pi Y_l^{*m} Y_{l'}^{m'} \sin\theta d\theta d\varphi = \delta_{ll'}\delta_{mm'}, \tag{A.5}$$

where i is the imaginary unit and $P_l^m(x)$ is the associated Legendre polynomial. When $m = 0$, we have $P_l^m(x) = P_l(x)$. The first few spherical harmonics are

$$\begin{aligned} Y_0^0(\theta, \varphi) &= \frac{1}{\sqrt{4\pi}}, \\ Y_1^0(\theta, \varphi) &= \sqrt{\frac{3}{4\pi}} \cos\theta, \\ Y_2^0(\theta, \varphi) &= \sqrt{\frac{5}{16\pi}} \left(3\cos^2\theta - 1\right). \end{aligned} \tag{A.6}$$

Appendix B
Relaxation Rates of Maxwellian Molecules

Consider the Boltzmann equation for the Maxwellian molecules, in the homogeneous relaxation problem

$$\frac{\partial f}{\partial t} = \sqrt{\frac{2K}{m}} \iint B(\theta)[f(\boldsymbol{v}'_*)f(\boldsymbol{v}') - f(\boldsymbol{v}_*)f(\boldsymbol{v})]d\Omega d\boldsymbol{v}_*. \tag{B.1}$$

Relaxation of shear stress—On multiplying Eq. (B.1) by $mc_{\langle i}c_{j\rangle}$ and integrating the resultant equation with respect to $\boldsymbol{v}$, we have

$$\begin{aligned}
\frac{\partial \sigma_{ij}}{\partial t} =& m\sqrt{\frac{2K}{m}} \iiint B(\theta)(c'_{\langle i}c'_{j\rangle} - c_{\langle i}c_{j\rangle})f(\boldsymbol{v}_*)f(\boldsymbol{v})d\Omega d\boldsymbol{v}_* d\boldsymbol{v} \\
=& m\sqrt{\frac{2K}{m}} \iint (v_r^2 I_{\langle ij\rangle} - 2v_r c_{\langle i} I_{j\rangle})f(\boldsymbol{v}_*)f(\boldsymbol{v})d\boldsymbol{v}_* d\boldsymbol{v},
\end{aligned} \tag{B.2}$$

where we have used the relation between the post- and pre-collision velocities (2.5), i.e., $c'_i = c_i - v_r \cos\theta_m k_i$ with $v_{ri}k_i = v_r \cos\theta_m$ and

$$\begin{aligned}
I_i =& \int_0^{2\pi}\int_0^{\pi} k_i \cos\theta_m B(\theta)\sin\theta d\phi d\theta \\
=& 8\pi \int_0^{\pi/2} k_i \cos^2\theta_m \sin\theta_m B(\pi - 2\theta_m)d\theta_m, \\
I_{ij} =& \int_0^{2\pi}\int_0^{\pi} k_i k_j \cos^2\theta_m B(\theta)\sin\theta d\phi d\theta \\
=& 8\pi \int_0^{\pi/2} k_i k_j \cos^3\theta_m \sin\theta_m B(\pi - 2\theta_m)d\theta_m.
\end{aligned} \tag{B.3}$$

It is clear that I_i must be of the form $I_i = a_1 v_{ri}/v_r$ [1]. On multiplying the first equation of Eq. (B.3) by v_{ri}/v_r, we have $a_1 = \chi^{(2,1)}$ with

L. Wu, *Rarefied Gas Dynamics*,
https://doi.org/10.1007/978-981-19-2872-7

$$\chi^{(r,s)} = 8\pi \int_0^{\pi/2} \cos^{r+1}\theta_m \sin^s\theta_m B(\pi - 2\theta_m) d\theta_m. \tag{B.4}$$

Similarly, I_{ij} must be of the form $I_{ij} = a_2 v_{ri} v_{rj}/v_r^2 + a_3\delta_{ij}$. On multiplying the second equation of Eq. (B.3) by $v_{ri}v_{rj}/v_r^2$ and δ_{ij}, respectively, we have $a_2 + a_3 = \chi^{(4,1)}$ and $a_2 + 3a_3 = \chi^{(2,1)}$ so that $a_2 = \chi^{(2,1)} - 3\chi^{(2,3)}/2$ and $a_3 = (\chi^{(2,1)} - \chi^{(4,1)})/2 = \chi^{(2,3)}/2$. Thus, Eq. (B.2) becomes

$$\begin{aligned}\frac{\partial \sigma_{ij}}{\partial t} &= -3\sqrt{\frac{2K}{m}}\chi^{(2,3)} n\sigma_{ij} \\ &= -\frac{3}{4}2\pi\sqrt{\frac{2K}{m}}\int_0^{\pi} d\theta \sin^3\theta B(\theta) n\sigma_{ij} = n\lambda_{02}\sigma_{ij},\end{aligned} \tag{B.5}$$

where λ_{02} is defined in Eq. (2.46). It can be shown that the shear viscosity (3.27), obtained from the Chapman-Enskog expansion of the Boltzmann equation, is

$$\mu = -\frac{k_B T}{\lambda_{02}}, \tag{B.6}$$

so the relaxation of shear stress satisfies

$$\frac{\partial \sigma_{ij}}{\partial t} = -\frac{p}{\mu}\sigma_{ij}. \tag{B.7}$$

Relaxation of heat flux—On multiplying Eq. (B.1) by $(m/2)c^2 c_i$ and integrate the resultant equation with respect to $\boldsymbol{v}$, we have

$$\begin{aligned}\frac{\partial q_i}{\partial t} =& \frac{m}{2}\sqrt{\frac{2K}{m}}\iint B(\theta)(c'^2 c'_i - c^2 c_i) f(\boldsymbol{v}_*) f(\boldsymbol{v}) d\boldsymbol{v}_* d\Omega d\boldsymbol{v} \\ =& \frac{m}{2}\sqrt{\frac{2K}{m}}\iint (-2v_r c_i c_k I_k - v_r c^2 I_i + v_r^2 c_i I_{kk} + 2v_r^2 c_k I_{ki} - v_r^3 I_{ikk}) \\ & \times f(\boldsymbol{v}_*) f(\boldsymbol{v}) d\boldsymbol{v}_* d\boldsymbol{v} \\ =& -\frac{2}{3}\frac{p}{\mu} q_i.\end{aligned} \tag{B.8}$$

Note that in deriving the above equation, we have used the following identity:

$$\begin{aligned}I_{ijk} =& \int_0^{2\pi}\int_0^{\pi} k_i k_j k_k \cos^3\theta_m B(\theta)\sin\theta d\phi d\theta \\ =& 8\pi\int_0^{\pi/2} k_i k_j k_k \cos^4\theta_m \sin\theta_m B(\pi - 2\theta_m) d\theta_m \\ =& a\frac{v_{ri}v_{rj}v_{rk}}{v_r^3} + \frac{b}{3v_r}(\delta_{ij}v_{rk} + \delta_{ik}v_{rj} + \delta_{jk}v_{ri}),\end{aligned} \tag{B.9}$$

with

$$a + \frac{5}{3}b = \chi^{(4,1)}, \quad a + b = \chi^{(6,1)}. \tag{B.10}$$

Reference

1. Struchtrup H (2005) Macroscopic transport equations for rarefied gas fows: approximation methods in kinetic theory. Springer, Heidelberg, Germany

Appendix C
Numerical Quadratures

C.1 Gauss-Legendre Quadrature

The integral $\int_a^b f(x)dx$ is approximated numerically as $\sum_{i=1}^{n} f(x_i)w_i$, where x_i and w_i are the i-th abscissa and weight in the Gauss-Legendre quadrature of order n. Note that this code is written by Greg von Winckel on 25 February 2004.

```
function [x,w]=lgwt(N,a,b)

% This script is for computing definite integrals using
% Legendre-Gauss Quadrature. Computes the Legendre-Gauss
% nodes and weights on an interval [a,b] with truncation order N

% Suppose you have a continuous function f(x) defined on [a,b]
% Then compute the definite integral using sum(f.*w);

% Written by Greg von Winckel - 02/25/2004

N=N-1;  N1=N+1; N2=N+2; xu=linspace(-1,1,N1)';

% Initial guess
y=cos((2*(0:N)'+1)*pi/(2*N+2)) + (0.27/N1)*sin(pi*xu*N/N2);

L=zeros(N1,N2); % Legendre-Gauss Vandermonde Matrix
Lp=zeros(N1,N2); % Derivative of LGVM

% Compute the zeros of the N+1 Legendre Polynomial
% using the recursion relation and the Newton-Raphson method
y0=2;

% Iterate until new points are uniformly within epsilon of old points
while max(abs(y-y0))>eps
```


L. Wu, *Rarefied Gas Dynamics*,
https://doi.org/10.1007/978-981-19-2872-7

```
    L(:,1)=1;    L(:,2)=y;
    for k=2:N1
        L(:,k+1)=( (2*k-1)*y.*L(:,k)-(k-1)*L(:,k-1) )/k;
    end
    Lp=(N2)*( L(:,N1)-y.*L(:,N2) )./(1-y.^2);
    y0=y;
    y=y0-L(:,N2)./Lp;
end
x=(a*(1-y)+b*(1+y))/2;     % Linear map from[-1,1] to [a,b]
w=(b-a)./((1-y.^2).*Lp.^2)*(N2/N1)^2;  % Compute the weights
```

C.2 Gauss-Hermite Quadrature

The integral $\int_{-\infty}^{\infty} f(x)dx$ is approximated numerically as $\sum_{i=1}^{n} f(x_i)w_i$, where x_i and w_i are the i-th abscissa and weight in the Gauss-Hermite quadrature of order n. Note that this code is written by Geert Van Damme on 21 February 2010.

```
function [x, w] = GaussHermite(n)

% This function determines the abscisas (x) and weights (w)
% for the Gauss-Hermite quadrature of order n>1,
% on the interval [-INF, +INF].

% This function is valid for any degree n>=2,
% as the companion matrix (of the n'th degree Hermite polynomial)
% is constructed as a symmetrical matrix,
% guaranteeing that all the eigenvalues (roots) will be real.

% Geert Van Damme
% geert@vandamme-iliano.be
% February 21, 2010

% Building the companion matrix CM
% CM is such that det(xI-CM)=L_n(x), with L_n the Hermite polynomial
% under consideration. Moreover, CM will be constructed in such a way
% that it is symmetrical.

i   = 1:n-1;
a   = sqrt(i/2);
CM  = diag(a,1) + diag(a,-1);

% Determining the abscissas (x) and weights (w)
% - since det(xI-CM)=L_n(x), the abscissas are the roots of the
```

```
%   characteristic polynomial, i.d. the eigenvalues of CM;
% - the weights can be derived from the corresponding eigenvectors.

[V L]     = eig(CM);
[x ind] = sort(diag(L));
V         = V(:,ind)';
w         = sqrt(pi) * V(:,1).^2;

w=w./exp(-x.^2);
```

Appendix D
Implementation of Fast Spectral Method

D.1 Algorithm 1: Zero-Padding

This pseudo-code calculates the Boltzmann collision operator by the zero-padding technique. Note that F stands for the FFT in MATLAB, S represents the MATLAB function FFTSHIFT that shifts the zero-frequency component to the center of spectrum, F^{-1} is the inverse FFT, and the function $\Re$ gets the real parts of complex numbers.

step 1. $\hat{f} = S\{\mathrm{F}^{-1}[S(f)]\}$

step 2. $\hat{Q}^{+} = 0$

For $\theta_p = (1, 2, \cdots, M-1)\frac{\pi}{M}$ and $\varphi_q = (1, 2, \cdots, M)\frac{\pi}{M}$

$t_1 = \hat{f} \cdot \phi_{\alpha+\gamma}(\xi_l, \theta_p, \varphi_q)$; $t_2 = \hat{f} \cdot \psi_{\gamma}(\xi_m, \theta_p, \varphi_q)$

zero-padding t_1, t_2 to the dimension $\geq \frac{3N_1}{2} \times \frac{3N_2}{2} \times \frac{3N_3}{2}$

$\hat{Q}^{+} = \hat{Q}^{+} + F(t_1) \cdot F(t_2) \cdot \sin\theta_p$

End

step 3. $t_1 = \hat{f}$; $t_2 = \hat{f} \cdot \phi_{loss}$

zero-padding t_1, t_2 to the dimension $\geq \frac{3N_1}{2} \times \frac{3N_2}{2} \times \frac{3N_3}{2}$

$\hat{Q}^{-} = F(t_1) \cdot F(t_2)$

step 4. $\hat{Q} = F^{-1}(\hat{Q}^{+} - \hat{Q}^{-})$; delete the redundant data in $\hat{Q}$

step 5. $Q = \frac{4\pi^2}{\mathrm{Kn}' M^2} S\{\Re[F[S(\hat{Q})]]\}$

D.2 Algorithm 2: No Zero-Padding

Note that if N is large enough, the zero-padding is not necessary, since the aliasing error in the FFT-based convolution is negligible.

step 1. $\hat{f} = S\{F^{-1}[S(f)]\}$

step 2. $Q^{+} = 0$

L. Wu, *Rarefied Gas Dynamics*,
https://doi.org/10.1007/978-981-19-2872-7

For $\theta_p = (1, 2, \cdots, M-1)\frac{\pi}{M}$ and $\varphi_q = (1, 2, \cdots, M)\frac{\pi}{M}$

$t_1 = \hat{f} \cdot \phi_{\alpha+\gamma}(\xi_l, \theta_p, \varphi_q)$; $t_2 = \hat{f} \cdot \psi_\gamma(\xi_m, \theta_p, \varphi_q)$

$Q^+ = Q^+ + F(t_1) \cdot F(t_2) \cdot \sin\theta_p$

End

$Q^+ = \frac{4\pi^2}{\text{Kn}' M^2} S[\Re(Q^+)]$

step 3. $\nu = \frac{4\pi^2}{\text{Kn}' M^2} S\{\Re[F[S(\hat{f} \cdot \phi_{loss})]]\}$; $Q^- = \nu f$

step 4. $Q = Q^+ - Q^-$

D.3 Algorithm 3: Collision Frequency

One can get an accurate collision frequency by enlarging the velocity domain by a factor of two. Details are given below.

step 1. get ϕ_{loss}^{ex} from Eq. (4.33) with $L^{ex} = 2L$ and $R^{ex} = \sqrt{2}L$

step 2. create a zero-value array f^{ex} of size $2N_1 \times 2N_2 \times 2N_3$

step 3. copy the value of f to the middle of f^{ex}

step 4. $\hat{f}^{ex} = S\{F[S(f^{ex})]\}$

step 5. $\nu^{ex} = \frac{4\pi^2}{\text{Kn}' M^2} S\{\Re[F^{-1}[S(\hat{f}^{ex} \cdot \phi_{loss}^{ex})]]\}$

step 6. copy the middle region value of ν^{ex} to ν.

Appendix E
MATLAB Code for Normal Shock Wave

This MATLAB code finds the structure of normal shock wave by solving the Boltzmann equation, using the fast spectral method for the Boltzmann collision operator and the second-order upwind finite difference method for the streaming operator. The 3D molecular velocity space is discretized uniformly, while the 1D spatial coordinate is discretized non-uniformly, with refined grids in the center of normal shock wave where the spatial gradient is large. The symmetry is considered so that the number of solid angle discretization is halved.

With proper modifications, the code can be used to simulate the 0D homogeneous relaxation problem, and the 1D Fourier and Couette flows, as well as 2D thermal transpiration.

Note that it uses the Gauss-Legendre quadrature to calculate the functions in Eqs. (4.26) and (4.28), and to discretize the solid angle in Eq. (4.27), so it calls the "lgwt" subroutine in Appendix C.1.

```
clear; clc;
%% type of interatomic potential
omega=0.5; % viscosity index
% parameters in the collision kernel B(theta,v_r)
alpha=2*(1-omega); Gama=0;

%% Rankine-Hugoniot relation for the shock wave
r=5/3; % specific heat ratio
Ma=3; % Mach number
ul=sqrt(r/2)*Ma; % upstream velocity
nr=(r+1)*Ma^2/(2+(r-1)*Ma^2); % downstream density
ur=ul/nr; % downstream velocity
% downstream temperature
Tr=2*(r-1)*(1+(r-1)/2*Ma^2)*(2*r*Ma^2/(r-1)-1)/(r+1)^2/Ma^2;

%% spataal non-uniform grids
y_max=8;      Ny=50;
ratio=5; % ratio of the maximum and minimum step size
NN=(Ny-1)/2;
d1=5*y_max/(NN)/(ratio+4); % minimum step size
d2=ratio*d1;    % maximum step size
y=d1*(-NN:NN)+0.2*(d2-d1)*(-NN:NN).^5/(NN)^4;

```


L. Wu, *Rarefied Gas Dynamics*,
https://doi.org/10.1007/978-981-19-2872-7

```
%% uniform velocity and frequency space
L=12; % 6*sqrt( max(T) )
Lx=L;  Nvx=42;    vx=linspace(-Lx,Lx,Nvx);    delta_vx=vx(2)-vx(1);
Ly=L;  Nvy=42;    vy=linspace(-Ly,Ly,Nvy);    delta_vy=vy(2)-vy(1);
Lz=L;  Nvz=42;    vz=linspace(-Lz,Lz,Nvz);    delta_vz=vz(2)-vz(1);
% three-dimensional frequency space
fre_vx=(-Nvx/2:Nvx/2-1)*2*pi/(Nvx*2*Lx/(Nvx-1));
fre_vy=(-Nvy/2:Nvy/2-1)*2*pi/(Nvy*2*Ly/(Nvy-1));
fre_vz=(-Nvz/2:Nvz/2-1)*2*pi/(Nvz*2*Lz/(Nvz-1));
[Fre_vx,Fre_vy,Fre_vz]=ndgrid(fre_vx,fre_vy,fre_vz);

%% calculate kernel modes
method=1; % 1: trapezoidal rule; 2: Gauss-Legendre quadrature
% discretization of solid angle
M=3; M2=2; % symmetry taken into account
R0=2*Lx/(2+sqrt(2));
R=sqrt(2)*R0;
[abscissa,weight]=lgwt(600,0,R); % Gauss-Legendre quadrature
ss=linspace(0,-sqrt(3)*min([fre_vx(1),fre_vy(1),fre_vz(1)]),8000);
ss_y1=zeros(size(ss));  ss_y2=zeros(size(ss));
for loop=1:8000
    ss_y1(loop)=2*sum(abscissa.^(alpha+Gama) ...
                .*cos(abscissa*ss(loop)).*weight);
    ss_y2(loop)=2*pi*sum(abscissa.^(1-Gama)...
                .*besselj(0,abscissa*ss(loop)).*weight);
end

if method==1 % trapezoidal rule
    integral_loop=M*M2;
    alpha1=zeros(Nvx,Nvy,Nvz,integral_loop);
    alpha2=alpha1; alpha3=zeros(Nvx,Nvy,Nvz);
    index=0;
    for loop=1:M2
    theta=loop*pi/(2*M2+1);
    for loop2=1:M
        index=index+1;
        theta2=(loop2-1)*pi/(2*M-1);
        s=Fre_vx*sin(theta)*cos(theta2) ...
         +Fre_vy*sin(theta)*sin(theta2)+Fre_vz*cos(theta);
        s(s==0)=eps;  s=abs(s);
        alpha1(:,:,:,index)=interp1(ss,ss_y1,s,'PCHIP', ...
                            'extrap')*sin(theta);
        s=Fre_vx.^2+Fre_vy.^2+Fre_vz.^2-s.^2;
        s(s<=0)=eps;  s=sqrt(s);
        if loop2==1
            alpha2(:,:,:,index)=interp1(ss,ss_y2,s,'PCHIP', ...
                               'extrap')*pi/(2*M-1)*pi/(2*M2+1)*1/2;
        else
            alpha2(:,:,:,index)=interp1(ss,ss_y2,s,'PCHIP', ...
                                  'extrap')*pi/(2*M-1)*pi/(2*M2+1);
        end
        alpha3=alpha3+alpha1(:,:,:,index).*alpha2(:,:,:,index);
    end
    end
else % Gauss-Legendre quadrature
    M=3;
    integral_loop=M*M;
    alpha1=zeros(Nvx,Nvy,Nvz,integral_loop);
    alpha2=alpha1; alpha3=zeros(Nvx,Nvy,Nvz);
    [jd,wt]=lgwt(2*M,0,pi);
```

```
      index=0;
      for loop=1:M
      theta=jd(loop);
      for loop2=1:M
          index=index+1;
          theta2=jd(loop2);
          s=Fre_vx*sin(theta)*cos(theta2)...
           +Fre_vy*sin(theta)*sin(theta2)+Fre_vz*cos(theta);
          s(s==0)=eps; s=abs(s);
          alpha1(:,:,:,index)=interp1(ss,ss_y1,s,'PCHIP','extrap')...
                                   *sin(theta);
          s=Fre_vx.^2+Fre_vy.^2+Fre_vz.^2-s.^2;
          s(s<=0)=eps;   s=sqrt(s);
          alpha2(:,:,:,index)=interp1(ss,ss_y2,s,'PCHIP','extrap')...
                                   *wt(loop)*wt(loop2);
          alpha3=alpha3+alpha1(:,:,:,index).*alpha2(:,:,:,index);
      end
      end
end

% symmetry condition permits to reduce the storage
vx=vx(1:Nvx/2);         Nvx=Nvx/2;
vz=vz(1:Nvz/2);         Nvz=Nvz/2;
[Vx,Vy,Vz]=ndgrid(vx,vy,vz);   % three-dimensional velocity space
% weights in the integration with respect to velocity space
dV=4*delta_vx*delta_vy*delta_vz;

%% Kundsen number
Kn=5*pi/16;
Kn_prime=Kn*64/5*2^(alpha/2)*gamma((alpha+3+Gama)/2)*gamma(2-Gama
    /2);

%% initial VDF
f=zeros(Nvx,Nvy,Nvz,Ny);
for ly=1:Ny
if y(ly)<=0
    f(:,:,:,ly)=1/pi^1.5*exp(-((Vy-ul).^2+Vx.^2+Vz.^2));
else
    f(:,:,:,ly)=nr/(pi*Tr)^1.5*exp(-((Vy-ur).^2+Vx.^2+Vz.^2)/Tr);
end
end

% macroscopic quantities
for ly=1:Ny
    rho(ly)=sum(sum(sum(f(:,:,:,ly))))*dV;
    uy(ly)=sum(sum(sum(f(:,:,:,ly).*Vy)))*dV/rho(ly);
    c2=Vx.^2+(Vy-uy(ly)).^2+Vz.^2;
    T(ly)=2*sum(sum(sum(f(:,:,:,ly).*c2)))*dV/rho(ly)/3;
    qy(ly)=sum(sum(sum( f(:,:,:,ly).*c2.*(Vy-uy(ly)) )))*dV;
end

%% main
for iteration_step=1:300
    iteration_step
%% collision calculated by FSM
for ly=3:Ny-2
    % VDF in the full velocity space and its spectrum
    ft(1:Nvx,:,1:Nvz)=f(:,:,:,ly);
    ft(Nvx+1:2*Nvx,:,1:Nvz)=f(end:-1:1,:,:,ly);
    ft(:,:,Nvz+1:2*Nvz)=ft(:,:,Nvz:-1:1);
```

```
    f_spec=fftshift(ifftn(fftshift( ft )));
    % gain term
    gain=0;
    for loop=1:integral_loop
        gain= gain+fftn(f_spec.*alpha1(:,:,:,loop)) ...
              .*fftn(f_spec.*alpha2(:,:,:,loop));
    end
    gain=gain+gain(:,:,end:-1:1); % symmetry in vz direction
    gain=gain+gain(end:-1:1,:,:); % symmetry in vx direction
    gain=fftshift(gain);
    % collision frequency
    temp=fftshift(fftn(fftshift( f_spec.*alpha3 )));
    temp=temp+temp(:,:,end:-1:1); % symmetry in vz direction
    temp=temp+temp(end:-1:1,:,:); % symmetry in vx direction
    %
    f(:,:,:,ly) =4/Kn_prime*real( gain(1:Nvx,:,1:Nvz) );
    mu(:,:,:,ly)=4/Kn_prime*real( temp(1:Nvx,:,1:Nvz) );
end

%% sweep of positive velocity
index=Nvy/2+1:Nvy;
for ly=3:Ny-2 % 2nd-order upwind
    a=(y(ly-2)-y(ly))*(y(ly-1)-y(ly))*(y(ly-2)-y(ly-1));
    coe1=Vy(:,index,:)*((y(ly-1)-y(ly))^2-(y(ly-2)-y(ly))^2) /a;
    b   =Vy(:,index,:)*(y(ly-2)-y(ly))^2                      /a;
    c   =Vy(:,index,:)*(y(ly-1)-y(ly))^2                      /a;
    f(:,index,:,ly)=( f(:,index,:,ly)-b.*f(:,index,:,ly-1) ...
                    +c.*f(:,index,:,ly-2))./(coe1+mu(:,index,:,ly));
end

%% sweep of negative velocity
index=1:Nvy/2;
for ly=Ny-2:-1:3 % 2nd-order upwind
    a=(y(ly+2)-y(ly))*(y(ly+1)-y(ly))*(y(ly+2)-y(ly+1));
    coe2=Vy(:,index,:)*((y(ly+1)-y(ly))^2-(y(ly+2)-y(ly))^2) /a;
    b   =Vy(:,index,:)*(y(ly+2)-y(ly))^2                      /a;
    c   =Vy(:,index,:)*(y(ly+1)-y(ly))^2                      /a;
    f(:,index,:,ly)=( f(:,index,:,ly)-b.*f(:,index,:,ly+1)...
                    +c.*f(:,index,:,ly+2))./(coe2+mu(:,index,:,ly));
end

%% macrosocpic quantities
for ly=1:Ny
    rho(ly)=sum(sum(sum(f(:,:,:,ly))))*dV; % density
    uy(ly)=sum(sum(sum(f(:,:,:,ly).*Vy)))*dV/rho(ly); % velocity
    c2=Vx.^2+( Vy-uy(ly) ).^2+Vz.^2;
    % temperature
    T(ly)=2*sum(sum(sum(f(:,:,:,ly).*c2)))*dV/rho(ly)/3;
    % normal stress
    sigma_yy(ly)=2*sum(sum(sum( f(:,:,:,ly) ...
              .*( (Vy-uy(ly)).^2-c2/3 ))))*dV;
    qy(ly)=sum(sum(sum(f(:,:,:,ly).*c2.*(Vy-uy(ly)))))*dV; % heat flux
end

end
figure(11); plot(y,rho,y,uy,y,T); hold on
figure(22); plot(y,sigma_yy,y,qy); hold on
```

Appendix F
MATLAB Code for Poiseuille Flow and Thermal Transpiration

This MATLAB code simulates the linearized Poiseuille flow and thermal transpiration between two parallel plates for the linearized Boltzmann equation, based on the modeled collision kernel (4.6) for inverse power-law potentials. The shear viscosity, however, is accurately calculated by the fast spectral method presented in Sect. 5.5.2. The algorithm is also powered by the GSIS in Chap. 8 so that it finds the converged solution within dozens of iterations.

The 3D molecular velocity space is discretized non-uniformly, to capture the discontinuities and "over-concentration" in the velocity distribution function, which are typical in wall-bounded rarefied gas flows at large Knudsen numbers. The frequency space, however, is always discretized uniformly so that the FFT-based convolution can be applied. The symmetry is considered so that the number of solid angle discretization is halved.

The 1D spatial coordinate is discretized non-uniformly, with refined grids near the solid walls, so that the Knudsen layer structure is resolved.

By changing the source term around the line 220, the Poiseuille or thermal transpiration can be simulated.

Note that it uses the Gauss-Legendre quadrature to calculate the functions in Eqs. (4.26) and (4.28), so it calls the "lgwt" subroutine in Appendix C.1.

```
1  clear;
2  
3  Kn=1; % Knudsen number
4  delta_rp=sqrt(pi)/2/Kn; % rarefaction parameter
5  
6  %% input parameters
7  accom=1; % Accommodation coefficient in Maxwell's BC
8  omega=1; % viscosity index
9  Nvx=24; Nvz=24; % unifrom discretization of velocity space
10 
11 Nvy=32; % non-unifrom discretization
12 imath=3;% odd
13 
14 Nvy2=32; % uniform frequency components
15 Ny=30;  % nonuniform spatial grids
16 
```


L. Wu, *Rarefied Gas Dynamics*,
https://doi.org/10.1007/978-981-19-2872-7

```
%% nonuniform spatial grids
idx=(0:Ny-1)/2/(Ny-1);
y=idx.^3.*(10-15*idx+6*idx.^2);
% weight
dy=zeros(1,Ny);
dy(1)=(y(2)-y(1))/2;
dy(Ny)=(y(Ny)-y(Ny-1))/2;
dy(2:Ny-1)=(y(3:Ny)-y(1:Ny-2))/2;

%% uniform velocity and frequency space
L_v=6;     % maximum velocity
vx=linspace(-L_v,L_v,Nvx);     delta_vx=vx(2)-vx(1);
vz=linspace(-L_v,L_v,Nvz);     delta_vz=vz(2)-vz(1);
%
id=-Nvy/2+1/2:Nvy/2-1/2;

vy=id.^imath*L_v/id(end)^imath;
wt_vy=imath*id.^(imath-1)*L_v/id(end)^imath;
%
fre_vx=(-Nvx/2:Nvx/2-1)*2*pi/(Nvx*2*L_v/(Nvx-1));
fre_vz=(-Nvz/2:Nvz/2-1)*2*pi/(Nvz*2*L_v/(Nvz-1));
fre_vy=(-Nvy2/2:Nvy2/2-1)*2*pi/(Nvy*2*L_v/(Nvy-1));
% three-dimensional frequency space
[Fre_vx,Fre_vy,Fre_vz]=ndgrid(fre_vx,fre_vy,fre_vz);

%% calculate kernel modes
alpha=2*(1-omega); Gamma=(1-alpha)/2; % collision kernel
% truncation collision operator
R0=2*L_v/(2+sqrt(2)); R=sqrt(2)*R0;
[abscissa,weight]=lgwt(64,0,R);
ss=linspace(0,-sqrt(3)*min([fre_vx(1),fre_vy(1),fre_vz(1)]),8000);
ss_y1=zeros(size(ss));  ss_y2=zeros(size(ss));
for loop=1:8000
    ss_y1(loop)=2*sum(abscissa.^(alpha+Gamma) ...
                .*cos(abscissa*ss(loop)).*weight);
    ss_y2(loop)=2*pi*sum(abscissa.^(1-Gamma) ...
           .*besselj(0,abscissa*ss(loop)).*weight);
end

M=5;  % discretization of solid angle
integral_loop=M*M;
alpha1=zeros(Nvx,Nvy2,Nvz,integral_loop);
alpha2=alpha1;
alpha3=zeros(Nvx,Nvy2,Nvz);
[jd,wt]=lgwt(2*M,0,pi);
index=0;% generate alpha_p
for loop=1:M
    theta=jd(loop);
    for loop2=1:M
        index=index+1;
        theta2=jd(loop2);
        s=Fre_vx*sin(theta)*cos(theta2) ...
         +Fre_vy*sin(theta)*sin(theta2)+Fre_vz*cos(theta);
        s(s==0)=eps;  s=abs(s);
        alpha1(:,:,:,index)=interp1(ss,ss_y1,s,'PCHIP', ...
                                    'extrap') *sin(theta);
        s=Fre_vx.^2+Fre_vy.^2+Fre_vz.^2-s.^2;
        s(s<=0)=eps;  s=sqrt(s);
        alpha2(:,:,:,index)=interp1(ss,ss_y2,s,'PCHIP',...
```

```
                                        'extrap')*wt(loop)*wt(loop2);
            alpha3=alpha3+alpha1(:,:,:,index).*alpha2(:,:,:,index);
        end
end

%% symmetry f(vz)=f(-vz)
vz=vz(1:Nvz/2);
Nvz=Nvz/2;
[Vx,Vy,Vz]=ndgrid(vx,vy,vz);   % 3D velocity space

%%
feq=1/(pi)^1.5*exp( -(Vx.^2+Vy.^2+Vz.^2) );
fgain1=zeros(Nvx,Nvy2,2*Nvz,integral_loop);
fgain2=zeros(Nvx,Nvy2,2*Nvz,integral_loop);

%% the spectrum of the equilibrium distribution function
f_spec=zeros(Nvx,Nvy2,Nvz); ft=zeros(Nvx,Nvy2,2*Nvz);
% vy direction, nonuniform fft
dim=2;
f_spec=0*f_spec;
for lf=1:Nvy2
    for lv=1:Nvy
        f_spec(:,lf,:)=f_spec(:,lf,:)+wt_vy(lv)*feq(:,lv,:) ...
                                *exp(1i*fre_vy(lf)*vy(lv));
    end
end
f_spec=f_spec/L_v/2*(Nvy-1)/Nvy;
% vx direction; unifrom fft
dim=1;
f_spec=fftshift( ifft(fftshift(f_spec,dim),[],dim),dim );
% vz direction, symmetry
ft(:,:,1:Nvz)=f_spec(:,:,:);
ft(:,:,Nvz+1:2*Nvz)=f_spec(:,:,end:-1:1);
dim=3;
f_spec2=fftshift( ifft(fftshift(ft,dim),[],dim),dim );
f_spec_eq=f_spec2;
%% loss
mu=zeros(Nvx,Nvy,Nvz);
dim=3;
mu_temp=fftshift( fft(fftshift(f_spec2.*alpha3,dim),[],dim),dim );
mu_temp=mu_temp+mu_temp(:,:,end:-1:1);
mu_temp2=mu_temp(:,:,1:Nvz);
dim=1;
f_spec=fftshift( fft(fftshift(mu_temp2,dim),[],dim),dim );
%
dim=2;
for lv=1:Nvy
    temp=0;
    for lf=1:Nvy2
        temp=temp+f_spec(:,lf,:)*exp(-1i*fre_vy(lf)*vy(lv));
    end
    mu(:,lv,:)=real(temp);
end
mu=mu+mu(end:-1:1,:,:);

%%
for loop=1:integral_loop
    fgain1(:,:,:,loop)=fftn(f_spec2.*alpha1(:,:,:,loop));
    fgain2(:,:,:,loop)=fftn(f_spec2.*alpha2(:,:,:,loop));
```

```
end

%% initial velocity distribution function
vx=vx(1:Nvx/2); % anti-symmetry: f(vx)=-f(-vx)
Nvx=Nvx/2;
feq=feq(1:Nvx,:,:);
mu=mu(1:Nvx,:,:);
[Vx,Vy,Vz]=ndgrid(vx,vy,vz);  %three-dimensional velocity space
[~,Wt_vy,~]=ndgrid(vx,wt_vy,vz);
Wt_vy=Wt_vy*4*delta_vx*delta_vz;

f=zeros(Nvx,Nvy,Nvz,Ny);
ux=zeros(1,Ny); ux2=ux;

for ly=1:Ny
    f(:,:,:,ly)=0*feq;
    ux(ly)=sum(sum(sum( Vx.*f(:,:,:,ly).*Wt_vy )));
    c2=Vx.^2+Vy.^2+Vz.^2-2.5;
    qx(ly)=sum(sum(sum( f(:,:,:,ly).*Vx.*c2.*Wt_vy )));
end

%% find the exact shear viscosity
g=f(:,:,:,1);
Shear_prime=1; % initial gauss
error=1;
while (error>1e-5)
    % generate the spectrum
    dim=2; % vy direction, nonuniform fft
    f_spec=0*f_spec;
    for lf=1:Nvy2
        temp=0;
        for lv=1:Nvy
            temp=temp+wt_vy(lv)*g(:,lv,:)*exp(1i*fre_vy(lf)*vy(lv)
    );
        end
        f_spec(1:Nvx,lf,:)=temp;
        f_spec(Nvx+1:2*Nvx,lf,:)=-temp(end:-1:1,:,:);
    end
    f_spec=f_spec/L_v/2*(Nvy-1)/Nvy;
    % vx direction; unifrom fft
    dim=1;
    f_spec=fftshift( ifft(fftshift(f_spec,dim),[],dim),dim );
    % vz direction, symmetry
    ft(:,:,1:Nvz)=f_spec(:,:,:);
    ft(:,:,Nvz+1:2*Nvz)=f_spec(:,:,end:-1:1);
    dim=3;
    f_spec2=fftshift( ifft(fftshift(ft,dim),[],dim),dim );

    %% gain
    f_temp2=0;
    for loop=1:integral_loop
        f_temp2= f_temp2+2*fgain1(:,:,:,loop) ...
                 .*fftn(f_spec2.*alpha2(:,:,:,loop));
        % or
        % f_temp2= f_temp2+2*fgain2(:,:,:,loop) ...
        %          .*fftn(f_spec2.*alpha1(:,:,:,loop)) ;
    end
    f_temp2=f_temp2-fftn(f_spec_eq).*fftn(f_spec2.*alpha3);
    f_spec2=fftshift( ifftn(f_temp2) );
```

```

    dim=3;
    mu_temp=fftshift( fft(fftshift(f_spec2,dim),[],dim),dim );
    mu_temp=mu_temp+mu_temp(:,:,end:-1:1);
    mu_temp2=mu_temp(:,:,1:Nvz);
    dim=1;
    f_spec=fftshift( fft(fftshift(mu_temp2,dim),[],dim),dim );
    f_spec=f_spec(1:Nvx,:,:)-f_spec(end:-1:Nvx+1,:,:);
    %
    dim=2;
    for lv=1:Nvy
        temp=0;
        for lf=1:Nvy2
            temp=temp+f_spec(:,lf,:)*exp(-1i*fre_vy(lf)*vy(lv));
        end
        Coll_g(:,lv,:)=real(temp);
    end
    %%
    g=(Coll_g+2*feq.*Vx.*Vy)./mu;
    Shear=sum(sum(sum(g.*Vx.*Vy.*Wt_vy)));

    %% convergence criterion
    error=abs( (Shear-Shear_prime)./Shear_prime );
    Shear_prime=Shear;
end

%% main
source=feq.*Vx; % Poiseuille flow
source=-feq.*Vx.*(Vx.^2+Vy.^2+Vz.^2-2.5); % thermal creep

%% if the hard sphere model is compared to Kyoto's group
%  please uncomment the following 7 lines
% k=0.8
% Kn=k*( 5*sqrt(pi)/8 );

Kn_prime=Kn*64*(sqrt(2))^alpha/5*gamma((alpha+Gamma+3)/2)*gamma(2-
    Gamma/2);
% 2*Shear_prime*delta_rp should become 4/Kn_prime, therefore
Shear_prime=4/Kn_prime/( 2*delta_rp );

%%
muBGK=delta_rp;
iteration_step=0;   error=1;

while (error>1e-5)
iteration_step=iteration_step+1
fboundary=f(:,1:Nvy/2,:,1);  % boundary

%% collision;
for ly=1:Ny
    %% spectrum
    % vy direction, nonuniform fft
    dim=2;
    f_spec=0*f_spec;
    for lf=1:Nvy2
        temp=0;
        for lv=1:Nvy
            temp=temp+wt_vy(lv)*f(:,lv,:,ly)*exp(1i*fre_vy(lf)*vy(
    lv));
```

```
            end
            f_spec(1:Nvx,lf,:)=temp;
            f_spec(Nvx+1:2*Nvx,lf,:)=-temp(end:-1:1,:,:);
        end
        f_spec=f_spec/L_v/2*(Nvy-1)/Nvy;
        % vx direction; unifrom fft
        dim=1;
        f_spec=fftshift( ifft(fftshift(f_spec,dim),[],dim),dim );
        % vz direction, symmetry
        ft(:,:,1:Nvz)=f_spec(:,:,:);
        ft(:,:,Nvz+1:2*Nvz)=f_spec(:,:,end:-1:1);
        dim=3;
        f_spec2=fftshift( ifft(fftshift(ft,dim),[],dim),dim );
        %% gain
        f_temp2=0;
        for loop=1:integral_loop
            f_temp2=f_temp2+2*fgain1(:,:,:,loop) ...
                                .*fftn(f_spec2.*alpha2(:,:,:,loop)) ;
        end
        f_temp2=f_temp2-fftn(f_spec_eq).*fftn(f_spec2.*alpha3);
        f_spec2=fftshift( ifftn(f_temp2) );

        dim=3;
        mu_temp=fftshift( fft(fftshift(f_spec2,dim),[],dim),dim );
        mu_temp=mu_temp+mu_temp(:,:,end:-1:1);
        mu_temp2=2*Shear_prime*delta_rp*mu_temp(:,:,1:Nvz);
        dim=1;
        f_spec=fftshift( fft(fftshift(mu_temp2,dim),[],dim),dim );
        f_spec=f_spec(1:Nvx,:,:)-f_spec(end:-1:Nvx+1,:,:);
        %
        dim=2;
        for lv=1:Nvy
            temp=0;
            for lf=1:Nvy2
                temp=temp+f_spec(:,lf,:)*exp(-1i*fre_vy(lf)*vy(lv));
            end
            Coll_g(:,lv,:)=real(temp);
        end
        Coll(:,:,:,ly)=Coll_g-2*Shear_prime*delta_rp*mu.*f(:,:,:,ly);
        Coll_BGK(:,:,:,ly)=muBGK*( feq.*(2*Vx*ux(ly) )-f(:,:,:,ly));
        %
        f(:,:,:,ly)=Coll_g+source;
end

%% boundary condition
index=Nvy/2+1:Nvy;
f(:,index,:,1)=(1-accom)*fboundary(:,end:-1:1,:);

%% stream of positive velocity
index=Nvy/2+1:Nvy;
ly=2;
coe1=Vy(:,index,:)/(y(ly)-y(ly-1));
f(:,index,:,ly)=( f(:,index,:,ly)+coe1.*f(:,index,:,ly-1) ) ...
                    ./( coe1+2*Shear_prime*delta_rp*mu(:,index,:) );
for ly=3:Ny
    a=(y(ly-2)-y(ly))*(y(ly-1)-y(ly))*(y(ly-2)-y(ly-1));
    coe1=(Vy(:,index,:))*((y(ly-1)-y(ly))^2-(y(ly-2)-y(ly))^2) /a;
    b   =(Vy(:,index,:))*(y(ly-2)-y(ly))^2                     /a;
    c   =(Vy(:,index,:))*(y(ly-1)-y(ly))^2                     /a;
```

```
    f(:,index,:,ly)=( f(:,index,:,ly)-b.*f(:,index,:,ly-1) ...
                    +c.*f(:,index,:,ly-2)) ...
                ./(coe1+2*Shear_prime*delta_rp*mu(:,index,:));
end

%% stream of negative velocity
index=1:Nvy/2;
f(:,1:Nvy/2,:,Ny)=f(:,Nvy:-1:Nvy/2+1,:,Ny); % symmetry
ly=Ny-1;
coe2=-Vy(:,index,:)/( y(Ny)-y(Ny-1) );
f(:,index,:,ly)=( f(:,index,:,ly)+coe2.*( 2*f(:,index,:,ly+1) ...
                -0.5*f(:,end:-1:Nvy/2+1,:,ly) ) ) ...
            ./( 1.5*coe2+2*Shear_prime*delta_rp*mu(:,index,:)
    );

for ly=Ny-2:-1:1
    a=(y(ly+2)-y(ly))*(y(ly+1)-y(ly))*(y(ly+2)-y(ly+1));
    coe2=(Vy(:,index,:))*((y(ly+1)-y(ly))^2-(y(ly+2)-y(ly))^2) /a;
    b   =(Vy(:,index,:))*(y(ly+2)-y(ly))^2                    /a;
    c   =(Vy(:,index,:))*(y(ly+1)-y(ly))^2                    /a;
    f(:,index,:,ly)=( f(:,index,:,ly)-b.*f(:,index,:,ly+1) ...
                   +c.*f(:,index,:,ly+2)) ...
                ./(coe2+2*Shear_prime*delta_rp*mu(:,index,:));
end

%% macroscopic quantities
for ly=1:Ny
    ux2(ly)=sum(sum(sum( Vx.*f(:,:,:,ly).*Wt_vy )));
    pxy(ly)=sum(sum(sum( 2*Vx.*Vy.*f(:,:,:,ly).*Wt_vy )));
    c2=Vx.^2+Vy.^2+Vz.^2-2.5;
    qx(ly)=sum(sum(sum( f(:,:,:,ly).*Vx.*c2.*Wt_vy )));
    % used in General Synthetic Iterative Scheme
    A1(ly)=sum(sum(sum( (Coll(:,:,:,ly) ...
                -Coll_BGK(:,:,:,ly)).*(2*Vx.*Vy).*Wt_vy )));
    A2(ly)=sum(sum(sum( f(:,:,:,ly).*(2*Vy.^2-1).*Vx.*Wt_vy )));
end

%% General Synthetic Iterative Scheme
Csource=sum(sum(sum( 2*Vx.*source.*Wt_vy )));
Pxy(Ny)=pxy(Ny);
for ly=Ny-1:-1:1
    Pxy(ly)=Pxy(ly+1)-Csource *( y(ly+1)-y(ly) );
end
% correction of ux in VDF
Pu(Ny)=A1(Ny)-muBGK*Pxy(Ny);
Pu(2)=A1(2)-muBGK*Pxy(2)-( A2(2)-A2(1) ) / ( y(2)-y(1) );
ux3=ux2;
ly=2;
ux3(ly)=ux3(ly-1)+Pu(ly) *( y(2)-y(1) );
f(:,:,:,ly)=f(:,:,:,ly)+( ux3(ly)-ux2(ly) )*Vx.*feq*2;
for ly=3:Ny
    a=(y(ly-2)-y(ly))*(y(ly-1)-y(ly))*(y(ly-2)-y(ly-1));
    coe1=((y(ly-1)-y(ly))^2-(y(ly-2)-y(ly))^2) /a;
    b   =(y(ly-2)-y(ly))^2 /a;
    c   =(y(ly-1)-y(ly))^2 /a;
    Pu(ly)=A1(ly)-muBGK*Pxy(ly) ...
        -( coe1*A2(ly)+b*A2(ly-1)-c*A2(ly-2) );
    ux3(ly)=( Pu(ly)-b*ux3(ly-1)+c*ux3(ly-2) ) / coe1;
    f(:,:,:,ly)=f(:,:,:,ly)+( ux3(ly)-ux2(ly) )*Vx.*feq*2;
```

```
end
ux2=ux3;

%%
error=sum( abs(ux2./ux-1).*dy )
ux=ux2;

end

%
% plot(y,ux,'r.')
% hold on
% plot(y,-qx,'b*')

G=2*sum(ux.*dy) % mass flow rate
Q=2*sum(qx.*dy) % mass flow rate


% plot the velocity distributon function
h2=sum(sum(f(:,:,:,end),3),1)*delta_vx*delta_vz;

h2=sum(sum(f(:,:,:,end).*(c2),3),1)*delta_vx*delta_vz;

h2=reshape(h2(1,:,1,:),1,Nvy);

figure(1)
plot(y-0.5,ux,'ko')
hold on

figure(2)
plot(y-0.5,qx,'ko')
hold on

% plot(Kn*vy,h2/Kn,'b','linewidth',2) % the VDF at y(end)
% hold on
% axis tight
% axis([-4 4 -0.2 0.1])
```

Index

L. Wu, *Rarefied Gas Dynamics*,
https://doi.org/10.1007/978-981-19-2872-7

The manufacturer's authorised representative in the EU is Springer Nature Customer Service Centre GmbH, Europaplatz 3, 69115 Heidelberg, Germany. If you have any concerns regarding our products, please contact ProductSafety@springernature.com

Printed and bound by CPI Group (UK) Ltd, Croydon, CR0 4YY
31/07/2026
02182943-0001